HIGH-DOSE DOSIMETRY

The following States are Members of the International Atomic Energy Agency:

AFGHANISTAN	HAITI	PARAGUAY
ALBANIA	HOLY SEE	PERU
ALGERIA	HUNGARY	PHILIPPINES
ARGENTINA	ICELAND	POLAND
AUSTRALIA	INDIA	PORTUGAL
AUSTRIA	INDONESIA	QATAR
BANGLADESH	IRAN, ISLAMIC REPUBLIC OF	ROMANIA
BELGIUM	IRAQ	SAUDI ARABIA
BOLIVIA	IRELAND	SENEGAL
BRAZIL	ISRAEL	SIERRA LEONE
BULGARIA	ITALY	SINGAPORE
BURMA	IVORY COAST	SOUTH AFRICA
BYELORUSSIAN SOVIET SOCIALIST REPUBLIC	JAMAICA	SPAIN
	JAPAN	SRI LANKA
CAMEROON	JORDAN	SUDAN
CANADA	KENYA	SWEDEN
CHILE	KOREA, REPUBLIC OF	SWITZERLAND
CHINA	KUWAIT	SYRIAN ARAB REPUBLIC
COLOMBIA	LEBANON	THAILAND
COSTA RICA	LIBERIA	TUNISIA
CUBA	LIBYAN ARAB JAMAHIRIYA	TURKEY
CYPRUS	LIECHTENSTEIN	UGANDA
CZECHOSLOVAKIA	LUXEMBOURG	UKRAINIAN SOVIET SOCIALIST REPUBLIC
DEMOCRATIC KAMPUCHEA	MADAGASCAR	
DEMOCRATIC PEOPLE'S REPUBLIC OF KOREA	MALAYSIA	UNION OF SOVIET SOCIALIST REPUBLICS
	MALI	
DENMARK	MAURITIUS	UNITED ARAB EMIRATES
DOMINICAN REPUBLIC	MEXICO	UNITED KINGDOM OF GREAT BRITAIN AND NORTHERN IRELAND
ECUADOR	MONACO	
EGYPT	MONGOLIA	
EL SALVADOR	MOROCCO	UNITED REPUBLIC OF TANZANIA
ETHIOPIA	NAMIBIA	
FINLAND	NETHERLANDS	UNITED STATES OF AMERICA
FRANCE	NEW ZEALAND	URUGUAY
GABON	NICARAGUA	VENEZUELA
GERMAN DEMOCRATIC REPUBLIC	NIGER	VIET NAM
GERMANY, FEDERAL REPUBLIC OF	NIGERIA	YUGOSLAVIA
GHANA	NORWAY	ZAIRE
GREECE	PAKISTAN	ZAMBIA
GUATEMALA	PANAMA	

The Agency's Statute was approved on 23 October 1956 by the Conference on the Statute of the IAEA held at United Nations Headquarters, New York; it entered into force on 29 July 1957. The Headquarters of the Agency are situated in Vienna. Its principal objective is "to accelerate and enlarge the contribution of atomic energy to peace, health and prosperity throughout the world".

© IAEA, 1985

Permission to reproduce or translate the information contained in this publication may be obtained by writing to the International Atomic Energy Agency, Wagramerstrasse 5, P.O. Box 100, A-1400 Vienna, Austria.

Printed by the IAEA in Austria
February 1985

PROCEEDINGS SERIES

HIGH-DOSE DOSIMETRY

PROCEEDINGS OF AN INTERNATIONAL SYMPOSIUM
ON HIGH-DOSE DOSIMETRY
ORGANIZED BY THE
INTERNATIONAL ATOMIC ENERGY AGENCY
AND HELD IN VIENNA,
8–12 OCTOBER 1984

INTERNATIONAL ATOMIC ENERGY AGENCY
VIENNA, 1985

HIGH-DOSE DOSIMETRY
IAEA, VIENNA, 1985
STI/PUB/671
ISBN 92—0—010085—6

FOREWORD

The International Symposium on High-Dose Dosimetry was organized by the International Atomic Energy Agency and held in Vienna from 8 to 12 October 1984. Over 70 participants from 30 Member States and 2 international organizations attended. The purpose of the symposium was to provide an international forum for the exchange of technical information on the latest developments in this fast moving field.

Interest in the application of radiation in industry is being shown world wide as it offers potential technological advantages as well as safety and economy in the following fields: sterilization of medical supplies, irradiation of food, treatment of wastes, and production of plastics and a variety of other products widely used in modern society.

Radiation dosimetry provides reliable quality control of radiation processes and is the basis for regulatory acceptance of irradiated products. This symposium, the first of its kind to be held, was therefore designed to be of particular interest to scientists developing dosimetry and dose assurance techniques for both research and applied industry.

Papers presented at the meeting discussed the expectations of industry, the problems facing dosimetry, the concern of industrialists and their activities, thoughts and hopes for the future. However, it was noted that most of the dosimetry systems in use need further improvement in order to better meet the essential characteristics of routine dosimeters and thus facilitate the operation of irradiation facilities.

It was recognized that reliable dosimetry could be a unique tool for good irradiation practice and that by standardizing dosimetry the quality assurance of irradiated products may cease to be a problem in the future. The symposium clearly aroused wide interest and it is hoped that it will prove to be a milestone for the future development of high-dose assurance in radiation-applied science and industry.

EDITORIAL NOTE

The papers and discussions have been edited by the editorial staff of the International Atomic Energy Agency to the extent considered necessary for the reader's assistance. The views expressed and the general style adopted remain, however, the responsibility of the named authors or participants. In addition, the views are not necessarily those of the governments of the nominating Member States or of the nominating organizations.

Where papers have been incorporated into these Proceedings without resetting by the Agency, this has been done with the knowledge of the authors and their government authorities, and their cooperation is gratefully acknowledged. The Proceedings have been printed by composition typing and photo-offset lithography. Within the limitations imposed by this method, every effort has been made to maintain a high editorial standard, in particular to achieve, wherever practicable, consistency of units and symbols and conformity to the standards recommended by competent international bodies.

The use in these Proceedings of particular designations of countries or territories does not imply any judgement by the publisher, the IAEA, as to the legal status of such countries or territories, of their authorities and institutions or of the delimitation of their boundaries.

The mention of specific companies or of their products or brand names does not imply any endorsement or recommendation on the part of the IAEA.

Authors are themselves responsible for obtaining the necessary permission to reproduce copyright material from other sources.

CONTENTS

GENERAL ASPECTS (Session 1)

Dosimétrie et exploitation d'installations d'irradiation (IAEA-SM-272/42) 3
   P.E. Vidal
Quality control through dosimetry at a contract radiation processing facility
(IAEA-SM-272/3) 13
   T.A. Du Plessis, A.H.A. Roediger
Dosimetry practice for irradiation of the Mediterranean fruit fly
Ceratitis capitata (Wied.) (IAEA-SM-272/36) 23
   J.L. Zavala, M.M. Fierro, A.J. Schwarz, D.H. Orozco, M. Guerra
Status and prospects of high-dose metrology in the Philippines
(IAEA-SM-272/7) 31
   E.M. Valdezco, E.G. Cabalfin, L.M. Ascaño, C.C. Singson, Q.O. Navarro
Statistical and metrological aspects of 20 years' experience of radiation
processing in Poland (IAEA-SM-272/26) 47
   P.P. Panta, Z. Bułhak
Irradiations gamma au Centre d'étude de l'énergie nucléaire de Belgique
(IAEA-SM-272/28) 61
   W. Boeykens, L. Ghoos
Dosimetry practice in the GOU-1 gamma irradiator (IAEA-SM-272/29) 69
   M.G. Hristova, V. Stenger, A. Kovács
Experimental chemical dosimetry and computer calculations: Combined
application in a ^{60}Co research unit (IAEA-SM-272/24) 79
   H.M. Bär, J. Reinhardt, M. Remer
Electron beam dosimetry in the ranges 1 to 10 MeV and 0.1 to 10 kW
(IAEA-SM-272/31) 85
   I. Kálmán

DOSIMETRY TECHNIQUES AND SYSTEMS (Session 2)

Dosimetry and quality control in the processing of polymers
(IAEA-SM-272/16) 95
   L. Wiesner
LiF thermoluminescence dosimetry for mapping absorbed dose distributions
in the gamma ray disinfection of machine-baled sheep wool
(IAEA-SM-272/23) 105
   Dexi Jiang
Measurement of high doses near metal and ceramic interfaces
(IAEA-SM-272/10) 109
   W.L. McLaughlin, J.C. Humphreys, M. Farahani, A. Miller

Evaluation of irradiated ethanol-monochlorobenzene dosimeters by the
conductivity method (IAEA-SM-272/33) 135
    A. Kovács, V. Stenger, G. Földiák, L. Legeza
Consistency of ethanol-chlorobenzene dosimetry (IAEA-SM-272/13) 143
    D. Ražem, L. Anđelić, I. Dvornik
Radiation dosimetry applications and their effects on glass optical fibres
(IAEA-SM-272/25) 157
    P.P. Panta, R. Romaniuk, K.Jędrzejewski
Geiger-Müller gamma detectors operated at up to 1000 G/h
(IAEA-SM-272/34) 165
    P.L. Lecuyer, P.M. Chaise
Silver dichromate as a routine dosimeter in the range 1 to 12 kGy
(IAEA-SM-272/35) 171
    J. Thomassen
Рабочие химические детекторы и индикаторы поглощенной дозы и
их метрологическое обеспечение (IAEA-SM-272/46) 183
    С.В. Климов, Б.М. Ванюшкин, Н.Г. Коньков,
    С.М. Николаев, В.В. Генералова, М.Н. Гурский,
    В.К. Амбросимов, Б.В. Толкачев, М.Н. Гринев
(Operational chemical detectors and indicators of absorbed dose and the associated metrology: S.V. Klimov et al.)
Dosímetro de nitrato/nitrito de potasio para altas dosis (IAEA-SM-272/1) 193
    E.M. Dorda, S.S. Muñoz
Standard measurement of processing level gamma ray dose rates with a
parallel-plate ionization chamber (IAEA-SM-272/17) 203
    R. Tanaka, H. Kaneko, N. Tamura, A. Katoh, Y. Moriuchi
Progress in alanine/ESR transfer dosimetry (IAEA-SM-272/39) 221
    D.F. Regulla, U. Deffner
Dosimetry of electron and gamma radiation with alanine/ESR spectroscopy
(IAEA-SM-272/12) 237
    M.K.H. Schneider, M. Krystek, C.C.J. Schneider
Applications of alanine-based dosimetry (IAEA-SM-272/41) 245
    A. Bartolotta, B.Caccia, P.L. Indovina, S. Onori, A. Rosati
Application of commercial silicon diodes for dose rate measurements
(IAEA-SM-272/30) 255
    M.S.I. Rageh, A.Z. El-Behay, F.A.S. Soliman
Rapid determination of optimal irradiation conditions on large radionuclide
sources (IAEA-SM-272/2) 263
    M. Pešek
Investigations of the use of LiF crystals for routine high-level dosimetry at
CERN (IAEA-SM-272/11) 275
    B. Baeyens, F. Coninckx, P. Maier, H. Schönbacher

Use of 'memory effect' of Al_2O_3 TL detectors in high exposure dosimetry
(IAEA-SM-272/32) 285
M. Osvay, F. Golder

DOSE STANDARDIZATION AND CALIBRATION, PHYSICAL ASPECTS
(Session 3)

Some parameters affecting the radiation response and post-irradiation
stability of red 4034 Perspex dosimeters (IAEA-SM-272/5) 293
B. Whittaker, M.F. Watts, S. Mellor, M. Heneghan
Dosimetry methods applied to irradiation with Tesla-4 MeV linear electron
accelerators (IAEA-SM-272/37) 307
I. Janovský
Methods for measuring dose and beam profiles of processing electron
accelerators (IAEA-SM-272/18) 317
R. Tanaka, H. Sunaga, T. Agematsu
Oxygen effects in cellulose triacetate dosimetry (IAEA-SM-272/38) 333
P. Gehringer, E. Proksch, H. Eschweiler
Response of radiochromic film dosimeters to electron beams in different
atmospheres (IAEA-SM-272/22) 345
Wenxiu Chen, Haishen Jia, W.L. McLaughlin
Standardization of high-dose measurement of electron and gamma ray
absorbed doses and dose rates (IAEA-SM-272/44) 357
W.L. McLaughlin
Calibration and intercomparison of red 4034 Perspex dosimeters
(IAEA-SM-272/6) 373
K.M. Glover, M. King, M.F. Watts
Energy dependence of radiochromic dosimeter response to X- and γ-rays
(IAEA-SM-272/9) 397
W.L. McLaughlin, A. Miller, R.M. Uribe, S. Kronenberg, C.R. Siebentritt
Calculation of energy dependence of some commonly used dosimeters
(IAEA-SM-272/27) 425
A. Miller

List of Chairmen of Sessions and Secretariat of the Symposium 437
List of Participants 439
Author Index 447
Transliteration Index 449
Index of Papers by Number 449

GENERAL ASPECTS

(Session 1)

Chairmen

D.F. REGULLA
Federal Republic of Germany

S. ONORI
Italy

Mémoire présenté sur demande

DOSIMETRIE ET EXPLOITATION D'INSTALLATIONS D'IRRADIATION

P.E. VIDAL
Association internationale d'irradiation
    industrielle (AIII),
Charbonnières-les-Bains

Abstract–Résumé

DOSIMETRY AND OPERATION OF IRRADIATION FACILITIES.
    The industrial use of ionizing radiation has required, from the very first, the measurement of delivered and absorbed doses; hence the necessity of providing dosimetric systems. Laboratories, scientists, industries and potential equipment manufacturers have all collaborated in this new field of activity. Dosimetric intercomparisons have been made by each industry at their own facilities and in collaboration with specialists, national organizations and the IAEA. Dosimetry has become a way of ensuring that treatment by irradiation has been carried out in accordance with the rules. It has become in effect assurance of quality. Routine dosimetry should determine a maximum and minimum dose. Numerous factors play a part in dosimetry. Industry is currently in possession of routine dosimetric systems that are sufficiently accurate, fairly easy to handle and reasonable in cost, thereby satisfying all the requirements of industry and the need for control. Dosimetry is important in the process of marketing irradiated products. The operator of an industrial irradiation facility bases his dosimetry on comparison with reference systems. Research aimed at simplifying the practice of routine dosimetry should be continued. New physical and chemical techniques will be incorporated into systems already in use. The introduction of microcomputers into the operation of radiation facilities has increased the value of dosimetry and made the conditions of treatment more widespread. Stress should be placed on research in several areas apart from reference systems, for example: dosimetric systems at temperatures from $+8°C$ to $-45°C$, over the dose range 100 krad to a little more than 1 Mrad, liquids and fluidized solids carried at high speed through ducts, thin-film liquids circulating at a high flow rate, and various other problems.

DOSIMETRIE ET EXPLOITATION D'INSTALLATIONS D'IRRADIATION.
    L'utilisation industrielle des rayonnements ionisants a, dès le début, nécessité la mesure des doses de rayonnements délivrées et absorbées, d'où le problème de la création de systèmes dosimétriques. Laboratoires, scientifiques, fabricants potentiels d'équipements et industriels ont collaboré dans ce nouveau champ d'activité. Des intercomparaisons dosimétriques ont été réalisées par chaque industriel dans ses installations et en collaboration avec les spécialistes, les organismes nationaux et l'AIEA. Finalement, la dosimétrie est devenue l'assurance que le traitement par irradiation a été réalisé dans toutes les règles de l'art. Elle est devenue une assurance-qualité. La dosimétrie de routine doit déterminer une dose minima et une dose maxima. De nombreux facteurs interviennent dans la dosimétrie.

Actuellement, l'industrie dispose de systèmes dosimétriques de routine suffisamment précis, assez pratiques à utiliser, et d'un coût raisonnable donnant satisfaction aux impératifs industriels et de contrôle. La dosimétrie est importante dans le processus de commercialisation des produits irradiés. L'exploitant d'installation industrielle d'irradiation rapporte ses systèmes dosimétriques de routine à des systèmes de référence. Il convient de poursuivre les travaux destinés à simplifier la pratique de la dosimétrie de routine. De nouvelles techniques physiques ou chimiques complètent les systèmes déjà utilisés. L'utilisation de micro-ordinateurs dans l'exploitation des installations d'irradiation a accru l'intérêt de la dosimétrie ainsi que la diffusion des conditions de traitement. Il est nécessaire d'accentuer les recherches dans plusieurs domaines outre les systèmes de référence: systèmes dosimétriques aux températures de $+8°C$ à $-45°C$, dans la bande des doses de 100 kilorads à un peu plus de 1 mégarad, liquides et solides fluidisés transportés à grande vitesse dans des conduites, liquides sous film de faible épaisseur circulant à grande vitesse ainsi que d'autres problèmes.

1. INTRODUCTION

L'utilisation industrielle des rayonnements ionisants (faisceaux d'électrons, gamma) a été envisagée il y a environ 35 ans. Il n'existait alors pratiquement pas de moyens d'irradiation. Ce n'est qu'en 1953, pour ma part, que j'ai suggéré et utilisé les barreaux d'uranium en cours de désactivation du réacteur français EL2 pour étudier les possibilités d'utiliser les rayonnements ionisants dans l'industrie et l'agriculture. Immédiatement s'est posé le problème de la mesure des doses délivrées et absorbées. Le CEA conçut alors un appareil dont la précision n'était pas la grande caractéristique. Quoiqu'il en fut, comme il n'existait rien, ni aucune connaissance sur l'irradiation, cet appareil nous a servi à déterminer des repères. Malgré tout, nous avons obtenu quelques résultats encourageants qui nous ont bien fait augurer de l'avenir.

Cette introduction confirme, une fois de plus:
— que l'on trouve d'abord et que l'on cherche ensuite pour perfectionner et acquérir des connaissances;
— que, dès l'origine, tous les travaux de découverte et de recherche nécessitent une référence de mesures et des procédés de mesure de routine, ce qui paraît évident mais n'est pas toujours facile à réaliser.

Rapidement, de nombreux scientifiques, laboratoires et industriels intéressés par les problèmes de mesures des doses délivrées et des doses absorbées de rayonnements ionisants ont cherché, conçu et fabriqué des dosimètres et les équipements et procédés de lecture. Les industriels, qui soit désiraient apprécier les répercussions éventuelles de ces techniques sur leurs activités présentes et futures, soit entrevoyaient les développements futurs, ont compris la nécessité de la dosimétrie. En conséquence, dans de nombreux pays, scientifiques, laboratoires et fabricants potentiels ont collaboré pour faire avancer les méthodes de dosimétrie.

Par suite, on a vu apparaître les travaux d'un certain nombre de scientifiques de haute qualité dont je ne citerai pas les noms de peur d'en oublier, mais dont beaucoup sont présents. Pour la plupart, ils n'ont jamais oublié dans leurs études que le but principal était la dosimétrie industrielle, sans pour cela négliger leur rôle scientifique. L'AIEA a joué un rôle important, en collaboration avec des organismes nationaux, en servant de pôle d'attraction pour ces spécialistes. De ce fait, il s'est établi une excellente coordination entre ces spécialistes, les laboratoires nationaux et les industriels. Les résultats acquis en montrent l'efficacité.

Nous allons examiner ce que l'industrie attend de la dosimétrie et des caractéristiques que doivent présenter les dosimètres et les équipements de lecture, le rôle que jouent les doses dans l'industrie et, finalement, quelques uns des problèmes à résoudre actuellement.

2. PROBLEMES POSES A LA DOSIMETRIE

Les industriels se sont trouvés en face de trois problèmes liés à la nécessité de mesurer les doses délivrées et les doses absorbées. Ces problèmes sont importants, car il est tout de suite apparu que la mesure des doses délivrées apporterait un contrôle du bon fonctionnement de l'installation et une assurance de résultats et de qualité pour les produits traités. Il fallait donc:

1) disposer d'une ou plusieurs références pour étalonner, contrôler le ou les systèmes dosimétriques utilisés en routine;
2) disposer de systèmes dosimétriques de routine dont nous verrons plus loin les critères qu'ils doivent présenter;
3) compte tenu des échanges nationaux et internationaux des produits, matériaux et marchandises irradiées, pour tous les utilisateurs, disposer d'un langage dosimétrique similaire, quel que soit le système dosimétrique de routine utilisé.

Ces problèmes n'ont pas échappé aux spécialistes en dosimétrie, à l'AIEA et aux exploitants d'installations d'irradiation. Aussi, très tôt, ces derniers ont effectué des intercomparaisons et tout d'abord dans leurs propres installations en utilisant plusieurs systèmes dosimétriques (au début, en se référant au sulfate ferreux). Puis ils ont réalisé entre eux des intercomparaisons qui ont donné lieu à des discussions et à des échanges de vues avec les fournisseurs de dosimètres. Le résultat de ces travaux, auxquels ont collaboré des scientifiques spécialisés, a été de faire progresser la dosimétrie.

Ces intercomparaisons sont excellentes, car elles permettent de remettre systématiquement en question la dosimétrie pratiquée et d'étudier les nouveaux systèmes dosimétriques. Tout cela est un facteur de progrès.

Dans l'économie d'échange de produits, d'échange de matériaux irradiés à tous les stades de fabrication ou de commercialisation, la dosimétrie apporte une assurance-qualité garantissant que les produits et les matériaux traités l'ont été dans toutes les règles de l'art. Elle informe le possesseur temporaire ou l'utilisateur final du traitement qui a été effectué. Ainsi, dans la stérilisation des matériels médicaux et chirurgicaux à usage unique, la routine a cherché à imposer, à la stérilisation par rayonnements ionisants, les méthodes de contrôle microbiologiques appliquées aux autres procédés de stérilisation. Or, il est maintenant reconnu que ces méthodes ne sont pas fiables et que les dosimétries physiques ou chimiques donnent une meilleure assurance de la réalisation de la stérilité. Il en est de même pour les denrées alimentaires irradiées où les doses sont les éléments de base (Codex Alimentarius).

3. PARTICULARITES DE LA DOSIMETRIE INDUSTRIELLE DE ROUTINE

Nous laisserons de côté le spectre du rayonnement utilisé, car il est déterminé par la source de rayonnements et la dégradation du spectre dans les produits, les matériels et les marchandises.

L'installation ayant été réglée pour travailler industriellement, il convient pour l'exploitant de déterminer le spectre des doses absorbées et d'établir une carte des doses absorbées.

Les sources de rayonnements composés d'électrons ou de photons ne délivrent pas, en chaque micropoint du produit, une dose identique. En pratique, sur une petite surface ou volume, il n'existe pas une grande différence, et les dosimètres intègrent les doses, donc donnent un excellent aperçu de la dose absorbée.

L'hétérogénéité de l'émission de la source de rayonnements, les caractéristiques du rayonnement utilisé, les distances de la source par rapport aux différentes parties du volume des produits traités et la structure du produit ne permettent pas d'obtenir une dose identique dans tout le volume du produit. En conséquence, il existe un minimum et un maximum des doses absorbées. Industriellement, ces deux doses absorbées sont très importantes quant aux résultats de l'opération d'irradiation. Si la dose minima n'est pas suffisante, on peut ne pas atteindre le but recherché. Si la dose maxima est trop élevée, on peut obtenir des effets d'altération des produits traités.

On détermine donc, soit le ratio $\dfrac{\text{Dose maxima}}{\text{Dose minima}}$, soit le ratio $\dfrac{\text{Dose minima}}{\text{Dose maxima}}$

Parfois, on porte attention à la dose moyenne. Cette valeur convient pour des exposés ou des présentations, mais n'est pas toujours suffisante industriellement. Quand on veut réaliser un traitement par rayonnements ionisants, on doit donner:
— soit les doses minima et maxima,
— soit la dose minima et le ratio.

4. FACTEURS INTERVENANT DANS LA MESURE DES DOSES ABSORBEES

Nous laisserons de côté les dispositifs, l'autorisation d'exploitation des installations d'irradiation et les dispositifs pratiques décrits pour déterminer ces doses minima et maxima. De nombreuses recommandations existent à ce sujet.

Sans entrer dans les détails, nous rappellerons que:

1) les caractéristiques des dosimètres peuvent être influencées par la régularité des dimensions (épaisseur en particulier), l'homogénéité de la matière utilisée, la pureté du liquide chimique utilisé, l'emballage, la température, l'humidité, la stabilité, le fading, les variations en cours de stockage avant et après irradiation, etc;
2) les soins pris au cours de la préparation d'un dosimètre ou d'un lot de dosimètres peuvent avoir une influence;
3) la fiabilité des équipements d'analyse ou de lecture et leur contrôle et réétalonnage à certains intervalles sont à prendre en considération;
4) dans certains cas, le choix des paramètres de lecture est d'une grande importance.

Cette énumération montre que le système dosimétrique idéal est difficile à réaliser. Il existe un grand nombre de matériaux, de produits sensibles aux rayonnements (films plastiques, systèmes colorés, produits, solutions chimiques, verres, etc.) qui peuvent être la base de systèmes dosimétriques.

5. CARACTERISTIQUES DES DOSIMETRES DE ROUTINE

Nous résumons ci-dessous les principales caractéristiques des dosimètres de routine:

— ils doivent être faciles à utiliser dans les processus industriels;
— ils doivent être aussi précis que possible, et il convient de remarquer que, pour la plupart, leur précision est bonne; en effet, à part quelques cas particuliers, la précision est très supérieure à celle relevant de la fourchette des doses entre la dose minima nécessaire pour obtenir l'effet désiré et la dose maxima susceptible de faire subir au produit des débuts de dommage par irradiation;
— leurs dimensions doivent être aussi restreintes que possible;
— ils doivent donner une bonne reproductibilité;
— ils doivent être stables dans le temps: il s'agit de la stabilité lors du stockage des dosimètres avant leur utilisation et de celle après irradiation; cette dernière est importante et doit être considérée avec attention car les dosimètres irradiés fournissent une preuve tangible complétant les documents établis pour montrer que les doses absorbées ont été celles prévues;

— ils doivent pouvoir être utilisés selon des procédures d'analyse ou de lecture simples et rapides;
— ils doivent couvrir une large bande de doses;
— leur coût et surtout celui des équipements doivent être raisonnables.

6. DOSIMETRIE ET COMMERCIALISATION DES PRODUITS IRRADIES

Pour une installation à source de cobalt 60 ou césium 137, les résultats de la dosimétrie démontrent que la source a fonctionné normalement, en bonne position, et que la vitesse du convoyeur est restée inchangée durant toute la durée des opérations d'irradiation. Pour un accélérateur, les résultats de la dosimétrie montrent que le courant moyen, le système de balayage du faisceau d'électron et la vitesse du convoyeur n'ont pas changé durant les opérations d'irradiation.

6.1. La dosimétrie est donc un moyen pratique et simple de vérifier le bon fonctionnement de tous les éléments concourant aux impératifs fixés pour obtenir le résultat désiré.

La dosimétrie est l'élément de base concourrant à la bonne réalisation de l'irradiation des produits, matériaux et marchandises. Elle s'ajoute donc aux autres critères donnant l'assurance de la qualité d'un produit. La dosimétrie, en définitive, est un condensé montrant que les opérations d'irradiation ont été faites dans toutes les règles de l'art. Elle fait partie des éléments constituants les bonnes pratiques de fabrication.

C'est pour ces raisons que, dès sa création en 1970, l'AIII a établi son «Règlement professionnal des exploitants d'installations industrielles d'irradiation» décrivant les responsabilités des exploitants. L'article 5 stipule que «les exploitants sont responsables des doses délivrées...» et l'article 7 est rédigé ainsi: «*Dosimétrie* — L'exploitant tient un registre où sont inscrits les relevés des mesures dosimétriques et les caractéristiques principales de l'installation d'irradiation au moment des mesures. Ces mesures dosimétriques sont obligatoirement faites par l'exploitant pour chaque lot, avec une fréquence fixée par les conditions particulières de chaque exploitant ou conformes aux règlements s'il en existe.»

De même, en 1975, dans son «Code of Good Practice in and Control of Radiation Facilities for Radiosterilizations of Medical Products», l'article 2.3 *Dosimetry* stipule: «(...) le but de la dosimétrie est d'assurer et d'affirmer que chaque produit élémentaire a reçu au moins la dose minima requise et que des doses excessives pouvant endommager le produit n'ont pas été délivrées (...)».

6.2. L'activité d'irradiation industrielle peut dans une vue d'ensemble se répartir en trois groupes:

6.2.1. Les exploitants utilisant une installation d'irradiation leur appartenant dans leurs processus de fabrication de produits qu'ils commercialisent.

La dosimétrie constitue un contrôle assurant à tous les services que l'irradiation a été effectuée dans les conditions requises ou imposées. Les relevés dosimétriques servent également de justificatifs s'il apparait que la qualité des fabrications peut être mise en cause et que l'irradiation est susceptible d'en être responsable.

6.2.2. Les exploitants utilisant leurs propres installations d'irradiation dans le processus de préparation de produits de base ou semi-ouvrés devant être ultérieurement transformés ou terminés par une autre société.

Outre les intérêts présentés au paragraphe 6.2.1, la dosimétrie a de plus pour objectif de porter à la connaissance du transformateur ou fabricant final les doses déjà appliquées aux produits qu'il reçoit. Cette connaissance est absolument nécessaire, les doses étant cumulatives. Si le transformateur ou fabricant final doit à nouveau irradier ses produits, il faut éviter de les détériorer par un surdosage. Ce surdosage peut modifier ou détériorer les caractéristiques au point de rendre les produits terminés invendables.

6.2.3. Les exploitants d'une installation d'irradiation traitant à façon des produits, matériaux et marchandises ne leur appartenant pas.

L'exploitant est responsable des doses absorbées comprises entre un minimum et un maximum demandés par son client. Il doit donc fournir à ce dernier la preuve que l'irradiation a été effectuée conformément à sa demande. Les relevés dosimétriques apportent cette preuve.

6.3. Litiges commerciaux

Dans les trois cas de figure ci-dessus, il peut exister un litige concernant le produit finalement commercialisé. L'irradiation étant encore insuffisamment connue et étant pour beaucoup mystérieuse, donc dangereuse, la tendance sera de rendre l'irradiation responsable de tout litige concernant le produit final.
Les connaissances mondialement acquises en dosimétrie et le «Règlement professionnel» représentent des arguments de défense si l'exploitant n'a pas commis de faute et a respecté les règles de bonne pratique.
Tout ceci montre l'importance industrielle et commerciale de la dosimétrie.

Actuellement, l'introduction des micro-ordinateurs dans l'exploitation des installations industrielles d'irradiation facilite et optimise les mesures dosimétriques et les contrôles de toutes sortes. Elle facilite leur diffusion et, en particulier, le contrôle des doses absorbées et le pilotage des installations.

Etant donné le développement et l'extension des échanges de produits dans l'économie, le langage dosimétrique doit faciliter les relations entre les détenteurs successifs des produits irradiés.

Tous doivent être convaincus que le langage dosimétrique, bien qu'ayant utilisé des systèmes différents, est le même. Pour cela, les utilisateurs de dosimètres de routine se rapportent à une référence commune ou plusieurs références reconnues et acceptées.

7. REFERENCES DOSIMETRIQUES

Il y a deux possibilités:

7.1. Déclarer qu'une institution ou un laboratoire est la référence et détient la «vérité».

Cette première solution ne semble pas devoir être retenue car elle figera les problèmes de dosimétrie et limitera les possibilités de nouveauté. Par ailleurs, pourquoi choisir un organisme plutôt qu'un autre? Cela ne peut que créer des affrontements stériles sans intérêt. Enfin, on n'obtiendra pas des références compatibles avec le libre échange des produits irradiés.

7.2. Définir un ou plusieurs systèmes dosimétriques de référence, les plus fiables possible, couvrant la plus large bande de doses possible, bien déterminés et fixés quant à leur conception et leur mise en œuvre, et permettant facilement des comparaisons.

Cette deuxième solution, si elle n'est pas parfaite, paraît la meilleure. Elle doit être volontaire et non imposée. Le ou les systèmes de référence retenus, le ou les plus aptes à un moment donné pourra ou pourront être substitués par de nouveaux plus fiables. Commercialement, les dosimètres de routine assureront que les doses absorbées étaient bien celles requises. L'exploitant irradiateur, outre ses intercomparaisons personnelles, aura toujours une base de référence qui sera le témoin impartial, vis-à-vis des tiers. Cette solution est indépendante de motivations sentimentales, psychologiques ou nationalistes. Elle est l'objectif des industriels professionnels qui désirent toujours progresser tout en assurant des opérations d'irradiation comparables et de qualité.

Maintenant il y a lieu d'aborder les problèmes pratiques à résoudre dans un proche avenir.

8. QUELQUES PRIORITES A CONSIDERER

Les doses utilisées industriellement vont de plusieurs milliers de rads (dizaines de Gy) à quelques dizaines de millions de rads (centaines de kGy). Pour des doses plus élevées (100 à 200 millions de rads), concernant par exemple les tenues de matériaux sous rayonnement, les méthodes consistant à cumuler les doses relevées ou à multiplier une dose de base donnent une approximation suffisante, en attendant mieux.

8.1. Dosimétrie de référence

Pour les faisceaux d'électrons, les mesures calorimétriques sont les plus utilisées. Pour les rayonnements gamma, les études pour le choix de dosimètres de référence se poursuivent. Les innovateurs et spécialistes cherchent à mettre à notre disposition ou à perfectionner des systèmes dosimétriques permettant d'améliorer constamment le contrôle et la comparaison des dosimètres de routine.

Bien qu'il existe peu de spécialistes dans la microcalorimétrie, cette technique permettrait d'étudier et d'affiner les systèmes actuels.

L'AIII, dans le domaine des rayonnements gamma, compte dans un avenir proche exposer un choix. Ce choix semble le meilleur pour le moment et pourra être modifié en fonction d'innovations qu'il convient de souhaiter.

8.2. Dosimétrie de routine

Plusieurs systèmes dosimétriques de routine donnent satisfaction actuellement pour les applications industrielles. Pour la plupart, l'influence de certains facteurs complique la dosimétrie. En conséquence, des travaux sont nécessaires pour essayer de simplifier la pratique des mesures dosimétriques de routine en assurant, par exemple, une continuité de fabrication et de caractéristiques des dosimètres, une simplification des mesures et une grande fiabilité des équipements de lecture.

8.3. Autres priorités

1) Adaptation ou mise au point des systèmes dosimétriques de routine dans le domaine du traitement par rayonnements ionisants des produits et matériaux à des températures voisines de 0°C ou à des températures de −15°C à −45°C (par exemple les denrées alimentaires réfrigérées et surgelées).
2) Systèmes dosimétriques adaptés à l'irradiation sous un gaz déterminé, éventuellement sous pression.

3) Dosimétrie d'un liquide, de produits solides de faibles dimensions fluidisés s'écoulant à grande vitesse dans des tuyauteries, d'un liquide s'écoulant à grande vitesse sous forme d'un film de faible épaisseur.
4) Systèmes dosimétriques pour des doses comprises entre 100 kilorads (1 kGy) et un peu plus de 1 mégarad (10 kGy), principalement pour les denrées alimentaires. Plusieurs systèmes dosimétriques sont proposés et commencent à être mis à l'épreuve industrielle. Des solutions doivent intervenir rapidement dans ce domaine.

Il existe encore des efforts importants à réaliser par les spécialistes scientifiques en dosimétrie, par les concepteurs et par les fabricants de dosimètres et d'équipements de lecture.

On a essayé ici de résumer les préoccupations des industriels dans les domaines de la dosimétrie, de leurs activités et de leurs projets.

9. CONCLUSION

Les techniques d'irradiation dans tous les domaines de leur utilisation sont synonymes avant tout d'innovation, de conservation ou d'amélioration de la qualité ou des caractéristiques des produits, matériaux et marchandises. La dosimétrie assure et garantit que ces avantages sont obtenus pratiquement et industriellement.

C'est là que réside le succès de nos activités.

IAEA-SM-272/3

QUALITY CONTROL THROUGH DOSIMETRY AT A CONTRACT RADIATION PROCESSING FACILITY

T.A. DU PLESSIS, A.H.A. ROEDIGER
Iso-Ster (Pty) Ltd,
Kempton Park, South Africa

Abstract

QUALITY CONTROL THROUGH DOSIMETRY AT A CONTRACT RADIATION PROCESSING FACILITY.

Reliable dosimetry procedures constitute a very important part of process control and quality assurance at a contract gamma radiation processing facility that caters for a large variety of different radiation applications. The choice, calibration and routine intercalibration of the dosimetry systems employed form the basis of a sound dosimetry policy in radiation processing. With the dosimetric procedures established, detailed dosimetric mapping of the irradiator upon commissioning (and whenever source modifications take place) is carried out to determine the radiation processing characteristics and performance of the plant. Having established the irradiator parameters, routine dosimetry procedures, being part of the overall quality control measures, are employed. In addition to routine dosimetry, independent monitoring of routine dosimetry is performed on a bi-monthly basis and the results indicate a variation of better than 3%. On an annual basis the dosimetry systems are intercalibrated through at least one primary standard dosimetry laboratory and to date a variation of better than 5% has been experienced. The company also participates in the Pilot Dose Assurance Service of the International Atomic Energy Agency, using the alanine/ESR dosimetry system. Routine calibration of the instrumentation employed is carried out on a regular basis. Detailed permanent records are compiled on all dosimetric and instrumentation calibrations, and the routine dosimetry employed at the plant. Certificates indicating the measured absorbed radiation doses are issued on request and in many cases are used for the dosimetric release of sterilized medical and pharmaceutical products. These procedures, used by Iso-Ster at its industrial gamma radiation facility, as well as the experience built up over a number of years using radiation dosimetry for process control and quality assurance are discussed.

1. INTRODUCTION

In the radiation processing industry reliable dosimetry procedures constitute a very important part of process control and quality assurance. Radiation dosimetry is the only acceptable method to guarantee that the irradiated product has undergone the correct radiation treatment. Radiation processes are developed through research and are defined by certain fixed absorbed radiation dose levels as well as dose uniformities. Quality control of these radiation processes must be based on the assurance that the product has been given the correct absorbed radiation dose. This implies that quality control in radiation processing is associated with accurate in-plant dosimetry, especially so in the case of contract radiation processing. Comprehensive procedures for plant radiation dosimetry at the Iso-Ster radiation facility were compiled in accordance with Section 5 of the South African Bureau of Standards (SABS) Code of Practice [1]. These procedures are also equally important in polymer modifications and in the treatment of foodstuffs (radurization).

2. DOSIMETRY SYSTEMS EMPLOYED

2.1. Choice of dosimetry systems

Choice of the dosimetry systems employed by Iso-Ster was based on experience gained over more than a decade at the South African Atomic Energy Board (AEB). Ease of carrying out the dosimetry and ruggedness, especially for intercalibration purposes, were prime requirements for the selection of these systems. This choice is also in agreement with the recommendations of the SABS Code of Practice [1]. The ceric sulphate dosimeter was chosen as the primary dosimetry system for calibration of the plant both at commissioning and whenever plant processing parameters are changed, for intercalibration purposes, as well as for routine monitoring of the process dosimetry. In addition, the ferrous ammonium sulphate dosimeter (Fricke dosimeter) was employed for calibrating both the fixed-geometry Gammacell-220 and the Gammabeam-650 research irradiators (AEB) used in the calibration of both red and amber acrylic dosimeters. For continuous process dosimetry both the red acrylic dosimeter and the amber acrylic dosimeter supplied by the Irradiation and Dosimetry Services, Atomic Energy Research Establishment, Harwell, United Kingdom, are employed as the secondary dosimetry system.

2.2. Calibration of dosimetry systems

The second step in the chain to establish accurate dosimetry is calibration of the chosen dosimetry systems. For both the ferrous ammonium sulphate dosimeter and the ceric sulphate dosimeter, primary dosimetry solutions were made up in triplicate using accepted procedures [2]. In making up these solutions, special attention was paid to the purity of the chemicals and the water used (using a specially designed distillation system), as well as the cleanliness of the glassware employed. Both these primary dosimeters were irradiated in fixed positions for specific irradiation times at ambient temperature in the two AEB research irradiators mentioned in subsection 2.1 in order to calibrate these research irradiators. In the case of the ceric sulphate dosimeter, the dosimeter solution giving results corresponding closest to published data was sealed in glass vials for future use. The secondary dosimetry systems were, in turn, calibrated

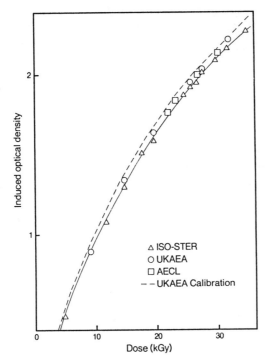

FIG.1. *Intercomparison of the red acrylic dosimetry system employed by Iso-Ster with those of the UKAEA and AECL.*

by irradiating sets of five dosimeters to irradiation doses from 1 to 35 kGy at intervals not exceeding 5 kGy and determining the dosimeter response. It was found that the red acrylic dosimeter gives a reproducibility of better than 3.5% and that of the amber acrylic dosimeter was found to be better than 5% [3, 4]. In addition, red acrylic dosimeters were pre-irradiated to fixed irradiation doses at Harwell, employing three dosimeters at each dose, covering the same dose range from 5 to 35 kGy.

These dosimeters were subsequently read by Iso-Ster after their return from the United Kingdom, allowing for fading during shipment. Similarly, at the commissioning of the plant it was possible to carry out an intercalibration using the red acrylic dosimeter supplied by the Atomic Energy of Canada Limited (AECL). By comparing the Iso-Ster calibration, the dosimeters pre-irradiated at Harwell and the results obtained from the AECL dosimeters a close correlation was found, as indicated in Fig.1. As follows from the figure, the calibration curve for the red acrylic dosimeter supplied by Harwell shows an unacceptable deviation from the three calibrated systems, especially at the higher dose range. This forms the basis for the Iso-Ster policy not to rely only on the calibration

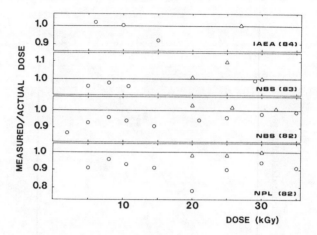

FIG.2. Results of intercalibrations carried out through two primary standards dosimetry laboratories and the IAEA Pilot Dose Assurance Service.

curves provided by the suppliers of dosimeters but to carry out its own calibration on every batch of dosimeters supplied.

2.3. Intercalibration of dosimetry systems

To verify and maintain a high level of dosimetry, it is considered imperative that the developed dosimetry systems be intercalibrated on a regular basis with those of reputable laboratories overseas. As was discussed in subsection 2.2, this was carried out at commissioning of the Iso-Ster plant, making use of data from both Harwell and AECL. A procedure has also been established to carry out intercalibrations of all Iso-Ster dosimetry on an annual basis through the following two organizations: Centre for Radiation Research, National Bureau of Standards (NBS), United States of America, and Division of Radiation Science, National Physical Laboratory (NPL), United Kingdom.

In addition, Iso-Ster was invited to participate in the Pilot Dose Assurance Service of the IAEA which commenced in 1983. The results obtained from the various intercalibrations carried out since 1982 are presented in Fig.2. From this figure it follows that a high measure of agreement between the Iso-Ster dosimetry system and those of the two primary standard dosimetry laboratories was obtained. Furthermore, this was confirmed by the gratifying agreement between the doses measured according to the Iso-Ster dosimetry system and that measured through the alanine/ESR dosimetry system of the Pilot Dose Assurance Service of the IAEA. These results indicate that the Iso-Ster dosimetry system currently has a precision that compares very favourably with that of the analytical techniques employed.

2.4. Tertiary radiation monitoring system

On arrival at the pre-irradiation storage area of the plant, every plant package to be radiation processed is provided with a visual radiation indicator (PVC/dimethyl yellow system) which turns from yellow to red upon receiving an irradiation dose in excess of 10 kGy; for the lower dose range a visual radiation indicator which turns from brown to purple is used. These systems act as a tertiary monitoring system to provide a visual means of determining whether a plant package was radiation processed.

3. CHOICE AND CALIBRATION OF DOSIMETRY INSTRUMENTATION

Coupled with the choice of dosimeter systems is the choice of readout instrumentation. Considering the importance of instrumentation in ensuring an accurate dosimetry procedure, it was decided to acquire only instrumentation of high quality designed for continuous reliable operation backed up by a similar service. To minimize instrumental error, all routine dosimetry is carried out with the same instrumentation used in the calibration and intercalibration of the dosimetry systems employed. Furthermore, it is essential that the operator of the instrumentation understands the physical or chemical principles involved and that regular calibration and maintenance of all instrumentation is carried out. Accordingly, Iso-Ster has highly trained personnel to monitor all dosimetry procedures and has a fixed schedule for maintaining and calibrating all instrumentation associated with dosimetry and the cycle time system of the irradiator, which is in compliance with subsection 6.7 of the SABS Code of Practice [1].

4. CALIBRATION OF RADIATION FACILITY AT COMMISSIONING

The personnel at AECL, suppliers of the Iso-Ster IR-113 Industrial Irradiator, carried out a detailed determination of the absorbed dose distribution and all relevant process parameters before commissioning of the facility. These calibration procedures were carried out by using dummy plant packages of a constant bulk density of 226 kg/m^3 — being the maximum bulk product density handled by the plant. At the same time, primary and secondary dosimeters were introduced by Iso-Ster to confirm AECL dosimetry before any actual radiation processing was carried out. Similar recalibrations are repeated whenever source augmentations or modifications are carried out. The calibration at commissioning was carried out in compliance with subsection 5.5 of the SABS Code of Practice [1].

5. ROUTINE DOSIMETRY PROCEDURES

5.1. Continuous process dosimetry

A secondary red acrylic dosimeter or amber acrylic dosimeter is provided for each separate batch of product received from a manufacturer for radiation processing. Such secondary dosimeters are placed in the product carriers in that position which receives the minimum absorbed radiation dose in the product carrier as determined empirically at plant commissioning, this position being the top centre and bottom centre of the product carrier. Whenever a batch of product received from a manufacturer exceeds ten carriers, an additional secondary dosimeter is used for every additional ten carriers or part thereof. A total of 37 carriers within the biological shield means that at any point in time there will be at least three secondary dosimeters within the shield. Routine dosimetry at Iso-Ster is carried out in compliance with subsection 5.6 of the SABS Code of Practice [1]. To minimize human error in dosimetry calculations, all dosimetry calibration has been computerized to eliminate the use of calibration curves to determine the absorbed radiation dose. By doing this the sophistication of the dosimetry procedure is further enhanced.

5.2. Monitoring of process dosimetry

Iso-Ster introduced an internal procedure for monitoring the continuous dosimetry carried out by the plant process controllers. This monitoring entails an independent bi-monthly check of routine dosimetry by the Research and Development Department of the company. A procedure is followed by means of which both primary and secondary dosimeters are used together with the additional secondary dosimeters during normal running of the plant and then carrying out an intercomparison. Whenever an overall difference of more than 3% is found between the two systems, the cause of such discrepancies is investigated.

6. PRODUCT IDENTIFICATION AND ACCOUNTABILITY

Each plant package entering the pre-irradiation storage area of the irradiation plant is uniquely and permanently marked by means of a self-adhesive label, as shown in Fig.3. Two different labels are used for gamma sterilization and for food radurization, the latter also having a 'Radura' emblem indicating that the particular foodstuff has been radiation treated. To uniquely identify the product the following information is recorded on the label:

(1) The name of Iso-Ster
(2) The date on which the product is received at the plant

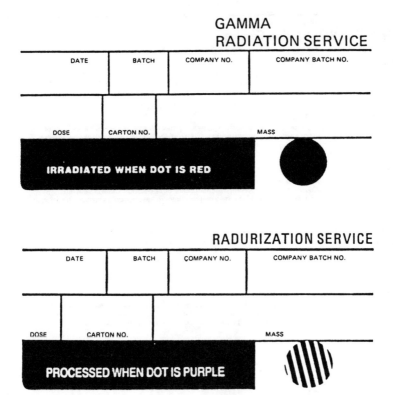

FIG.3. *Self-adhesive labels attached to every plant package received at Iso-Ster for radiation processing. The top label is used for sterilization purposes and the bottom one for food irradiation.*

(3) The Iso-Ster batch number for that particular day
(4) A code number identifying the manufacturer of the product
(5) The batch number of the manufacturer
(6) The requested radiation dose
(7) The number of the particular plant package in the consignment
(8) The gross mass of the particular plant package.

In addition, the tertiary radiation indicator is attached to every label. After weighing the particular consignment, an additional label is printed stating the total mass of the consignment and the number of plant packages in the consignment, together with the above information. This label is attached to the quality control sheet which accompanies the consignment to the point of dispatch.

Radiation Sterilization Service

Sertifikaat van Bestralingsdosis

Hierby word gesertifiseer dat die produk in die houers met onderstaande verwysings-/kontrolenommers ondervermelde gemiddelde minimum bestralingsdosis ontvang het. Hierdie dosis is deur rooi poli(metielmetakrilaat)-dosismeting bepaal en deur chemiese seriumdosismeting bevestig.

Certificate of Irradiation Dose

It is hereby certified that the products in the containers bearing the undermentioned reference/control numbers have received the average minimum irradiation dose stated below. This dose was determined by red poly (methyl methacrylate) dosimetry and verified by cerium chemical dosimetry.

Certificat de Dose D'Irradiation

La présente certifie que le produit dans les boîtes marquées avec les numéros de référence/contrôle mentionnés ci-dèssous a reçu la dose moyenne minimum declarée ci-dessous. Cette dose a été déterminée à l'aide d'un dosimètre au poly (méthyle méthacrylate) rouge ainsi qu'un dosimètre chimique au cérium.

Bescheinigung der Bestrahlungsdosis

Hiermit wird bescheinigt, dasz die Produkte in den Behältern mit den unten erwähnten Geschäfts/Kontrollnummern mit der genannten mittleren Mindestdosis bestrahlt wurden. Diese Dosis ist mit Rotpoly-(Methylmethacrylat)-Dosierung bestimmt und mit zeriumchemischer Dosierung überprüft worden.

Certificado de Dosis de Irradiación

Certifico que el producto dentro los envases con los números de referencia/control señalados mas abajo ha recibido el promedio minimo de dosis de irradiación indicado mas adelante. Esta dosis se determinó por el dosimetro rojo poly (metilo metacrilato) y se verificó por dosimetro quimico de cerio.

Maatskappy Company Societe Gesellschaft Empresa	Maatskappy se verwysingsnommer Company reference number Numero donné par la Société Gesellschaftsnummer Numero de referencia de la Empresa	Bestralingsdatum Irradiation date Date d'irradiation Bestrahlungsdatum Fecha de la irradiacion

Datum
Date
Fecha

Iso-Ster-kontrolenommers Iso-Ster control numbers Iso-Ster numéros de contrôle Iso-Ster Kontrollnummern Iso-Ster números de control	Gemiddelde minimum dosis ontvang Average minimum dose received Dose moyenne minimum reçue Mittlere Mindestdosis erhalten Promedio minimo de dosis recibida
	kGy

qq Iso-Ster
pp Iso-Ster

FIG.4. Certificate of measured minimum absorbed gamma radiation dose based on red or amber acrylic dosimetry as issued by Iso-Ster.

7. QUALITY CONTROL DOCUMENTATION

Detailed permanent records pertaining to the following dosimetry and quality control procedures are compiled and kept by Iso-Ster in compliance with Section 7 of the SABS Code of Practice [1]:

(1) Calibration of the dosimetry system
(2) Intercalibration of the dosimetry system
(3) Calibration of dosimetry instrumentation

(4) Calibration of the radiation facility at commissioning
(5) Continuous routine dosimetry
(6) Monitoring of the routine process dosimetry
(7) Radiation certificates.

A certificate in five languages stating the actual measured minimum absorbed radiation dose(s) for a specific product batch is available on request to every user of the radiation service, see Fig.4. For gamma sterilization, most health authorities accept such a certificate as sufficient proof of sterility, obviating the need for costly sterility testing.

8. INSPECTION OF THE RADIATION PLANT AND QUALITY CONTROL PROCEDURES

Iso-Ster encourages manufacturers, officials from the Department of Health and all other persons having a bona fide interest in radiation processing to inspect the operations and quality control procedures employed at its facilities.

REFERENCES

[1] SOUTH AFRICAN BUREAU OF STANDARDS, Code of Practice for the Industrial Sterilization of Medical Products. Part I. Gamma Radiation Facilities, Rep. SABS 0185, Part I, SABS, Pretoria (1982).
[2] ANTONIADES, M.T., "Dosimetry calibration", Proc. 2nd Gamma Processing Seminar, Ottawa, Atomic Energy of Canada Ltd, Technical Paper TP 13-01 (1980).
[3] BARRETT, J.H., Int. J. Appl. Radiat. Isot. 33 (1982) 1177.
[4] DOYLE, Y., "The red acrylic dosimetry system", Proc. 2nd Gamma Processing Seminar, Ottawa, Atomic Energy of Canada Ltd, Technical Paper TP 09-01 (1980).

IAEA-SM-272/36

DOSIMETRY PRACTICE FOR IRRADIATION OF THE MEDITERRANEAN FRUIT FLY
Ceratitis capitata (Wied.)

J.L. ZAVALA, M.M. FIERRO, A.J. SCHWARZ,
D.H. OROZCO, M. GUERRA
Dirección General de Sanidad Vegetal,
Secretaría de Agricultura
  y Recursos Hidráulicos,
Metapa, Chiapas, Mexico

Abstract

DOSIMETRY PRACTICE FOR IRRADIATION OF THE MEDITERRANEAN FRUIT FLY *Ceratitis capitata* (Wied.).
  In a sterile insect technique (SIT) programme the sterility of mass-reared insects, in our case Mediterranean fruit flies, is of primary importance. Mediterranean fruit fly pupae are irradiated in an AECL-CP-JS-7400 irradiator. Originally the capacity was 31 300 Ci, but because of the natural decay of cobalt, the actual source strength is 14 836 Ci. Thus, the dose with which the pupae are irradiated is 14.5 ± 1 krad (145 ± 10 Gy). A great risk in the daily release of sterile flies is that some batches of fertile flies may also be released. To ensure that this does not occur continuous dosimetric check-ups have to routinely be carried out. Fricke dosimetry is ideal for this purpose because it has a range of response to doses of 4 to 40 krad (40 to 400 Gy) and because it is an economic and simple dosimetric system.

1. INTRODUCTION

Because the sterile insect technique (SIT) is one of the main components of integrated pest control it has been widely researched.

In 1977, after detection of the Mediterranean fruit fly in southern Mexico, a large SIT programme was developed with the objective of eradication by producing and releasing 500 million sterilized medflies per week [1].

The daily release of a great number of sterile flies is of great risk if some fertile fly batches are dispersed [2]; thus, a reliable method to determine the irradiation dose has to be used. There are a variety of dosimetry methods, some of which are Perspex (red and clear), radiochromic dye film, lyoluminescence, Fricke ferrous sulphate, etc. However, because the latter has a range of 4 to 40 krad, it is the method most recommended.[1]

---

[1] 1 rad = 1.00×10^{-2} Gy.

The main objective of our work has been to apply Fricke dosimetry to the irradiation technique used in the Mediterranean Fruit Fly Laboratory in Metapa, Chiapas, Mexico, and included:

(1) Calculation of the irradiation dose by determining the molar extinction coefficient
(2) Determination of the dose distribution in the irradiation box
(3) Determination of the dose distribution in the plastic container in which the pupae are irradiated.

2. MATERIAL AND METHODS

2.1. Calculation of the factor used to determine the irradiation dose

The Perkin Elmer model 552 spectrophotometer calibration and the molar extinction coefficient calculation were used to convert the absorbance reading into dose.

The hexahydrated ferric chloride (Fe Cl_3-6 H_2O) master solution was used (17.36 × 10^{-4} mol/L concentration); 2, 4, 6, 8 and 10 mL were diluted to 100 mL using 0.8 N sulphuric acid (H_2SO_4) to obtain solutions with iron(Fe^{+++}) concentrations of 3.59, 7.18, 10.78, 14.37 and 17.36 × 10^{-5} mol/L.

The room temperature during the calibration readings was 23 to 26°C and the solution temperatures in the spectrophotometer were 28 ± 1°C. The temperatures were obtained using a YSI model 42 telethermometer, allowing 1 min between readings for the temperature to stabilize.

The molar extinction coefficient in the spectrophotometer was 2249 at 28°C; this was obtained by calculating the slope obtained after drawing the different absorbance reading dilutions against iron (Fe^{+++}) molarity (Fig.1, Table I).

The molar extinction coefficient obtained at 28°C was corrected to 2202 at 25°C. The equation used in Refs [3–5] was used

$$D = \left(\frac{\Delta A \cdot N_A}{\rho \cdot G \cdot \epsilon \cdot d}\right)\frac{b}{k}$$

where D   is the absorbed dose
$\Delta A = A_i - A_0$ is the change in absorbance at 305 nm and 25°C, and is dimensionless
N_A is Avogadro's number
ρ is the density of the dosimeter solution
G   is the G-value (radiation yield of chemical substance)
ϵ is the molar extinction coefficient (at 305 nm and 25°C)

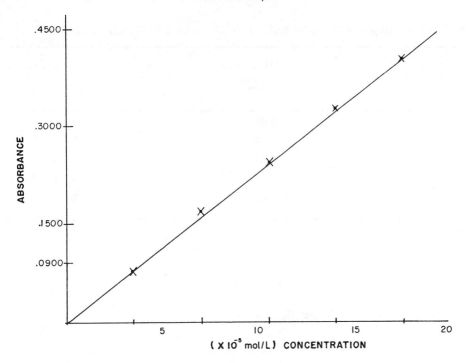

FIG.1. Calibration curve of the Perkin Elmer model 552 spectrophotometer.

d   is the optical path length (usually 1 cm)
b   is the energy conversion factor
k   is the volume conversion factor.

Substituting the values, we get

$$D = \frac{(6.022 \times 10^{23} \Delta A)}{1.024 \times 15.6 \times 10^{-2} \times 2202 \times 1} \times \frac{(1.602 \times 10^{-14})}{10^3}$$

$$D = (2.7425 \times 10^4) \Delta A$$

which, after correction at 25°C, is

$$D = \frac{(2.7425 \times 10^4)}{1 + 0.007 \, (T - 25°C)} \Delta A$$

The last formula was used to calculate the doses in the chemical dosimetry described in subsection 2.2.

TABLE I. ABSORBANCE VERSUS FERRIC ION (Fe^{+++}) MOLARITY

Fe^{+++} concentration $\times 10^{-5}$ mol/L	Absorbance readings	Temperature (°C)
3.59	0.0774	27.5
7.18	0.174	28.5
10.78	0.249	27.6
14.37	0.3298	28.5
17.36	0.4036	27.8

2.2. Dose distribution in the irradiation box

2.2.1. Preparation of dosimetric solution

The Fricke dosimetric solution was prepared in accordance with Ref.[3]; 0.392 g of ferrous ammonium sulphate ($Fe(NH_4)_2 SO_4 \cdot 6H_2O$) and 0.058 g of sodium chloride (NaCl) in 12.5 mL of 0.8 N sulphuric acid (H_2SO_4) were taken and then diluted to 1 L of 0.8 N sulphuric acid, which was prepared by dissolving 22.5 mL of sulphuric acid (1.84 g/cm³ density) in distilled water diluted to 1 L [3–5].

2.2.2. Dosimetry procedure to determine dose distribution in the empty irradiation box

The Fricke dosimetric solution was placed in glass vials with screw-top lids. Before closing, a piece of polyethylene was placed under the lid to prevent possible contamination.

The solution containers, henceforth called dosimeters, were placed on cardboard sheets (90 × 40 cm) which had previously been squared (10 cm²) so that a dosimeter was placed at every co-ordinate crossing.

All the sheets (55 for each of the three replicates) were placed in the centre of the empty aluminium boxes used for transport of the pupae-filled containers to the irradiator. Irradiation was done in an AECL-CP-JS-7400 irradiator containing a ^{60}Co source which had an initial power of 31 300 Ci; however, because of the natural decay of cobalt the actual strength is now 14 836 Ci (June 1984).[2] To determine the irradiation dose, once the dosimeters were

---

[2] 1 Ci = 3.70 × 10¹⁰ Bq.

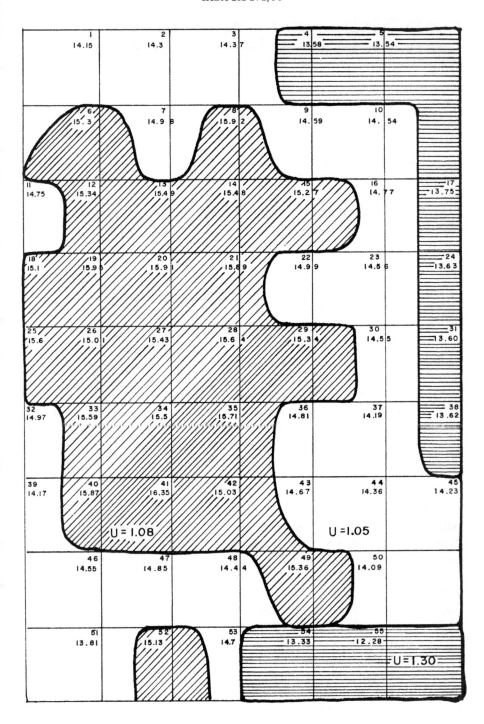

FIG.2. Dose distribution in aluminium boxes.

TABLE II. DOSE DISTRIBUTION IN THE ALUMINIUM BOX

	Mean dose (krad)	Dose uniformity ratio	Height (cm)
Upper	13.9	1.17	60–90
Medium	15.02	1.17	30–60
Lower	14.56	1.30	0–30

irradiated in the spectrophotometer, the optical density was obtained at an absorbance peak of 305 nm.

2.3. Dose distribution in plastic containers

The dosimetry procedures to determine dose distribution in the plastic containers are as follows.

The Fricke dosimetric solutions and dosimeters were prepared as described in subsections 2.2.1 and 2.2.2. Dosimeters were placed on the outer and inner surfaces of the containers at the four cardinal points and also in the centre; dosimeters in the containers filled with pupae were only placed at the northern and southern points. Only two levels were monitored. Four containers were used: container Nos 1 and 2 were irradiated at 13.5 krad, container No.3 at 12.5 krad and container No.4 at 10.5 krad.

3. RESULTS

Dosimetry in the empty box showed the following results (average of three replicates). By dividing the cardboard sheets into three levels (every 30 cm) the dose uniformity ratio was 1.30, 1.17 and 1.17 at the low, medium and upper levels, respectively (Fig.2, Table II). If the sheet is divided according to isodoses, the dose uniformity ratio is 1.08 for an isodose higher than 15 krad, 1.05 for a 14 to 15 krad isodose and 1.09 for an isodose below 14 krad (Fig.2, Table III). The dose uniformity ratio of the empty box was 1.30, with an average dose of 14.78 krad.

Dosimetry in the plastic container (with pupae) showed the following results (Table IV). For containers Nos 1 and 2 (13.5 krad expected dose) there was a difference between the levels of 0.92 and 0.88%, respectively, in relation to the dose uniformity ratio, with an average dose of 13.89 and 13.04 krad, respectively. Container No.3 (12.5 krad expected dose) showed a 2.58% difference between levels in the dose uniformity ratio and an average dose of 12.84 krad. Container

TABLE III. ISODOSES IN THE ALUMINIUM BOX

Isodose rate (krad)	Mean dose (krad)	Dose uniformity ratio
>15	15.53	1.08
14–15	14.52	1.05
<14	13.49	1.09

TABLE IV. DOSE UNIFORMITY RATIO (U) IN THE PLASTIC CONTAINERS

Plastic container No.	Level A (U)	Level B (U)	% difference (U)	Expected dose (krad)	Mean dose (krad)
1	1.08	1.07	0.92	13.5	13.89
2	1.13	1.12	0.88	13.5	13.04
3	1.16	1.13	2.58	12.5	12.80
4	1.20	1.18	1.66	10.5	10.68

No.4 (10.5 krad expected dose) showed a 1.76% difference between levels with respect to the dose uniformity ratio and an average dose of 10.68 krad.

4. CONCLUSIONS

Based on the results it can be concluded that:

(1) Because of minor differences in dose uniformity ratios, aluminium boxes can be utilized to irradiate pupae.
(2) The best position to irradiate biological material in the box is at the medium-level west side; this is because the dose uniformity ratio at this level and the isodose show the lowest differences.
(3) Even though the dose distribution and average dose can be determined by means of chemical dosimetry, it is necessary to combine this practice with biological dosimetry [6], which shows insect behaviour and, based on the fertility-sterility results, determines the final dose to be used (Table V). This not only reduces damage to the insect but also enhances the success of SIT.

TABLE V. COMPARISON BETWEEN MEAN DOSE AND PERCENTAGE STERILITY FROM DIFFERENT AGES OF MALES AND FEMALES

Mean dose (krad)	Age (d)	% sterility Males	Females
13.89	−1	99.61	100
13.04	−2	98.90	100
12.80	−2	98.76	100
10.68	−2	98.00	100

ACKNOWLEDGEMENT

The authors wish to thank J. Toledo for his help in obtaining the fertility-sterility results.

REFERENCES

[1] PATTON, T.P., "Mediterranean fruit fly eradication trial in Mexico", Proc. Symp. XIV Int. Congr. on Fruit Fly Problems, Kyoto and Naha, 1980.
[2] SCHWARZ, A.J., et al., Mass production of the mediterranean fruit fly in Metapa, Mexico (in preparation).
[3] INTERNATIONAL ATOMIC ENERGY AGENCY, Manual of Food Irradiation Dosimetry, Technical Reports Series No.178, IAEA, Vienna (1977) 30.
[4] CSERÉP, G., et al., Chemical Dosimetry Course on Laboratory Aid, Institute of Isotopes, Hungarian Academy of Sciences, Budapest (1971).
[5] McLAUGHLIN, W.L., Lecture on the Fundamentals of Dosimetry, IAEA Seminar on High-Dose Dosimetry in Industrial Radiation Processing, Risø National Laboratory, Roskilde (1982).
[6] OHINATA, K., et al., Mediterranean fruit flies: Sterility and sexual competitiveness in the laboratory after treatment with gamma irradiation in air, carbon dioxide, helium, nitrogen or partial vacuum, J. Econ. Entomol. **70** 2 (1977) 165.

STATUS AND PROSPECTS OF HIGH-DOSE METROLOGY IN THE PHILIPPINES

E.M. VALDEZCO, E.G. CABALFIN,
L.M. ASCAÑO, C.C. SINGSON,
Q.O. NAVARRO
Philippine Atomic Energy Commission,
Diliman, Quezon City,
Philippines

Abstract

STATUS AND PROSPECTS OF HIGH-DOSE METROLOGY IN THE PHILIPPINES.
　The Philippines has embarked on large-scale application of radiation processing with the construction of the first Philippine multi-purpose irradiation facility. Initially, this facility will be used for food preservation and will later be expanded to include other applications. The dosimetry activities carried out before commissioning of this facility are presented, including a critical review of various dosimetry studies being conducted locally using different chemical systems designed for high-dose measurements. The results of studies on the dose response characteristics of ferrous sulphate, modified Fricke, and dyed and undyed polymethyl methacrylate dosimeters are presented. Initial efforts to establish a dose quality assurance programme for the facility are described. The problems encountered and the prospects of high-dose metrology relative to the expansion programme of the facility to include applications other than food irradiation are outlined.

1. INTRODUCTION

　Remarkable growth and progress have been achieved in the use of large radiation sources for industrial processing over the last 20 years. This progress has resulted in a significant increase in radiation applications in industry. Radiation applications in established areas such as the polymer industry, the wire and cable industry, the rubber industry, the manufacture of disposable medical supplies, and the production and preservation of food continue to grow very rapidly.
　In the Philippines, however, there has been and still is some skepticism on the industrial application of radiation processing. Undoubtedly, the idea of large radiation sources as vital components of industrial endeavours is anathema to many industrialists. The main reasons for this attitude may be the uncertainties in calculation of the price of the product, fear of the potential hazards of radiation, or merely the considerable red tape that has to be overcome before an irradiation

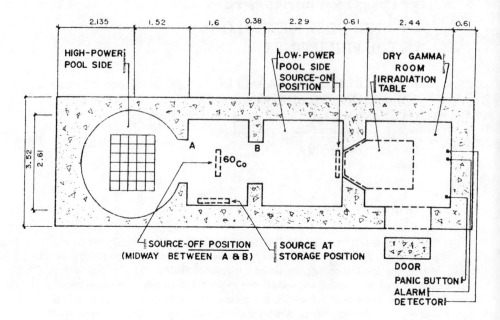

FIG.1. *Reactor pool plan view.*

unit is installed. Nevertheless, it is unrealistic to believe that radiation technology has no industrial future in the Philippines.

One of the main problems is the lack of an irradiation facility on a pilot-plant scale. The Philippine Atomic Energy Commission, therefore, embarked on a project to build a multi-purpose demonstration irradiation facility at its compound in Diliman, Quezon City. The main component of this facility, the irradiator, is supplied by the Atomic Energy of Canada Limited, through a technical assistance programme of the International Atomic Energy Agency.

When completed the irradiation plant will serve as a common facility for semi-pilot and pilot-scale studies to determine the economic viability of radiation sterilization of medical products and food irradiation. It will likewise be used to develop other applications of radiation processing.

On the other hand, application of dosimetry as a means of industrial radiation quality control has grown considerably because of increasing demands from users of large radiation sources.

This paper summarizes the results of activities related to the measurement of radiation absorbed dose in the high-dose range. The irradiation facility utilized for these studies is a one-sided ^{60}Co source plaque situated in the dry gamma room of the Philippine Research Reactor (PRR-1) of the Philippine Atomic Energy Commission. A Gammacell unit was used to irradiate dosimeters in conducting linearity tests and fading experiments.

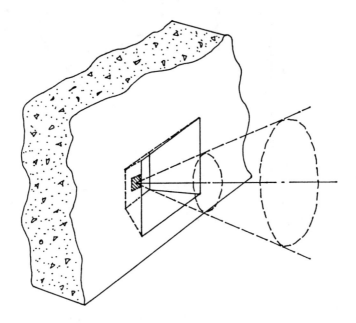

FIG.2. Dry gamma room window.

It is hoped that with the experience gained in running these experiments we will be able to perform routine dosimetry measurements in the new irradiation plant with a high level of confidence and to arrive at a decision on which type of dosimeter to use for dose quality assurance measurements during commercial operation of the irradiation plant.

2. DESCRIPTION OF IRRADIATION FACILITIES

2.1. Dry gamma room

The 1.94×10^{14} Bq ^{60}Co source is located in the PRR-1 pool. The source is made up of six ^{60}Co rods (1.25 cm in diameter and 32 cm in length) arranged parallel to each other at equal distances on a plane. It is mounted in a bracket which is provided with a hook catcher near the top of the ^{60}Co rods.

Irradiations are conducted in the dry gamma room adjacent to the low power pool section (Fig.1). The room measures approximately $2.4 \times 2.6 \times 3.0$ m. The window through which the ^{60}Co gamma rays pass flares out from about 2 ft² from the water side to 4 ft² at the dry gamma room side (Fig.2).[1]

---

[1] 1 ft² = 9.290×10^{-2} m².

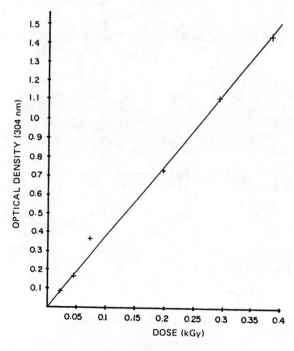

FIG.3. Fricke calibration curve.

For bulk irradiations, a rectangular work table measuring 2.4 m long by 1.2 m wide and 0.9 m from the floor was set up in the dry gamma room adjacent to the window. The effective areas for sample positioning are indicated on the table by dose rate curves which are calibrated periodically using Fricke dosimeters.

2.2. Gammacell irradiation unit

The Gammacell-220, a self-contained high intensity ^{60}Co irradiator, is equipped with a hollow, thin-walled cylindrical aluminium chamber measuring 15 cm in diameter and 20.6 cm in height. As of September 1984 the ^{60}Co source in the form of rods has an activity of 5.2×10^{13} Bq.

3. FRICKE DOSIMETRY STUDIES

The ferrous sulphate solution (Fricke dosimeter) has been used as a reference measurement system for the calibration and routine determination of the dose effect response of various dosimeters. At an absorption maximum of 304 nm using a Beckman Model 25 spectrophotometer with the solution held in silica window

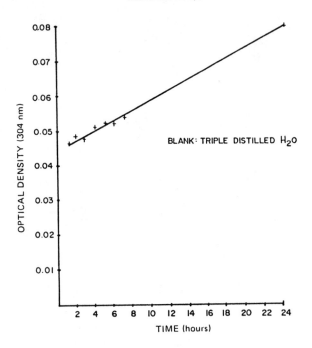

FIG.4. *Stability of modified Fricke solution.*

cuvettes, the molar extinction coefficient, ϵ_m(Fe III), was found to be 2206 L·mol^{-1}·cm^{-1} at 25°C.

The standard solution was irradiated in polyethylene sample vials using the Gammacell-220. The dose response is linear from 20 to 400 Gy, as shown in Fig.3.

For the modified Fricke, the ferrous sulphate and cupric sulphate solutions were prepared separately and freshly mixed just before irradiation. The stability of the dosimeter solution for the first 24 h was determined and was found to be linear, as shown in Fig.4. The dose response curve of the modified Fricke dosimeter is shown in Fig.5. More test runs are being conducted to determine temperature dependence, dose rate dependence, etc., which might affect the reproducibility of results.

4. POLYMETHYL METHACRYLATE (PMMA) DOSIMETRY STUDIES

Figures 6 and 7 show the optical density versus dose relationships of clear HX Perspex (1 mm thickness) and red (4034J) Perspex (3 mm thickness), respectively. These dosimeters were irradiated using the Gammacell-220 and read at 305 nm (for clear) and 640 nm (for red) using a Perkin Elmer 550 UV-Vis spectrophotometer.

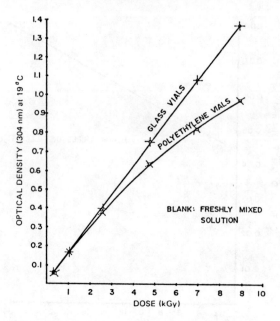

FIG.5. Dose response curve for modified Fricke solution.

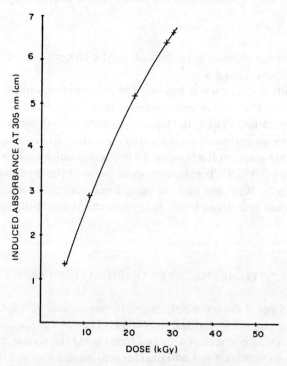

FIG.6. Calibration curve for clear Perspex HX (1 mm).

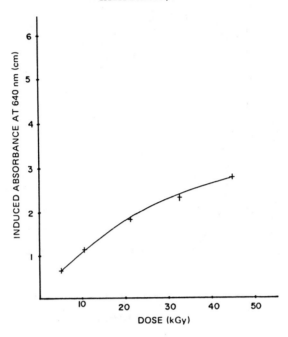

FIG. 7. Calibration curve for red Perspex 4034 (3 mm).

Fading tests were conducted for about 3 weeks for shorter time intervals and followed through until 100 days after irradiation; the results are shown in Figs 8 and 9.

5. DOSE DISTRIBUTION STUDIES

Dosimetry measurements using red Perspex 4034 were done on both dummy boxes and product boxes containing scalp vein sets. The latter were made of polyvinyl chloride with polyethylene double packaging, while dummy boxes contained styrofoam, simulating the product density of 0.1042 g/cm^3. The dimensions of the irradiation boxes were 48 × 20 × 15 cm.

The samples were irradiated to a dose of 25 kGy in the dry gamma room; the arrangement of the boxes is shown in Fig.10. Dosimeters were arranged in the boxes as shown in Fig.11.

Initially, three different conditions were tried during irradiation: (a) irradiation without turning, (b) irradiation with horizontal turning at half the irradiation time, and (c) irradiation with horizontal and vertical turning at half the irradiation time.

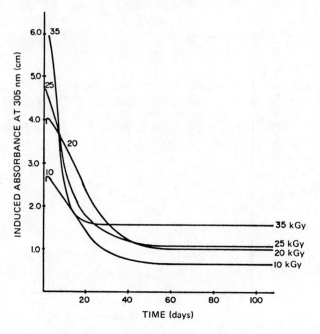

FIG.8. Fading of clear Perspex HX (1 mm).

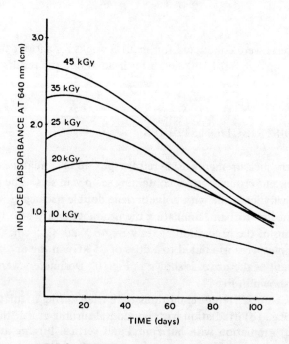

FIG.9. Fading of red Perspex 4034 (3 mm).

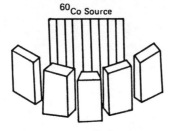

FIG.10. Position of boxes during irradiation in dry gamma room.

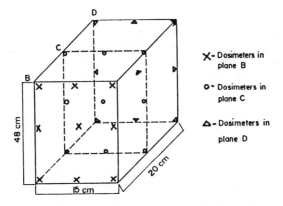

FIG.11. Box showing positions of dosimeters for depth-dose distribution (not drawn to scale).

Turning the boxes vertically and horizontally showed the absorbed dose to be more uniformly distributed, as shown in Figs 12 and 13. Likewise, the uniformity ratio went down from 3.12 to 1.13. The relative dose distribution in the irradiation boxes shown in Fig.14 indicates that radiation intensity is greater at the left side of the dry gamma room window. The uniformity ratio in the boxes ranged from 1.13 to 1.28.

The minimum absorbed dose was found to be at the top of the box along the midsection plane, while the maximum dose was found at the centre of the same midsection plane. Isodose curves along this midsection plane were determined by placing 72 dosimeters on cardboard which was positioned along this plane (Fig.15). The results of such determinations are shown in Fig.16.

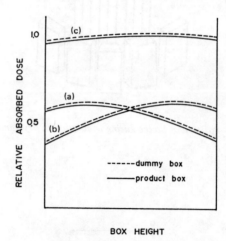

FIG.12. Dose distribution along the height of an irradiation box. (a) and (b) are relative absorbed doses during the first and second half of irradiation time, respectively; (c) is relative absorbed dose for full irradiation time.

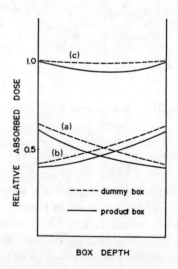

FIG.13. Dose distribution along the depth of an irradiation box. (a) and (b) are relative absorbed doses during the first and second half of irradiation time, respectively; (c) is relative absorbed dose for full irradiation time.

IAEA-SM-272/7

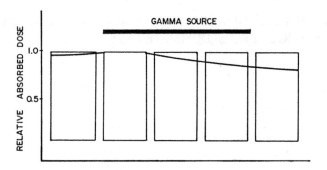

FIG.14. Relative dose distribution in irradiation boxes.

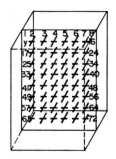

FIG.15. Dummy box showing positions of dosimeters for determination of isodose curves.

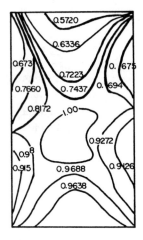

FIG.16. Isodose curves on a vertical plane parallel to the source plaque through the middle of a dummy product box. Results are given relative to maximum dose values.

TABLE I. RESULTS OF AAEC INTERCOMPARISON MEASUREMENTS WITH THE CERIC-CEROUS DOSIMETER SYSTEM

	Institute (Philippines)
Replicates	5, 6, 2
Nominated dose (ND)	20
Irradiation date(s)	17–27 July 1981
Irradiation time (h)	69.67

Sample position	Dose (kGy)			Mean
	5	6	2	
1	19.2	22.0	19.7	20.3
2	19.7	20.6	20.2	20.2
3	20.1	20.6	20.7	20.5
4	20.3	20.9	23.6	20.6
5	20.4	21.0	22.0	21.1
6	20.5	21.3	20.6	20.8
Mean	20.0	21.1	20.6	20.6
(SD)	(0.49)	(0.53)	(0.86)	(0.55)
ND	1.000	1.055	1.030	1.030
Dose rate ($kGy \cdot h^{-1}$)				0.296

Analysis of variance for Institute replicates:

Source of variation	df	SS	MS	F
Between replicates	2	4.5643	2.2822	2.66[ns]
Within replicates	15	12.8602	0.85735	
Total	17	17.4245		
Between positions	5	4.4844	0.89688	0.83[ns]
Within positions	12	12.9401	1.07834	
	17	17.4245		

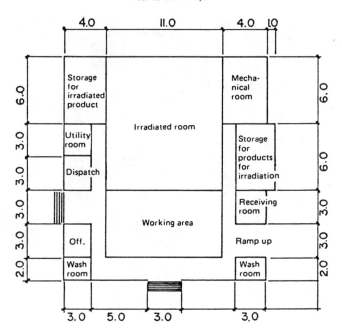

FIG.17. *Layout of facility (dimensions in metres).*

6. INTERCOMPARISON MEASUREMENTS

In view of the worldwide exchange of products arising from radiation processing, intercomparison programmes on dose measurements are organized to give the processor added confidence in his dosimetry technique.

The Philippines has participated in a programme organized by the Australian Atomic Energy Commission (AAEC) under IAEA Technical Contract 2821/TC on the Calibration of Some Asian and Pacific Radiation Facilities Used for Investigating Sterilization of Medical Products.

Ceric-cerous dosimeters prepared at the AAEC were forwarded to participating laboratories where three replicates, each of six samples, were irradiated at a dose of 20 kGy. The irradiated dosimeters and one unirradiated replicate were returned to the AAEC for measurement and comparison with an untravelled replicate.

The ceric-cerous dosimeters were irradiated in the dry gamma room irradiation facility at the required total dose of 20 kGy. The results are shown in Table I and are in good agreement (to within ±3%) with the required nominal dose.

7. FUTURE PLANS

The general layout of the new irradiation facility is shown in Fig.17. In addition to the irradiation cell, a mechanical room and storage rooms for products

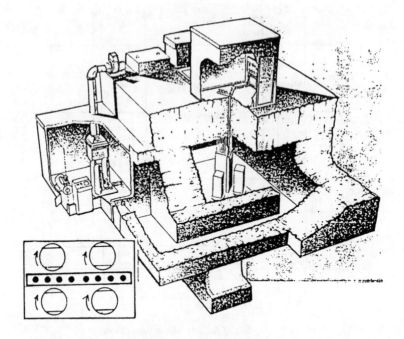

FIG.18. *Cut-out view of irradiator.*

before and after irradiation are included. The building has a total floor area of around 400 m^2.

The irradiation room is designed for a maximum ^{60}Co source activity of 7.5 × 10^{15} Bq. The design of the Gammabeam-651 PT (Figs 18 and 19), a batch-type irradiator, is proprietary to the AECL and includes the following components: ^{60}Co source and source mechanism, product turntables, monitoring devices, control console, and safety features.

Initially, the irradiator will be loaded with a 1 × 10^{15} Bq ^{60}Co source in three pencils installed in three source assemblies. The remaining source assemblies would provide capacity for additional source replenishment in the future.

Before commissioning, parallel experimental runs will be conducted using the same dosimeter systems to verify and/or determine additional parameters relevant to dosimetry. In addition, use of ethanol chlorobenzene as a routine dosimeter is being seriously considered in view of recent publications on the practicability and suitability of the dosimeter for the dose range covering food irradiation and medical sterilization applications.

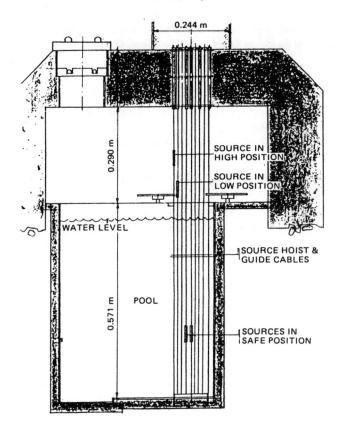

FIG.19. Cross-section of irradiation room.

ACKNOWLEDGEMENTS

The technical assistance of C. Carmona, who did the experimental runs using red and clear Perspex, is gratefully acknowledged. Likewise, the authors would like to thank L. Lopez, I. Pasion and D. Salom for their valuable assistance. Thanks are also due to A. del Rosario for her untiring effort in the preparation of the manuscript.

BIBLIOGRAPHY

CHADWICK, K.H., "Facility calibration, the commissioning of a process and routine monitoring practices", Radiosterilization of Medical Products (Proc. Symp. Bombay, 1974), IAEA, Vienna (1975) 69.

CHARLESBY, A., "Future prospects of industrial radiation processing", Industrial Application of Radioisotopes and Radiation Technology (Proc. Conf. Grenoble, 1981), IAEA, Vienna (1982) 105–115.

CHU, R.D.H., et al., "Use of ceric sulphate and Perspex dosimeters for the calibration of irradiation facilities", Radiosterilization of Medical Products (Proc. Symp. Bombay, 1974), IAEA, Vienna (1975) 83.

ETTINGER, K.V., et al., "Progress in high-dose radiation dosimetry", Biomedical Dosimetry: Physical Aspects, Instrumentation, Calibration (Proc. Symp. Paris, 1980), IAEA, Vienna (1981) 405–432.

INTERNATIONAL ATOMIC ENERGY AGENCY, High-Dose Measurements in Industrial Radiation Processing, Technical Reports Series No.205, IAEA, Vienna (1981).

INTERNATIONAL ATOMIC ENERGY AGENCY, Manual of Food Irradiation Dosimetry, Technical Reports Series No.178, IAEA, Vienna (1977).

NAM, J.W., Report on an IAEA Advisory Group Meeting on High-Dose Pilot Intercomparison, Food Irradiat. Newsl. 5 3 (1981) 16.

STATISTICAL AND METROLOGICAL ASPECTS OF 20 YEARS' EXPERIENCE OF RADIATION PROCESSING IN POLAND

P.P. PANTA, Z. BUŁHAK
Institute of Nuclear Chemistry
  and Technology*,
Warsaw, Poland

Abstract

STATISTICAL AND METROLOGICAL ASPECTS OF 20 YEARS' EXPERIENCE OF RADIATION PROCESSING IN POLAND.

A survey of work done in the field of radiation processing and high-dose dosimetry performed at the Institute of Nuclear Chemistry and Technology, Warsaw, is given in view of the worldwide trends that have taken place during the last 20 years. The status of radiation processing on linac LAE-13/9 at the Institute and an analysis of its technological capacity and operational problems are given. The metrological aspects of some commonly used dosimetric systems for routine and absolute uses are discussed. Some routine dosimetric systems are currently being tested and proposed for polymer and semiconductor radiation modification control. Attention is called to the statistical aspects of unexpected and very rare events connected with the unserviceability of the conveyor and scanning system which involved overdosing during irradiations. Some results of semiconductor irradiation illustrate the level of precision of radiation processing with reference to the reproducibility of electrical parameters depending on the level of industrial semiconductor technology.

1. INTRODUCTION

Worldwide radiation processing and high-dose radiation dosimetry have advanced from a euphoric beginning to the current mature stage [1–4] (Fig.1). Electron beam processing is now used by many industries for the cross-linking of plastic films, production of tubing or cable insulation, food preservation, rubber vulcanization, curing of coatings, semiconductor modifications, sterilization of medical supplies, etc.

The status of radiation facilities in Poland is as follows (Table I):

(1) One multipurpose semicommercial electron linac (LAE-13/9)
(2) One commercial electron accelerator (IŁU-6) for polymer modification
(3) One electron linac (LUE-6) mainly for research

---

* Formerly the Institute of Nuclear Research.

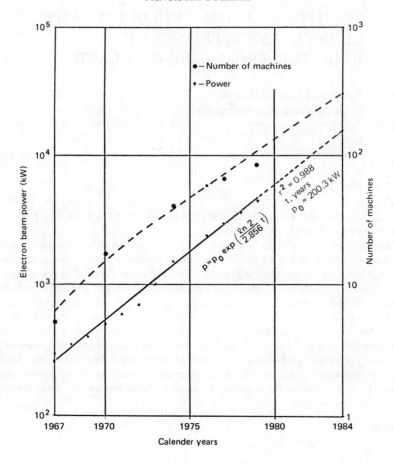

FIG.1. *Cumulative capacity of electron accelerators (only 46 countries).*

(4) A low energy accelerator under construction (EAK-400/100, ca.400 keV and 100 mA)
(5) Two laboratory-type ^{60}Co sources.

In 1985 a new multipurpose commercial-type electron accelerator (IŁU-6) will be installed at the Institute of Nuclear Chemistry and Technology (INCT) (Żerań centre, Warsaw). The establishment of this facility will definitely accelerate the development of radiation technology for the industry. The International Atomic Energy Agency is assisting the project through the supply of an accelerator and auxiliary equipment.

General conclusions based on an analysis of up-to-date experimental, pilot and full production radiation processing for industrial and medical applications are reported. Representative examples of radiation processing are:

TABLE I. MAIN RADIATION FACILITIES IN POLAND

Type	Manufacturer	Energy (MeV)	Power (kW)	Location
Semicommercial electron linac (LAE-13/9)	USSR (1971)	10 to 13	9	Warsaw (Żerań centre)
Commercial electron linac (IŁU-6)	USSR (1983)	0.5 to 2	20	Człuchów (north Poland)
IŁU-6 to be installed	USSR (1986)	0.5 to 2	20	Warsaw
Linac for research (LUE-6)	USSR (1983)	7 to 8	5	Łódź
Low energy linac under construction	Poland (1986)	0.4	40	Świerk (near Warsaw)
Laboratory gamma ray source	Poland (1967)	20 kCi (^{60}Co)		Łódź
Laboratory gamma ray source	USSR (1976)	16 kCi (^{60}Co)		Warsaw (Żerań centre)

(a) Radiosterilization of medical equipment and supplies and also, on a lesser scale, biological tissues (human bone grafts, cartilages, skin and aortic valves) [5–10].
(b) Cross-linking of polymers, mainly polyethylene in the form of tubes and tapes for heat-shrinkable products [11–14].
(c) Radiation modification of the dynamic parameters of semiconductor devices [15] such as thyristors and diodes, and also some studies of the radiation stability of integrated circuits [16–18].

Besides electron beam curing of surface coatings [19], food preservation [20] and radiation vulcanization of natural rubber latex [21] are also carried out but only on a small scale and only according to actual needs (Table II).

2. EXPERIMENTAL PROCEDURES

Radiosterilization of biological tissues, e.g. lyophilized human bone transplants, has been performed since 1964 using a core of the shutdown reactor EWA-10 [5, 6]

TABLE II. ESTABLISHED RADIATION TREATMENT BASED ON LINAC LAE-13/9 AT THE ŻERAŃ CENTRE

Radiation-induced process	Product	Remarks
Sterilization of biological tissues	Human bone grafts, cartilages and aortic valves	Spent fuel and ^{60}Co (1963 to 1971)
Sterilization of medical supplies and some biological tissues	PVC catheters, surgical disposable gloves, small containers for microbiological containers, disposable urine drainage bags and some bone and skin grafts	Electron beam (after 1972)
Cross-linking of polymer chains of polyethylene to impart heat-shrinking properties and increased heat resistance	Polyethylene pipes, tubes, cables, films and foils	Electron beam (after 1974)
Modification of electrical properties of semiconductors	Solid-state silicon diodes, thyristors and integrated circuits	Electron beam (after 1975)

and laboratory ^{60}Co sources (Gammacell-220 and over the next few years a 20 000 Ci source at Łódź).[1] For example, during 3 years of introductory activities over 6000 bone grafts have been sterilized at a dose of 3.3 ± 20% using gamma rays. All the grafts were found to be sterile according to bacteriological tests (*Bacillus subtilis*) and to be useful clinically [22, 23]. Some human and animal aortic valves were radiosterilized at very high gamma dose rates of up to 0.8 MGy/h (80 Mrad/h) [8] — then very close to the dose rate of an electron linac at the 5 to 10 kW power level (Fig.2).

The installation in 1972 of a Soviet manufactured 13 MeV/9 kW electron linac (LAE-13/9) [24] at INCT created a basis for industrial radiosterilization, polymer cross-linking and modification of the lifetime of minority carriers in solid-state silicon devices.

It is worth while adding that this work has reached the saturation level as it is limited by the capacity of LAE-13/9 and some research needs. Since December 1972 our plant has operated 12 hours a day (two shifts) and at least 5 days a week (approximately 2300 h annually), with the exception of planned shutdowns for maintenance and holiday periods. In almost 12 years our linac has been able to

---

[1] 1 Ci = 3.70 × 10^{10} Bq.

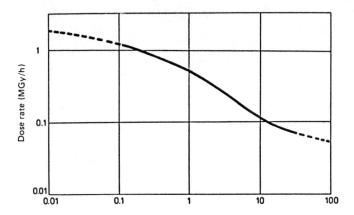

FIG.2. Gamma dose rate versus time after the EWA-10 reactor shutdown (core after 78 h at 7.2 MW).

operate at a utilization efficiency exceeding the 95% level. The irradiation capacity is shown in Table III and Fig.3.

The Department of Radiation Chemistry and Technology at INCT co-operates with many medical and industrial centres which send them products selected for radiation processing. For example, the department in co-operation with the ENERGOKABEL Research and Development Centre of Cable Industry developed and implemented in the Process Equipment Works at Człuchów full-scale processing of polyethylene thermo-shrinkable pipes using the electron linac IŁU-6 [14]. The production method and relating process equipment are covered by a Polish patent [25].

3. DOSIMETRIC AND METROLOGICAL ASPECTS

Traditional rigid thin PVC film (0.26 mm), produced by Kunststoffwerke Staufen, Federal Republic of Germany [9], is used routinely for absorbed dose control. We do not use radiochromic film dosimeters [26–32] even though they are commercially available because they are very expensive, are sensitive to UV light, etc. Nevertheless, some experiments with carefully synthesized leucocyanids of malachite green (MGCN) were successfully performed. At a MGCN concentration of 10% (in PVC matrix) even the linear relationship between dose and absorbance has been obtained [33]. Absolute PVC dosimeter calibration for the electron beam is carried out by means of a quasi-adiabatic water calorimeter with a thermistor temperature sensor and styrofoam thermal insulation (Fig.4). Such a simple water calorimeter was originally reported by Taimuty [34] in the 1950s and was improved and popularized in the 1960s by the team at the Risø National Laboratory, Roskilde, Denmark [35].

TABLE III. IRRADIATION CAPACITY[a] OF THE LINAC LAE-13/9 AT THE ŻERAŃ CENTRE

Products	Throughput (annual)	Dose (kGy)	Remarks
Catheters	950 000	28	–
Microbiological containers	2 500 000	35	–
Surgical gloves	200 000	35	–
Urine drainage bags	300 000	28	–
Bone and skin grafts	1 000	30	Sterilized in frozen state
Silicon diode structures, diodes and thyristors	350 000 to 500 000	Various	Throughput depends on structure size

[a] Total of up to 250 t or over 2 million US $ per year; for comparison, the commercial linac IŁU-6 enables irradiation (dose 150 kGY) of a total of 450 t of polyethylene pipes per year.

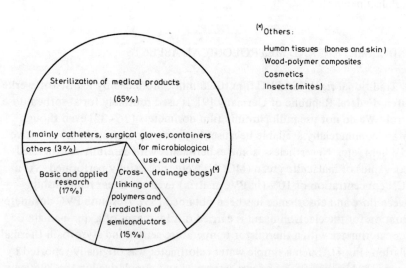

FIG.3. Radiation processing at the Żerań centre.

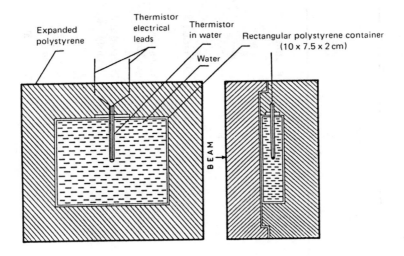

FIG.4. Cross-section of the quasi-adiabatic water calorimeter for electron beam dosimetry.

Analogically to other simple systems, such as the Fricke dosimeter, its operation can reveal some unexpected effects; these were recently discussed in detail by Fletcher [36] and Domen [37, 38].

The problems are connected with water alone (treated as a calorimetric body) and radiation 'water-like' polystyrene, which is commonly used as insulating material. Sometimes a modified calorimeter consisting of polystyrene discs embedded in water is proposed for reducing a heat defect below 1% [39, 40]. The precision of the water calorimeter is relatively good, about ±2%, but its overall accuracy is not better than ±8%, caused by systematic error. Thus, there exists some observable discrepancy between the response of the water calorimeter and the Fricke dosimeter treated as a secondary standard. This has been confirmed by some authors who suggest that certain radiation-induced reactions inside the water calorimeter result in up to 10% error [36]; however, this value seems to be significantly overestimated. On the other hand, a higher temperature rise of polystyrene thermal insulation can involve a small heat transfer to the water inside the calorimeter vessel and be responsible for an increase in dose readout. These effects reduce the water calorimeter accuracy under conditions of dose control during large-scale irradiations. For this reason improved calibration was done by using a graphite calorimeter with the same geometry and size as those of the water calorimeter (Table IV).

The graphite calorimeter is more sensitive, but needs a small correction owing to changes of specific heat as a function of temperature, which is practically negligible in water.

TABLE IV. SOME COMMON DOSIMETRIC SYSTEMS USED FOR RADIATION PROCESSING AT THE LAE-13/9 LINAC

System	Method of analysis	Approximate dose range (Gy)	Linearity correlation coefficient (r^2)	Precision (%)	Overall accuracy (%)	Remarks
PVC film	UV spectrophotometry (394 nm)	5×10^3 to 5×10^5	0.966	6	12	Routine
Water calorimeter	Resistance measurement	10^4 to 5×10^5	0.998	2	8	Calibration
Graphite calorimeter	Resistance measurement	10^3 to 10^5	0.99	2	5	Calibration
Ferrous sulphate	UV spectrophotometry (302 nm)	up to 4×10^2	0.999	2	3 to 5	Secondary standard
Polyethylene shrinkage after blow moulding	Relative change in length (calliper gauge)	3×10^4 to 4×10^5	0.993	5	10	Proposed and routinely tested
Silicon diode structures	Voltage drop measurement	10^3 to 10^4	0.99	5	10	Proposed and short-term tested
MGCN 10% in PVC matrix	VIS spectrophotometry (630 nm)	2×10^4 to 1.2×10^5	0.991	4	8	Sporadically used, low stability

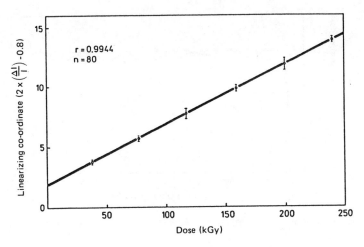

FIG.5. *Polyethylene 'shrink' dosimeter* [41].

Metrologically there are two necessary levels of accuracy of absorbed dose distribution during radiation processing:

(1) For massive processing (sterilization and cross-linking) with dose limits from ±10% up to ±30%
(2) For finer irradiations (bioproducts and semiconductors) with narrower uncertainty limits, i.e. better than ±10%.

It is the opinion of various authors that a routinely used PVC dosimeter enables an overall accuracy from about ±6% to ±12% [41]. For this reason, according to a generally accepted postulate of matching dosimeter material to an irradiated medium, some new dosimeter systems have been found and treated.

The first dosimeter consists of a polyethylene pipe segment in which the degree of shrinking after heat treatment is related to the absorbed dose degree of polyethylene cross-linking [41]. Practical performance comprises the following:

(a) Irradiation of the polyethylene target together with the dosimetric segment of polyethylene pipe (previously suitably blow moulded)
(b) Heating of the dosimetric segment to 150°C and then cooling to room temperature
(c) Measurement of the change of length of the dosimetric segment and determination of the absorbed dose from the calibration curve.

It should be emphasized that each new batch of the polyethylene material needs to be calibrated (by the calorimeter) separately. The calibration plot is a linearly fitted relationship between the absorbed dose and a quotient of length difference to initiate the length of the segment of pipe to a power of -0.8 (Fig.5).

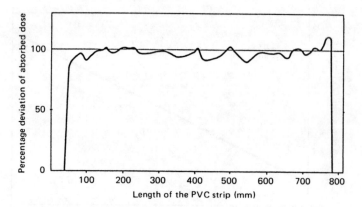

FIG.6. *Typical example of electron dose inhomogeneities along the diagonal of the irradiation container.*

A typical value of the coefficient for determination of the calibration plot lies between 0.990 and 0.994 and one can assume a relatively good fit (better than PVC film). An average accuracy is about ±5% within 20 to 250 kGy (2 to 25 Mrad) and approximately ±10% within 250 to 400 kGy (25 to 40 Mrad) [42]. This dosimeter has only batch-to-batch reproducibility.

The second new dosimeter, suitable for dose controlling during radiation modifications of semiconductors, consists of wafers made of high purity silicon monocrystals irradiated with the silicon dosimeters. Previously, an absorbance difference in the IR region was measured for controlling the density of radiation-induced recombination centres as a function of absorbed dose. However, spectrophotometric measurement needs a rather large wafer diameter (about 30 to 40 mm), which is very fragile and, moreover, requires surface polishing for the reduction of light loss due to reflection. Thus, improved versions of the silicon dosimeter only use small wafers (diameter of 5 to 7 mm) in which the absorbed dose is determined by measuring the voltage drop in a forward direction, which depends on the density of radiation-induced recombination centres in n-Si. Analogically to the polyethylene dosimeter, the silicon dosimeter reveals only the metrological batch-to-batch reproducibility of readouts. It is ready for re-use after annealing at 400 to 450°C for about 20 minutes. However, little experience has been gained with this dosimeter [43].

Independent of the dosimeters used, the electron beam over the horizontal conveyor is currently controlled by secondary emission of the beam current monitor, thus enabling a stable exposure dose to be obtained, usually from ±7 to ±8%. Such an absorbed dose level is easily attainable for thin, surface-type targets, for example thin plastic foils and films. However, non-uniformity of the absorbed dose inside the thick layers of irradiated materials can be up to ±25 to ±30%, as reported by Osipov [44]. Such extreme inhomogeneities of dose can

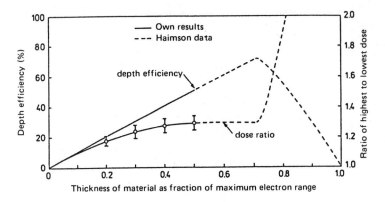

FIG.7. *Electron depth-dose efficiency characteristics.*

appear during irradiation of multilayer packages of PVC catheters or thick-walled polyethylene pipes and cables (Figs 6 and 7).

4. SOME STATISTICAL ASPECTS

During 12 years of routine linac irradiations some very rare and unexpected events have occurred.

The first event was related to the conveyor system and was caused by many years of friction with the blocking jib, with difficult accessibility. As a result, an erroneous readout of conveyor velocity appeared, which was responsible for a number of medical supplies being overdosed.

Other events related to overdosing were caused by the unserviceability of the beam scanning subsystem. This kind of incident occurs very seldom, i.e. less than one in 10 000 irradiation cases. However, the most dangerous incident occurred only once. It involved the self-ignition of a small sample of radiation-polymerized styrene; fortunately it caused little linac damage, i.e. one in about 5×10^5 irradiations. This self-ignition was caused by sparking as a result of space-charge accumulation and relaxation in styrene irradiated with high energy electrons.

On the other hand, about 0.5% of radiation-treated materials were only partly subirradiated because of an imperfection in the conveyor system (mutual friction between containers and the conveyor); such materials are re-irradiated as soon as possible. For radiosterilization, fractionation of the dose is in principle undesirable. However, Plester [45] reported that for practical purposes it can be assumed that the effects of resterilization are cumulative and additive. At present, a sterility level of 0.999999 is commonly obtained; this means that the non-sterile

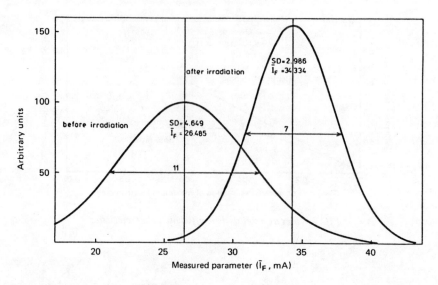

FIG.8. *Distribution of the electrical parameters of Si diodes before and after irradiation.*

rate is expected to be 10^{-6} microorganisms per article (i.e. it meets the World Health Organization sterility requirements). As reported by Czerniawski and Stolarczyk [46], typical contamination of microorganisms found on surfaces in the medical supplies factory is only about 2000 per single-use injection.
Thus, under normal technological conditions the sterility of irradiated medical supplies is maintained in the long term and can be controlled qualitatively by microbiological tests (i.e. an alternative — sterile or non-sterile). However, it is practically impossible to obtain full quantitative information for a large series of small irradiated products. Semiconducting diodes and thyristors are exceptions, since they need, at least after irradiation, individual detailed measurement of some electric parameters. A representative example could be a sample of 400 small diodes (BYP-350) measured carefully before and after an exposure of 13 MeV electrons (Fig.8).

Graphic analysis shows clearly that radiation processing is a more precise technique than industrial semiconductor technology, since it is possible to obtain a better mean value of the given parameter and also to attain a reduction in the standard deviation level (i.e. decrease of statistical spreading).

REFERENCES

[1] FOWLES, P., WALKER, J., EARWARKER, L.G., High intensity electron accelerators in radiation processing, J. Br. Nucl. Energy Soc. 16 2 (1977) 133.
[2] SILVERMAN, J., Current status of radiation processing, Radiat. Phys. Chem. 14 1–2 (1979) 17.

[3] MORGENSTERN, K.H., Industrial radiation processing: Present status and prospects, Radiat. Phys. Chem. **18** 1–2 (1981) 1.
[4] SANTAR, I., SCHWEINER, Z., Radiation processing in Czechoslovakia: Current state of development, Nukleonika **26** 7–8 (1981) 739.
[5] CZERNIEWSKI, M., PANTA, P., ZIELCZYNSKI, M., ŻAK, W., ŻARNOWIECKI, K., Bone tissue sterilization using reactor fuel gamma radiation. Nukleonica **10** 12 (1965) 791.
[6] CZERNIEWSKI, M., PANTA, P., ZIELCZYNSKI, M., ŻAK, W., ŻARNOWIECKI, K., Osseous tissue sterilization with gamma radiation from the 'EWA' reactor, Pol. Tyg. Lek. **20** 48 (1965) 1803.
[7] KOMENDER, J., MALCZEWSKA, H., BUŁHAK, Z., GOŁASZEWSKA, A., LESIAK-CYGANOWSKA, E., Radiation-sterilization of frozen biostatic grafts by means of an electron accelerator, Nukleonika **26** 1 (1981) 85.
[8] PANTA, P., Postepy Technol. Jadr. **22** (1978) 667.
[9] BUŁHAK, Z., Ph.D.Thesis, Institute of Nuclear Research, Warsaw, 1976.
[10] ZAGÓRSKI, Z.P., Sterylizacja Radiacyjna, PZWL, Warsaw (1981).
[11] KREJZLER, R., BUŁHAK, Z., ROBALEWSKI, A., Polish Patent No. 102,465, 1979.
[12] JAWORSKA, E., KAŁUSKA, I., STRZELCZAK, G., Isotopenpraxis **19** (1983) 305.
[13] JAWORSKA, E., KAŁUSKA, I., Radiation Chemistry (Proc. 5th Symp. Tihany, 1982), Akademia Kiado, Budapest (1983) 919.
[14] HEROPOLITANSKI, A., STACHOWIAK, R., Kontrahent **5** (1983) 5.
[15] DRABIK, L., PANTA, P., SZYJKO, J., Elektronizacja **4** (1981) 41.
[16] SZYJKO, J., BANY, B., BUŁHAK, Z., DRABIK, L., PANTA, P., STAMBULDZYS, A., SWIDERSKI, W., WINCEL, H., Polish Patent No. 116,059, 1982.
[17] PANTA, P. (in preparation).
[18] WISŁOWSKI, J., WRONSKI, W., Pr. ITE **8** (1983) 19.
[19] ACHMATOWICZ, T., JÓŹWIAK, H., ZIELIŃSKA, T., Nukleonika **24** 10–11 (1979) 1025.
[20] KROH, J. (Ed.), Technika Radiacyjna (Radiation Technology), WNT, Warsaw (1971).
[21] AMBROŻ, H., ZIELINSKI, W., JAWORSKA, E., Polim. Medycynie **3** 3 (1973) 181.
[22] OSTROWSKI, K., KOSSOWSKA, B., MOSKALEWSKI, S., KOMENDER, A., KURNATOWSKI, W., "Radiosterilization of tissues preserved for clinical purposes: Effect on tissue antigenicity", Radiosterilization of Medical Products (Proc. Symp. Budapest, 1967), IAEA, Vienna (1968) 139–143.
[23] STACHOWICZ, W., OSTROWSKI, K., DZIEDZIC-GOCŁAWSKA, A., KOMENDER, A., Nukleonika **15** (1970) 131.
[24] ZIMEK, Z., et al., Nukleonika **17** 1–2 (1972) 75.
[25] ROBALEWSKI, A., et al., Polish Patent No. 131,456, 1981.
[26] McLAUGHLIN, W.L., Int. J. Appl. Radiat. Isot. **17** (1966) 85.
[27] McLAUGHLIN, W.L., KOSANIC, M.M., Int. J. Appl. Radiat. Isot. **25** (1974) 249.
[28] McLAUGHLIN, W.L., MILLER, A., FIDAN, S., PEJTERSEN, K., BATSBERG-PEDERSEN, W., Radiat. Phys. Chem. **10** (1977) 117.
[29] KOSANIC, M.M., NENADOVIC, M.T., RADAK, B.B., MARKOVIC, V.M., McLAUGHLIN, W.L., Int. J. Appl. Radiat. Isot. **28** (1977) 313.
[30] McLAUGHLIN, W.L., HUMPHREYS, J.C., RADAK, B.B., MILLER, A., OLEJNIK, T.A., Radiat. Phys. Chem. **14** (1978) 535.
[31] GEHRINGER, P., ESCHWEILER, H., PROKSCH, E., Int. J. Appl. Radiat. Isot. **31** (1980) 595.
[32] GEHRINGER, P., PROKSCH, E., ESCHWEILER, H., Int. J. Appl. Radiat. Isot. **33** (1982) 1403.
[33] JÓŹWIAK, H., Private communication, 1984.
[34] TAIMUTY, S.I., Electron beam calorimetry, Nucleonics **15** 10 (1957) 182.

[35] BRYNJOLFSSON, A., HOLM, N.W., THARUP, G., SEHESTED, K., Industrial Sterilization at the Electron Linear-Accelerator Facility at Risø, Risø National Laboratory, Roskilde (1963).
[36] FLETCHER, J.W., "The absorbed dose water calorimeter: Use of radiation-chemical calculations to improve the accuracy and sensitivity", Radiation Research (Proc. 7th Congress Amsterdam, 1983), Amsterdam (1983) E2-14.
[37] DOMEN, S.R., Med. Phys. 7 (1980) 157.
[38] DOMEN, S.R., J. Res. Natl. Bur. Stand. 87 (1982) 211.
[39] DOMEN, S.R., A polystyrene-water calorimeter, Int. J. Appl. Radiat. Isot. 34 3 (1983).
[40] KUBO, H., Int. J. Appl. Radiat. Isot. 35 8 (1984) 808.
[41] KREJZLER, R., ROSZCZYNKO, W., BUŁHAK, Z., ROBALEWSKI, A., Polish Patent No. 96,561, 1979.
[42] KREJZLER, R., Private communication, 1984.
[43] PANTA, P. (in preparation).
[44] OSIPOV, B., et al., Sterilization by Ionizing Radiation (Proc. Int. Conf. Vienna, 1974) (GAUGHRAN, E.R.L., GOUDIE, A.J., Eds), Multiscience, Montreal (1974) 144.
[45] PLESTER, D.W., Industrial Sterilization (Proc. Int. Symp. Amsterdam, 1972) (PHILIPS, G. B., MILLER, W.S., Eds), Duke University Press, Durham (1973) 142.
[46] CZERNIAWSKI, E., STOLARCZYK, L., Acta Microbiol. Pol., Ser. B 23 4 (1974) 177.

IRRADIATIONS GAMMA AU CENTRE D'ETUDE DE L'ENERGIE NUCLEAIRE DE BELGIQUE

W. BOEYKENS
Département BR2,
CEN/SCK,
Mol

L. GHOOS
Contrôle-Radioprotection,
Mol

Belgique

Abstract–Résumé

GAMMA IRRADIATION AT THE BELGIAN NUCLEAR RESEARCH CENTRE.
 Gamma irradiation carried out at the Belgian Nuclear Research Centre (CEN/SCK) is geared to the study of the behaviour of different materials encountered in programmes of basic as well as applied research. Different irradiation units are described. Although the samples irradiated are usually small in size, one of the units permits the irradiation of instrumentation capsules or loops. Dosimetry is carried out by means of type red 4034 Perspex detectors and thermoluminescent aluminium-based detectors. Use and calibration of these detectors is discussed. Since the two systems have been developed separately, an intercomparison is made, showing good agreement.

IRRADIATIONS GAMMA AU CENTRE D'ETUDE DE L'ENERGIE NUCLEAIRE DE BELGIQUE.
 Les irradiations gamma effectuées au Centre d'étude de l'énergie nucléaire de Belgique (CEN/SCK) sont axées sur des études relatives au comportement de différents matériaux rencontrés dans les programmes de recherche aussi bien appliquée que fondamentale. Les différentes stations d'irradiation sont décrites. Bien qu'en général des échantillons de dimensions réduites soient irradiés, une des stations permet l'irradiation de capsules instrumentées ou de boucles. La dosimétrie se fait à l'aide de détecteurs Red 4034 Perspex et de détecteurs thermoluminescents à base d'alumine. L'utilisation de ces détecteurs est discutée ainsi que leur calibration. Comme les deux systèmes se sont développés séparément, une intercomparaison a été effectuée qui donne une bonne concordance.

Au CEN/SCK, différentes stations d'irradiation gamma sont installées: bien qu'en général des échantillons de dimensions réduites soient irradiés, une des stations permet l'irradiation de capsules instrumentées ou de boucles.

La dosimétrie se fait à l'aide de détecteurs Red 4034 Perspex et de détecteurs thermoluminescents à base d'Al_2O_3. Une intercomparaison entre les deux systèmes a donné de bons résultats.

1. IRRADIATIONS EXECUTEES

Les irradiations gamma effectuées au CEN/SCK sont axées sur des études telles que:

— le stockage à long terme des déchets radioactifs, par exemple l'étude des réactions physico-chimiques dans l'argile soumis à des irradiations, et l'examen de la résistance mécanique des verres utilisés pour le gainage des produits de fission;
— l'étude des dégâts dus aux rayonnements dans du matériel employé dans les installations nucléaires: câbles électriques, matériel d'isolation, lentilles de caméra, etc.;
— le comportement des sondes de mesure dans les champs à haut flux gamma et dans des conditions de température et d'environnement très variées;
— les essais de polymérisation ainsi que les irradiations de divers matériaux pour des études de l'état solide dans le cadre de la recherche fondamentale;
— le suivi, par le Service de radioprotection, de la dose reçue par les fenêtres à verre au plomb montées dans les parois des cellules chaudes.

2. STATIONS D'IRRADIATION

Compte tenu des demandes d'irradiation variées, différentes stations d'irradiation sont en exploitation.

2.1. **La première station d'irradiation gamma,** qui se trouve sous eau, utilise des éléments combustibles du réacteur BR2. Après chaque cycle de fonctionnement de 21 jours, une dizaine d'éléments combustibles sont déchargés du réacteur et transférés au GIF *(Gamma Irradiation Facility)* pour y être stockés temporairement en attente de leur retraitement (figure 1). Eventuellement, ils sont transportés directement dans la station d'irradiation γ équipée d'écrans en cadmium pour éviter tout risque de criticité. En variant les paramètres des éléments combustibles (burn-up, temps de refroidissement, nombre d'éléments combustibles), il est possible d'obtenir des flux gamma «à la carte».

L'implantation de cette station dans la partie CMF *(Core Mock-up Facility)* de la piscine du réacteur BR2 est donnée à la figure 1. La variation axiale du flux est indiquée à la figure 2 tandis que la variation du flux gamma en fonction du temps est donnée à la figure 3.

Cette station permet des irradiations à haute dose (10^5 à 10^8 Gy) dans des flux gamma intenses. Les échantillons de petit volume sont irradiés dans des tubes d'un diamètre maximum utile de 20 mm (flux gamma maximum: 4×10^5 Gy/h) introduits dans la cavité centrale d'un élément combustible BR2.

Des échantillons de volume moyen sont irradiés dans un tube d'un diamètre maximum utile de 78 mm, placé dans une cavité de la station d'irradiation,

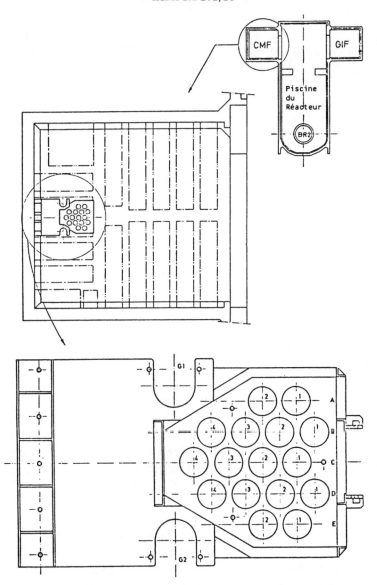

FIG.1. *Emplacement de la station d'irradiation CMF.*

entourée au maximum de 6 éléments combustibles (flux maximum: 10^5 Gy/h). Ces tubes ne sont pas instrumentés.

La station permet en plus l'irradiation simultanée de deux boucles d'irradiation instrumentées (air ou gaz inerte, analyse des gaz de radiolyse, contrôle de température, irradiation sous métal liquide, etc.). Un exemple de pareille boucle est donné à la figure 4. Le diamètre extérieur maximum est de 94 mm [1].

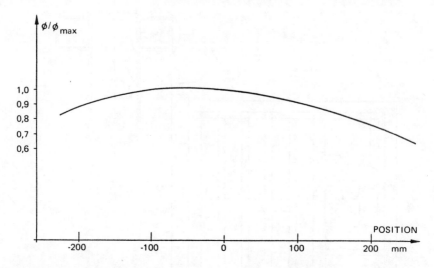

FIG.2. Distribution axiale du flux gamma dans l'installation d'irradiation utilisant les éléments combustibles BR2.

Le flux de neutrons pendant cette irradiation à été calculé. La valeur maxima au début d'un cycle d'irradiation vaut 2×10^5 n·cm^{-2}·s^{-1} en neutrons thermiques. La contribution au dépôt d'énergie est négligeable.

2.2. Une deuxième station, également située sous eau, est équipée de sources ^{60}Co avec une activité totale de 550 TBq. Ces sources se trouvent aux quatre coins d'un bac étanche en acier inox ayant les dimensions suivantes: $65 \times 50 \times 30$ cm. Le taux de dose actuel est de 9×10^2 Gy/h.

En général, des échantillons de volume réduit y sont irradiés à des doses ne dépassant pas 0,1 MGy. Parfois l'installation est utilisée pour stériliser des bouilons de culture pour l'élevage de micro-organismes.

2.3. Une troisième station est installée dans une cavité sphérique (diamètre: 1 m) se trouvant dans la colonne thermique verticale du réacteur BR1 (réacteur à uranium naturel, modéré au graphite).

Le système consiste en un tube en cadmium d'une épaisseur de 1 mm, d'une longueur de 50 mm et d'un diamètre extérieur de 54,5 mm: le cadmium est monté sur un tube en pyrex boraté à introduire dans l'axe vertical de la cavité. Le champ gamma est dû aux captures des neutrons thermiques dans le cadmium.

Le flux gamma est très homogène et son intensité est fonction de la puissance du réacteur avec un maximum de 0,15 kGy/h pour une puissance du réacteur de 4 MW.

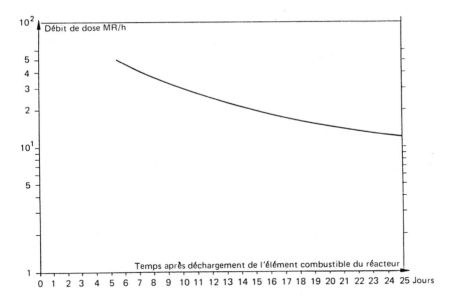

FIG.3. Variation du flux gamma pendant un cycle dans l'installation d'irradiation utilisant les éléments combustibles BR2 (1 R = 2.58 × 10⁻⁴ C/kg).

L'installation, bien que déjà utilisée telle quelle, est en plein développement: un flux gamma plus pur et plus intense est envisagé.

2.4. **La dernière station** considérée sert uniquement à la calibration des détecteurs thermoluminescents. Il s'agit d'un grand conteneur en plomb, de diamètre extérieur de 52 cm et dans lequel 18 sources de ^{60}Co sont rangées autour d'une enceinte cylindrique d'un diamètre de 6 cm.

Pour l'introduction des détecteurs dans cette enceinte, ceux-ci sont placées à l'intérieur d'un tube en acier inox, monté au bout d'une tige qui sert aussi de verrouillage. Ce tube a un diamètre intérieur de 2,5 cm et une longueur de 8 cm: l'épaisseur de la paroi est de 1,5 mm. Le taux de dose y est de 30 Gy/h et d'une distribution très homogène.

3. DOSIMETRIE: CALIBRATION ET UTILISATION DES DETECTEURS

3.1. **Les détecteurs Red 4034 S Perspex** (dimensions normales: 30 × 12 × 3 mm) sont d'emploi courant dans les deux premières stations. A chaque livraison de détecteurs, un jeu de détecteurs de référence est également commandé: ils ont été irradiés à des doses de 5 a 50 kGy, par intervalle de 5 kGy. Ces détecteurs de référence permettent de tracer une courbe d'étalonnage après lecture à 640 nm avec un spectrophotomètre Perkin Elmer type Hitachi 200.

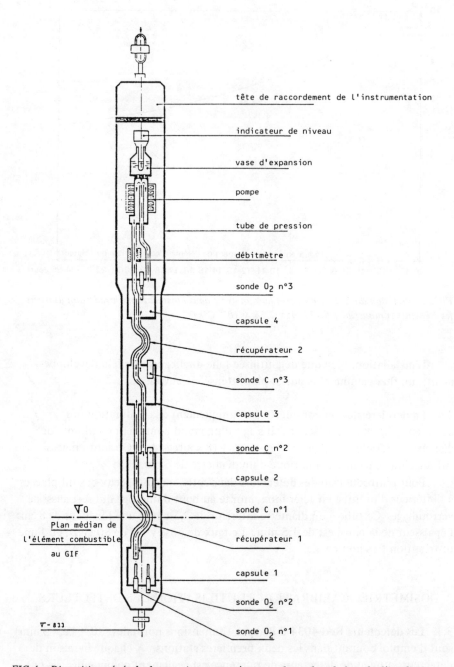

FIG.4. Disposition générale des moniteurs oxygène et carbone dans la boucle d'irradiation gamma.

Avant d'entamer une campagne d'irradiations gamma avec les éléments combustibles BR2 usés, le débit de dose au début du cycle est calculé en tenant compte des différents paramètres des éléments combustibles.

La dose gamma calculée est ensuite vérifiée par l'irradiation de détecteurs Perspex au début de la campagne d'irradiation. Cette irradiation se fait à l'intérieur d'un tube postiche. Les dosimètres Perspex sont fixés par du ruban adhésif sur une canne dosimétrique en aluminium avec des encoches correspondant à la position des échantillons à irradier.

Une mesure complémentaire à la fin du cycle d'irradiation est exécutée: la variation du flux gamma en fonction du temps après déchargement du réacteur a également été déterminée expérimentalement (figure 3).

Les doses étant mesurées avec un matériel léger, une petite correction doit être appliquée quand il s'agit d'une irradiation de matériel à densité élevée.

3.2. Le détecteur thermoluminescent

en usage est l'alumine (Al_2O_3). Il est utilisé, pour des raisons de radioprotection, à la surface interne des fenêtres en verre de plomb dans les parois des cellules chaudes. Jadis, les expériences traitées dans ces cellules n'étant jamais des sources de neutrons, du borate de lithium était utilisé.

La réponse de l'alumine est très peu influencée par la présence de neutrons, à condition que ces derniers ne dépassent pas 14 MeV en énergie. La réponse aux neutrons thermiques a été mesurée dans la cavité du réacteur BR1: 0,278 R équivalent pour 10^{10} n/cm² [2]. La réponse aux spectres de ^{252}Cf et de Pu-Be est du même ordre de grandeur [3].

Les détecteurs d'aluminium sont des pastilles frittées d'un diamètre de 7 mm, et d'une épaisseur de 0,9 mm. Pour certaines applications, l'alumine en poudre a été utilisé.

Lors de leur calibration, les détecteurs sont mis dans une ampoule en plexi dont la paroi a une épaisseur de 5 mm. Cette ampoule est chargée dans un tube en acier inox qui est introduit dans l'enceinte cylindrique entourée de source de ^{60}Co et dans laquelle le flux gamma varie de moins de 3%.

La calibration du flux gamma à l'intérieur de ce tube en acier inox a été effectuée à l'aide de détecteurs au fluorure de lithium calibrés dans notre installation de calibration où le taux de dose et la dose intégrée sont déterminés par une chambre d'ionisation du type Victoreen 555 Radocon II. Cette chambre est calibrée régulièrement dans un laboratoire primaire.

Les détecteurs alumine peuvent être réutilisés après un recuit à 600°C pendant 4 heures. Leur sensibilité n'est pas changée par ce traitement thermique.

Un fading à température ambiante de 40% par an a été constaté et le détecteur est utilisé à l'abri de la lumière.

Les détecteurs sont lus dans un appareil 7100 de Télédyne Isotopes.

TABLEAU I. INTERCOMPARAISON DES
DOSES DE CALIBRATION

Dose Perspex (Gy)	Dose Alumine (Gy)
215	240
430	450
860	870
1290	1250

3.3. **Une intercomparaison** entre les deux méthodes a été effectuée dans la mesure où les deux systèmes se sont développés séparément et où les calibrations se font de manière complètement indépendante.

Les détecteurs thermoluminescents ont été irradiés à des doses déterminées préalablement à l'aide des dosimètres Perspex.

Le tableau I résume les résultats.

REMERCIEMENTS

Beaucoup de personnes ont contribué aux travaux que nous venons de résumer. Nous remercions MM. R. Lenders et G. Balzer, du Département BR2, pour des informations qu'ils nous ont fournies ainsi que M. R. Menil, du Département Physique des Réacteurs. Nous remercions également M. C. Van Bosstraeten, de Contrôle-Radioprotection, qui a effectué le calcul du flux neutronique dans la station d'irradiation avec les éléments combustibles de BR2.

REFERENCES

[1] GANDOLFO, J.M., BOEYKENS, W., HEBEL, W., MOONS, F., SOENEN, M., «Gamma irradiations at the BR2-reactor», Euratom Irradiation Devices Working Group, Geesthaecht (8–10 octobre 1980).
[2] MENIL, R., communication privée.
[3] ABERHOFER, M., SCHARMAN, A., «Applied Thermoluminescence Dosimetry», Course held at the Joint Research Centre, Ispra (12–16 novembre 1980), publié par A. Hilger Ltd., Bristol (1981).

DOSIMETRY PRACTICE IN THE GOU-1 GAMMA IRRADIATOR

M.G. HRISTOVA
Institute of Nuclear Research and
   Nuclear Energy,
Bulgarian Academy of Sciences,
Sofia, Bulgaria

V. STENGER, A. KOVÁCS
Institute of Isotopes,
Hungarian Academy of Sciences,
Budapest, Hungary

Abstract

DOSIMETRY PRACTICE IN THE GOU-1 GAMMA IRRADIATOR.
   The GOU-1 research gamma irradiator at the Institute of Nuclear Research and Nuclear Energy, Sofia, was built and put into operation in 1975. Its large volume, high dose rate and special design features make it comparatively universal, with great possibilities for experimental work. The extent of coincidence between the calculated and obtained exposure dose rates in the irradiation chamber was checked by a Si(p) detector calibrated on the basis of the Fricke dosimeter, the oscillometric ethanol-monochlorobenzene dosimeter and the ionization chamber. The Si(p) detector was mainly used in the exploitation process; sometimes the two dosimeters already mentioned were also used. Reference points were chosen in the volume of the chamber where the exposure dose rate was periodically checked. Samples of varying sizes and density were irradiated after preliminary theoretical and experimental evaluations for the extent of field reduction in the material volume. The exposure dose rate distribution in the irradiation volumes of the chamber was also periodically checked.

1. INTRODUCTION

   The development and application of gamma radiation technology is connected with the construction of the necessary technical equipment. At present there are more than 80 industrial and several times more research irradiators world wide. Countries such as Canada, the United Kingdom, the USSR and France are already established manufacturers of gamma irradiators. In some countries, however, because of the specific conditions and requirements, gamma irradiators of single design with relatively greater possibilities for research and production have been built. Such an irradiator – the GOU-1 gamma irradiator – was designed and put into operation in 1975 at the Institute of Nuclear Research and Nuclear Energy, Sofia [1].

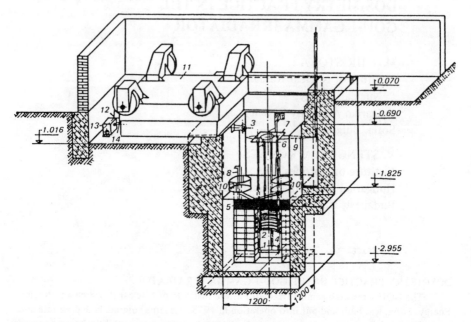

FIG.1. Axonometric projection of the GOU-1 gamma irradiator (1, 2, 4 – storage container; 3 – irradiation chamber; 5 – lead plate; 6,7 – source rack; 8 – balance weight; 9 – leverage; 10 – rotation platforms; 11 – shielding trapdoor; 12 – electric motor; 13 – reduction gear; 14 – drive screw).

2. GOU-1 GAMMA IRRADIATOR

Figure 1 gives the axonometric projection of the facility and Fig.2 shows an inside view of the irradiation chamber.

The gamma irradiator is of the ground type, with two zones situated one upon the other. The upper zone is the irradiation chamber with dimensions of 1200 × 1200 × 1100 mm. The lower zone is the storage compartment for the radioactive source assembly, which includes a hollow cylinder (with a diameter of 360 mm) on which the 16 linear source elements (with a total initial activity of 1.41×10^{16} Bq (38 kCi) ^{60}Co) are mounted. The length of the active elements is 320 mm. This design comprises two types of irradiation zones. The first is the inner space of the hollow cylinder. The gamma field in the cylinder is the same as that in the field obtained from a hollow cylindrical source with a homogeneous distribution of activity on the inner surface of the cylinder. To increase the axial homogeneity of the field, sources with a higher specific activity are loaded at both ends of the linear source elements. The second irradiation zone is the area around the source assembly. The exposure dose rate in this area decreases rapidly with the distance from the source. Because of the above-

FIG.2. Inside view of the GOU-1 irradiation chamber.

mentioned loading, the vertical homogeneity of the field is also increased. A graphical method was used to calculate the gamma field, thus enabling easy determination of the exposure dose rate distribution in the irradiation chamber [2].

3. IRRADIATION ACTIVITY

During 10 years of operation of GOU-1 thousands of samples from different research groups, institutes and plants have been irradiated, as well as small quantities of industrial products. The samples also differ in material, volume, quantity and required dose and, on this basis, they are irradiated in one of the following irradiation zones:

(1) *Central zone.* The free volume of the hollow source assembly cylinder; the dimensions are: height, 320 mm; diameter, 270 mm; inhomogeneity, 15%.

(2) *Zone surrounding the source assembly.* Taking into consideration the exposure dose rate distribution, this zone has been divided into irradiation volumes

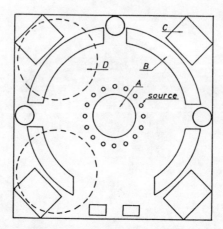

FIG.3. Location of the irradiation vessels inside the irradiation chamber (A — central zone; B — cylindrical sector cassettes with a radial width (co-linear with the gamma rays) of 100 mm; C — cassettes with dimensions of 250 X 200 X 500 mm; D — cylindrical cassettes with a height of 500 mm and diameters of 200, 300 and 400 mm placed on rotating platforms).

in which the inhomogeneity of the dose rate for materials with a density of 0.7 t/m^3 is within the limit of ± 10 to ± 20%. The configurations of these irradiation volumes are given in Fig.3.

4. DOSIMETRY PRACTICE

The great variety in samples, required doses and exposure dose rates, and the character of the gamma field in the irradiation chamber call for continuous optimization of the irradiation conditions for each sample. The exposure dose rates at the different reference points and volumes of the GOU-1 irradiation chamber are determined mainly by a Si(p) detector calibrated on the basis of two physical and two chemical methods (ionization chamber, calorimeter, Fricke dosimeter, ethanol-monochlorobenzene dosimeter) [3]. The necessary data from a number of points in the irradiation volume are easily and rapidly obtained since readings from the high sensitivity ammeter, with which signals from the Si(p) detector are measured, provide immediate exposure dose rates.

The calculated and measured exposure dose rates at some of the reference points in the irradiation chamber when GOU-1 was first put into operation

TABLE I. CALCULATED AND MEASURED EXPOSURE DOSE RATES IN GOU-1

Point X, Z[a] (cm)	Dose rate (kGy/h)				Difference (%)
	Calculated			Measured Dec. 1975	
	Certified activity (1.4×10^{15} Bq) Jun. 1974	Reduction from 4 mm steel	Dec. 1975		
X-0 Z-0	9.90	9.17	7.54	7.30	− 3.3
X-0 Z-100	9.50	8.80	7.17	6.98	− 2.7
X-0 Z-160	8.19	7.58	6.24	6.17	− 1.1
X-23 Z-0	6.93	6.42	5.37	4.50	− 16.2
X-48 Z-0	1.83	1.69	1.42	1.45	+ 2.1
X-73 Z-0	0.85	0.79	0.66	0.62	− 6.1
X-43 Z-44	1.62	1.50	1.26	1.15	− 8.7
X-69 Z-44	0.77	0.71	0.60	0.57	− 5.0

[a] The centre of the co-ordinate system is in the centre of the radioactive source assembly.

are shown in Table I. The significant difference between calculated and measured data outside the source assembly is due to improper positioning of the detectors.

In general, the dosimetry practice in GOU-1 is as follows:

(1) In the volume of chamber 3, reference points were chosen for checking the radioactive source assembly and its movement; the exposure dose rate is periodically measured by a Si(p) detector. Data for the central irradiation volumes are given in Table II. They show the stability of the radioactive source assembly and the accuracy of positioning of the source assembly.

(2) The exposure dose rate distribution in the irradiation volumes at the irradiation positions is periodically measured. Measurements done in June 1983 with Fricke and ethanol-monochlorobenzene dosimeters and the Si(p) detector are shown in Fig.4 [4, 5].

TABLE II. DOSE RATES AT REFERENCE POINTS IN THE CENTRAL IRRADIATION VOLUME

Date	Dose rate (kGy/h)								
	Apr. 1975	Oct. 1976	Mar. 1977	Oct. 1978	May 1979	Feb. 1980	Apr. 1981	Mar. 1982	Jun. 1983	Aug. 1984
Measured	8.20	6.65	6.32	5.14	4.76	4.21	3.5	3.22	2.63	2.32
Calculated (based on measurements made in Apr. 1975)	–	6.58	6.38	5.17	4.79	4.34	3.68	3.26	2.70	2.40

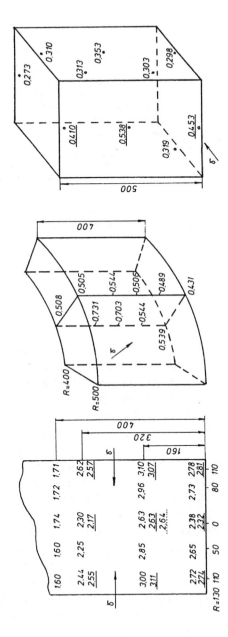

FIG.4. *Exposure dose rate distribution in the irradiation vessels, June 1983 (results in kGy·h⁻¹; distances in mm). The numerals underlined with an unbroken line were measured with the Fricke dosimeter; those with a broken line with the Si(p) detector; those not underlined with the ethanol-monochlorobenzene dosimeter.*

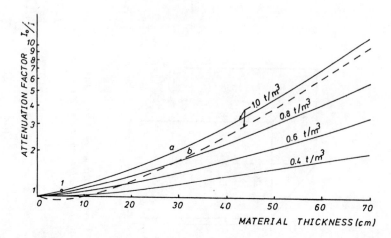

FIG.5. Radiation dose rate attenuation factors for material with densities of 0.8, 0.6 and 0.4 as a function of the water's attenuation factor (a — after Ref. [6]; b — after Ref. [7]). Point 1 — measured.

(3) Irradiation of the samples and exposure dose rate measurements in each specific case include: (a) Depending on the thickness of the material through which the gamma rays pass, the samples are divided into groups with thicknesses of up to 50, 100, 150, 250 and 350 mm. (b) Depending on the quantity of the samples, the required dose and the exposure dose rate in the irradiation chamber, the irradiation position is chosen. (c) Reduction of the exposure dose rate from the material is evaluated. If the material has a density of less than one, evaluation is done with the help of curves determined on the basis of the gamma ray reduction in the water curve (see Fig.5) [6, 7]. (d) The estimated dose rate then is checked at three to five chosen reference points by means of a Si(p) detector. (e) The necessary irradiation time is determined, taking into account the dose received during dose rate checking. The dose rate curves in air, the dose rate curves in the central volume full of water and the evaluated dose rate curves when the irradiation volumes are filled with material with a density 0.7 t/m^3 are shown in Fig.6. Data show that to obtain an inhomogeneity of 10% in the peripheral volumes when the material thickness exceeds 50 or 100 mm for the 400 or 500 mm corresponding distance from the centre of the radioactive source assembly continuous or 180° rotation of the material after half the irradiation time must be achieved.

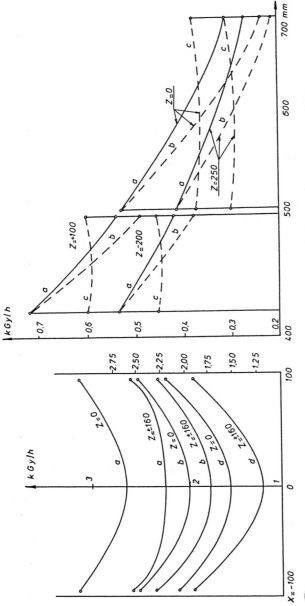

FIG.6. Dose rate curves in irradiation vessels (a — in air; b, c — in material with a density of $0.7 \ t/m^3$ without and with rotation respectively; d — in water-filled central volume).

The results of manifold research and practical activities obtained during the many years of operating GOU-1 have shown that the method chosen for irradiation and dosimetry is correct and that the required dose (maximum 10% inhomogeneity) has been received by the irradiated samples.

REFERENCES

[1] PANDEV, I.N., et al., Yad. Energ. 3 (1976) 68.
[2] HRISTOVA, M.G., et al., Yad. Energ. 5 (1977) 82.
[3] OSVAY, M., STENGER, V., FÖLDIÁK, G., "Silicon detectors for measurement of high exposure rate gamma rays", Biomedical Dosimetry (Proc. Symp. Vienna, 1975), IAEA, Vienna (1975) 623–632.
[4] DAVIS, J.V., LAW, J., Phys. Med. Biol. 8 (1963) 1.
[5] RAŽEM, D., DVORNIK, I., "Application of the ethanol-chlorobenzene dosimeter to electron-beam and gamma-radiation dosimetry. II. Cobalt-60 gamma rays", Dosimetry in Agriculture, Industry, Biology and Medicine (Proc. Symp. Vienna, 1972), IAEA, Vienna (1973) 405–419.
[6] KIMELL, L.P., MASHKOVICH, B.P., Zashchita ot Ioniziruyushchikh Izluchenii, Atomizdat, Moscow (1972).
[7] Engineering Compendium on Radiation Shielding, Vol. II, Shielding Materials, Springer-Verlag, New York (1975).

IAEA-SM-272/24

EXPERIMENTAL CHEMICAL DOSIMETRY AND COMPUTER CALCULATIONS
Combined application in a ^{60}Co research unit

H.M. BÄR, J. REINHARDT, M. REMER
Central Institute for Isotope and
 Radiation Research,
Academy of Sciences of the German
 Democratic Republic,
Leipzig, German Democratic Republic

Abstract

EXPERIMENTAL CHEMICAL DOSIMETRY AND COMPUTER CALCULATIONS: COMBINED APPLICATION IN A ^{60}Co RESEARCH UNIT.

Since 1976 a universal ^{60}Co irradiation unit of the PANORAMA type has been available at the Central Institute for Isotope and Radiation Research, Leipzig. The original activity of 1.2 PBq (32 kCi) ^{60}Co was distributed symmetrically to 12 revolving irradiation tubes. This permits irradiation from the outside and the inside as well as a combination of both variants. The first radiation field evaluation was made by means of chemical dosimetry (Fricke, chlorobenzene). Because of the variety of geometric positions of the radiation sources, the resulting radiation fields could not be determined precisely by chemical dosimetric experiments, or only with great difficulty. Therefore, a computer code was developed to describe these radiation fields. The efficiency of the program was increased step by step and was compared in all cases with the results of the experimental dosimetry. By taking into consideration the specific correction factors, the coincidence of the computational model and experimental data was attained. The basic model of the code DOSKMF2 contains computation of the exposure rate, which is dependent on a system of radiation sources (line and/or point sources) at one point of the radiation field. The geometric conditions are described by two co-ordinate systems. The program takes the shielding layers into consideration. Calculations of radiation fields in complex irradiation units are made by many organizational variants based on the basic model. The code was first used to compute different configurations of the sources within the irradiation tubes of the PANORAMA unit in order to determine the best way of carrying out the necessary charging and recharging procedures. A catalogue was also prepared with graphic representations of typical irradiation configurations. It contains radiation fields in the form of isolines and allows rapid determination of the necessary configurations, irradiation positions and irradiation times. The combination of experimental chemical dosimetry and computer calculations was also successfully used to develop chemical processes on an industrial scale.

Since 1976 a universal ^{60}Co gamma irradiation unit of the PANORAMA type has been available at the Central Institute of Isotope and Radiation Research, Leipzig. This 'open' irradiation unit, the largest of its kind for research purposes in the German Democratic Republic, was designed, constructed and completed in close co-operation with the All-Union Institute of Radiation Technology, Moscow.

The unit was installed in the centre of an irradiation room with labyrinth. The biological shielding was made of concrete (density, 2.3 t/m^3; thickness, 1.5 m). The cylindrical ^{60}Co sources (diameter, 11 mm; length, 16 mm), each with an activity of 18.5 TBq (500 Ci), have additionally been set in stainless-steel balls (diameter, 22.225 mm). These spherical sources are distributed to 12 revolving irradiation tubes. The total activity of PANORAMA is 1.2 PBq (32 kCi); each irradiation tube contained seven ^{60}Co sources when the unit was put into operation in 1976.

The length of the straight section of the irradiation tubes is about 400 mm. To enable use of the whole zone during irradiation processes, inactive smaller balls (diameter, 19.05 mm) are arranged between the sources. The sources are stored in a lead container consisting of five specially formed parts. The total weight is about 10 t.

The samples, products, vessels or reactors to be irradiated are arranged at the working table of PANORAMA and are surrounded by the 12 double-bent irradiation tubes. They can be turned by hand in steps of one degree. The diameter of the 'cage' which is formed by the 12 tubes varies between 120 and 800 mm. These facts allow the operator to choose a suitable dose rate by changing the distance between the object and the sources. The number of irradiation tubes to be fitted with sources can also be varied, e.g. it is possible to use one or two or five or each combination of the 12 tubes. Because of these possibilities the operator is able to solve nearly all the irradiation problems that arise. In special cases the bent irradiation tubes can also be exchanged for straight tubes.

Transport of the sources from the storage position into the irradiation position and vice versa is realized by means of inactive transport balls (diameter, 19.05 mm) which are moved by a worm gear. Microswitches signal on the control panel if the sources are in the irradiation (red light) or the storage position (green light). A very reliable safety system guarantees entrance into the irradiation room without any danger to staff members when the irradiation process is completed.

The geometric conditions of PANORAMA permit irradiation from the outside and the inside as well as a combinations of both variants.

After installation, the first radiation field evaluation was made by means of chemical dosimetry (Fricke, chlorobenzene). However, for several experiments different configurations of the tubes were needed. Because of the variety of geometric positions of the radiation sources, the resulting radiation fields could not be determined precisely by chemical dosimetric experiments, or only with great difficulty. Therefore, a computer code was developed to describe these radiation fields. The efficiency of the code was increased step by step and was compared in all cases with the results of the experimental dosimetry. By taking into consideration specific correction factors in relation to the irradiation unit, agreement between the computational model and experimental data was attained. For the characterization of complex radiation fields, the dose rate at many points of the field has to be determined. This means that the effect of all sources of the irradiation unit has to be calculated at each point of the field; this is done by the DOSKMF2 code [1, 2].

The basic model of the code contains computation of the exposure rate, which is dependent on a system of irradiation sources (line and/or point sources), at one point of the radiation field. The geometric conditions are described by two co-ordinate systems. The first is used to arrange the radiation sources and tubes; the second is used to describe the shielding layers in the form of concentric circles. The code realizes the connection of the two systems and computes all the geometric values, especially the distance between the source and the point of the field and the actual thickness of shielding layers. If shielding layers or the irradiated product are between a source and a point of the field we have to take into consideration not only the effect of absorption but also the effect of build-up. In the code the build-up factor is approached by Taylor coefficients; they are also used as a basis for multi-layer configurations. The effectiveness and bounds of error were checked regularly by means of experimental dosimetry. Calculations of radiation fields in complex irradiation units are made by many organizational variants based on the basic computational model. The points of the field are given by cylindrical or cartesian systems of co-ordinates. It is possible to choose an initial, a step and a final value for each co-ordinate and to calculate many points of the field in one run of the code.

Interpretation of the results is supported by statistical calculations such as mean and standard deviations, and the minimum and maximum values per plane and total system. This enables a rapid survey of the properties of the resulting radiation fields to be obtained. Another code was also developed which processes the data produced by DOSKMF2 on a magnetic tape and realizes a graphic output in the form of isolines on a plotter. This is a very good way of representing radiation fields. The codes are written in PL1 and FORTRAN. We use the codes on a computer ES 1040 with the operational system OS/ES.

After installation, experimental dosimetry was begun and a very simple computational model was used (for instance: only one layer with build-up factor, simple geometric conditions). However, in this early phase good agreement was not achieved. After detailed analyses were made the model was enlarged. It should be noted that not all the parameters needed for the description were known precisely. Therefore, the calculations were repeated with the enlarged model and better agreement was reached. After taking into consideration specific correction factors which represent the unknown parameters of the unit good conformity was achieved between the experimental and computed values of the dose rate, i.e. agreement in the mean value with a standard deviation of $\sim 5\%$. Using this model we started extensive calculations to obtain detailed information on the structure of the radiation field. In this phase we also compared the results of computation with experimental measurement and again achieved good conformity.

Thus it was possible to make the following calculations:

(1) Dose rate distributions for typical arrangements of the tubes, namely different diameters of the circle of tubes, different numbers of tubes, different diameters of the irradiated product and different media. The results were compiled in a catalogue in the form of isolines. This is very useful for practical work and allows rapid determination of the appropriate configurations of tubes, irradiation positions and irradiation times.
(2) Special arrangements for the particular requests of customers.
(3) Different configurations of the sources within the irradiation tubes in order to determine the best way of carrying out the necessary charging and recharging procedures, i.e. simulation of the procedures. Taking into consideration all the technological aspects, the best arrangement was chosen; then the dosimetric experiments were repeated.

The combination of experimental chemical dosimetry and computer calculations was also successful when used in connection with the development of chemical processes on an industrial scale [3–5]. Fundamental investigations were carried out in the PANORAMA unit. In developing radiation-induced chlorination of PVC, we used the code to calculate the radiation fields in a model reactor used in the PANORAMA unit. We then calculated some of the parameters of the pilot plant, especially the form of the reaction vessel, the configuration of the irradiation tubes, and the form of the radiation shielding.

After installation of the pilot plant we repeated the calculations under real conditions. This was necessary because some modifications had been made to the reaction vessel. Taking this into consideration, the new calculations and the dosimetric experiments reached good conformity. Thus, it is now possible to describe radiation fields by means of computer calculations.

To reduce the bacterial count of enzyme solutions, this method was used for the layout of a model reactor and experiments on a laboratory scale and for the design of an industrial plant, including the necessary biological shielding. It is important to note that throughout these applications we checked the calculations by means of experimental dosimetry. As a result, it was possible to test and enlarge the computational model.

The combination of experimental dosimetry and computer calculations for gamma irradiation facilities has the following advantages:

(1) On the basis of an adequate model it is possible to provide a good description of radiation fields; this is very important for the effective use of existing facilities.
(2) In the early design phase of new facilities it is possible to influence the economic aspects of, for example, sources and shieldings.

We effectively support the design of some partial systems of a gamma irradiation facility, e.g. arrangement of sources and tubes, source storage, reaction vessels, transport system and shieldings.

REFERENCES

[1] REMER, M., DOSKMF2, Code Description, Akademie der Wissenschaften der DDR, Zentralinstitut für Isotopen- und Strahlenforschung, Leipzig, Archiv-Nr. 6C34J1 (1982).

[2] REMER, M., DOSKMF2 — A Contribution to the Computer-Aided Design of Dose Rate Distributions, ZfI-Mitt. 71 (1983) 47–56.

[3] Patentschrift Nr. 148 188 (WP B 01 J/218069), 13.5.81, Reaktor für die Durchführung strahlenchemischer Reaktionen.

[4] BÄR, M., GRENSING, R., LANGGUTH, H., REINHARDT, J., SCHMIDT, P., SPERLING, U., "A pilot plant for the radiation-induced chlorination of PVC — experiences and results", Radioisotope Application and Radiation Processing in Industry (Proc. 2nd Working Meeting, Leipzig), Zentralinstitut für Isotopen- und Strahlenforschung, Leipzig (1982).

[5] ENFELD, G., HOFFMEISTER, H., REINHARDT, J., WÖLBING, M., "Reduction of bacterial count of enzyme solutions by gamma radiation", ibid.

IAEA-SM-272/31

ELECTRON BEAM DOSIMETRY
IN THE RANGES 1 TO 10 MeV
AND 0.1 TO 10 kW

I. KÁLMÁN
Research Institute for the Plastics
   Industry,
Budapest, Hungary

Abstract

ELECTRON BEAM DOSIMETRY IN THE RANGES 1 TO 10 MeV AND 0.1 TO 10 kW.
   In electron beam processing systems it is necessary to measure the beam power, either for the control of the accelerator or for dosimetry. For this purpose we developed an industrial comparative measuring method which has the following principles: a water pillow on a Plexiglas platter absorbes the electron beam, the water surface should be greater than the scanned beam, and the thickness of the pillow should be nearly twice the penetration depth of the beam. It is necessary to use an adjustable but extremely constant water flow rate. We measured the temperature difference between outlet and inlet water. To avoid difficulties in calorimetric methods, a heating spiral was built into the water pillow and heated electrically in order to measure the heating power. This method has proved to be useful for high-dose dosimetry in the range 100 to 1000 Gy/s and has been used in Hungary as a secondary etalon for measurement of beam power.

Various industrial technologies using electron beam processing are being developed world wide. The greatest use of electron irradiation is made in the production of heat-shrinkable materials. High beam capacities are used in other cross-linking plastic processes and in manufacturing thermally resistant endproducts such as ribbons, films, fibres and foams. For all these technologies electron beam processing is considered to be one of the most important techniques. This process is now almost exclusively performed by electron accelerators. In various industrial technologies a total beam power of about 10 to 15 MW is being used; this figure is increasing year by year. In all industrial production units, installation of such a radiation source should provide: continuous operation, steady parameters during operation and safety; it should also be economic.
   To attain optimum beam efficiency, a periodic check of the dose rate of the accelerator is required using an accurate and rapid dosimetry technique.
   The importance of correct dosimetry is supported by the fact that the minimum dose necessary for actual technology should strike the target, while the upper limit of the dose is determined by the degradation of the material. On the other hand, the levels and ratios of the different parameters of the accelerator

(energy, power, beam current, constants of the scanning device and the material-transporting machines) can only be determined by dosimetry.

The beam power of individual industrial radiation sources is, however, increasing, mainly for economic reasons. No industrial accelerator in service nowadays has a beam power below some kilowatts, the average power is 5 to 20 kW, while industrial accelerators of 50 to 100 kW are not infrequent. This development also makes further advances in radiation dosimetry possible.

The actual possibilities of industrial dosimetry are as follows:

(1) Calorimeters are used as primary etalons. They work well at a low beam power but difficulties arise in keeping adiabatic operation at beam powers of several kilowatts, which affects rapid and simple measurement.

(2) Chemical dosimeters (Fricke, cero, chlorobenzene, ethanol) are extensively and favourably used both in accelerator and gamma-source techniques. In the accelerator technique, direct application of some dosimetric methods (e.g. Fricke) is limited to dose rates of 1 to 10 Gy/s. The general trend is automation of measurement, a subject that will be covered in several papers at this Symposium. However, automated methods are mainly used in dosimetry of the product and are only suitable for checking the accelerator in indirect ways.

(3) Use of semiconductor dosimeters has recently been extended but they have not yet been accepted for general use.

(4) Film dosimeters are suitable for some practical work, especially where it is necessary to demonstrate that a predetermined minimum dose has been obtained by each target (e.g. in sterilization).

(5) The most widespread method of checking an electron beam is by electric measurement of the characteristic parameters of the radiation, namely radiation energy, radiation intensity or beam current, characteristics of the scanning system (such as location and width of the fan beam), and the speed of the target. Checking the characteristic parameters of the accelerator refers essentially to the stability of the electron beam treatment in time. Theoretically, a very careful check of the electric parameters permits the calculation of product dosimetry; however, in spite of this check a separate method is needed to ascertain if the technology has to be adjusted, if the machinery has to be maintained or repaired, or if some breakdown is observed.

The water-flow calorimetric technique discussed in this paper is suitable for this purpose.

In laboratory experimental accelerators, where beam power is in the range of milliwatts to watts, the temperature rise of the absorbing target is insignificant, thus adiabatic conditions can be approximated correctly. In industrial irradiations, however, owing to the economic higher beam powers applied, the target should be cooled. Because of this necessity, measurement of the flow rate and temperature

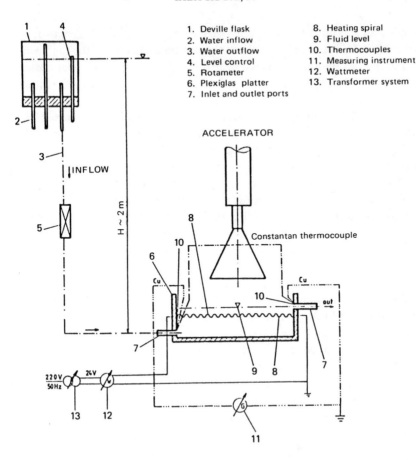

FIG.1. Control of accelerator power.

of the controlled stream of the cooling medium (preferably water) was suggested for determination of the absorbed beam power. Under practical conditions, i.e. at a beam power of 5 to 10 kW, the temperature difference is 4 to 5°C, thus the system is not considered to be adiabatic; for this reason a comparative technique was also used.

A heater (supplied from mains power at a known wattage) is immersed in the cooling medium. The water stream is alternatively heated by the electron beam and mains power until the same temperature gradient is attained by both energy supplies, i.e. when they both have an identical wattage. The equipment is shown in Fig.1.

The radiation target is a water layer poured into a tray with a depth of not less than double the penetration depth of the electron beam and with an area larger than that of the fan beam. The tray should be made of insulating material, preferably an acrylic sheet, to enable visual monitoring of the Cerenkov radiation of water by a TV set in order to check the positional failure of the fan beam.

The inlet of the tray is located near the bottom, while the upper outlet governs the height of the water level. This arrangement provides efficient rinsing.

Just below the water level, where the beam strikes the water surface, a kanthal heating filament is spanned at a depth of 0.5 $R\rho$ (half the penetration depth, R) so that its heating effect on the water layer is similar to that of the beam.

The heating filament should be gauged for a low-voltage supply of 24 V to provide the necessary electric power for the required accelerator energy (with more than one filament if needed).

The most critical factor of measurement is constancy of the water stream. The municipal water supply proved to be unsatisfactory; instead, a normal laboratory Deville flask level regulation was applied, with the stationary water stream controlled by a rotameter. Although this system is more complicated than a simple pumping circuit, the Deville flask also serves as a temperature balancing unit. Scattering of the experimental results was quite low.

The comparative indicator of measurement is the temperature difference between outlet and inlet water. It can be measured by thermistors or by differential switching of Fe-Constantan thermocouples, as was done in our experiments. Thermocouples should be shielded from the scattered electron beam at the outlet and inlet ports. The differential voltage (10 to 100 μV) of the thermocouples is either measured by a galvanometer or recorded by a compensation recorder. It can thus be concluded that the absolute values of the flow rate and the temperature difference have no significance; only the consistency of these values during measurement is crucial, thus calibration of the system is also very simple.

The electric heating is adjusted to the beam power by a regulating (e.g. a toroid) transformer from a 24 V mains transformer and is measured by a wattmeter.

Heating filaments proved to be more favourable than heating rods, since they have a lower time constant and promote consistency of the flow rate by preventing the possible formation of annular flow channels in the tray.

Measurement is carried out in the following stages:

(1) Assemble the measuring system
(2) Start the water flow, then wait for some minutes until the uniform water stream and the temperature balance are attained
(3) Switch on the electron beam; after a few minutes delay, the constant electron beam produces an unvaried temperature difference
(4) Switch off the electron beam and switch on the electric heating simultaneously, then adjust the latter so that the temperature rise will be the same as that of the electron beam
(5) If the adjustment is not immediately successful, switch over and over between the beam and the heating to iterate the correct adjustment.

If the water flow rate and the temperature difference is strictly constant, the wattage of the beam power will be identical to that of the heater.

According to practical experience, the consistency of flow and temperature conditions is attained in 3 to 5 min after opening the water stream if the beam power is between 0.4 and 8 kW.

After starting radiation, temperature equilibrium is achieved in about 3 min. With a correctly adjusted heating power, the re-equilibration time after switch-over to electric heating is well within 2 min. Iterated switch-overs take 1 min each.

The sensitivity of measurement can be demonstrated by the fact that even a singular breakdown is perceptible.

At lower beam powers, when lower flow rates should be adjusted, the duration of measurement is longer. Determination of a radiation of 100 W is still convenient although rather slow. Measurement of a 50 W beam is intricate. Theoretically, measurements can be performed above 10 kW as well, but the flow system of the medium should be modified since the necessary higher flow rate requires manifold inlets in order to avoid turbulence and internal stirring and so prevent any excessive evaporation due to local overheating which may impair the accuracy.

The radiation energy is not specifically limited. Very high energies, above 10 MV, are not practical because of the risk of activation. At an energy of 1 MV, the penetration depth of the electron beam in water is about 3 mm. At lower energies, location of the heating filament may be problematic at the half depth of penetration, i.e. less than 1.5 mm.

Measurement was compared with several dosimetric methods. The accuracy of these calibrations was better than ±7%.

It can be derived from the principle of this technique that accuracy is composed of the precision of temperature difference which can be easily sensed within ±1%, and the sensitivity of the rotameter used for controlling the consistency of the flow rate which is generally ±3 to 5%. These values support the accuracy of the comparative measurements obtained.

Another consequence of the principle of this method is that water is an absorbing medium which captures secondary electrons and those backscattered from the surface. However, the power of the bremsstrahlung is not sensed. If the method was used as an absolute measurement, it would cause a slight negative error.

If product dosimetry is intended, the geometry of the target will influence both the scattered radiation and the effect of secondary electrons. These uncertainties would result in even greater error.

If the surface of the water layer on the tray is larger than the area of the fan beam, i.e. the whole beam is absorbed, the total radiation power of the accelerator is measured. It is preferable to check the accelerator equipment by obtaining information on the status of the energy, and the focusing, current and scanning conditions. As the beam power is measured simultaneously with coulometric measurement of the current, it is possible to determine the radiation energy.

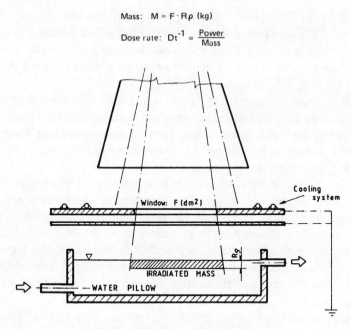

FIG.2. *Dose rate measurement.*

For product dosimetry, the procedure should be modified as follows. The water layer on the tray should be covered by a shield with a double aperture. The shield thickness should exceed the penetration depth of the electron beam. In high-performance accelerators, especially when the major proportion of the fan beam is shielded, the upper part of the shield should be cooled independently. At a power of 1 to 2 kW, if only a small part of the fan beam is covered, air cooling is satisfactory; at higher beam powers, the upper sheet should be equipped with a separate cooling coil.

The absorbed dose (in Gy) (see Fig.2) is defined as the absorbed radiation energy by mass M (in kg) of the target.

The dose rate is the quotient of the measured beam power to the target mass. The irradiated mass of the target is the product of the F area of the aperture (in dm^2) with penetration depth, R (in dm), multiplied by the density of the target (in kg/dm^3). Accordingly, the dose rate is

$$Dt^{-1} = \frac{Power}{M} \text{ (in Gy/s)}$$

The beam power can be determined by comparative measurement.

According to practical experience, this technique is favourable in the dose rate range from 100 to 1000 Gy/s but can readily be extended to between 10 Gy/s and 10 kGy/s.

As a secondary reference method this comparative water stream calorimetric method has proved to be effective in Hungary.

The technique has the following advantages: measurement is relatively simple with conventional equipment such as a wattmeter and a rotameter with normal accuracy as well as a low voltage circuit; assembling and performing measurement are rather rapid; it is independent of the type of accelerator; no special air conditioning is required, thus measurement can also be carried out under industrial conditions.

The disadvantages of the technique are that it only provides information on the average power of the radiation and that it requires a water delivery system in the target room.

DOSIMETRY TECHNIQUES AND SYSTEMS
(Session 2)

Chairmen

B. WHITTAKER
United Kingdom

H.M. BÄR
German Democratic Republic

W.L. McLAUGHLIN
United States of America

R. TANAKA
Japan

DOSIMETRY AND QUALITY CONTROL IN THE PROCESSING OF POLYMERS

L. WIESNER
BGS Beta-Gamma-Service,
Dr. Wiesner GmbH,
Wiehl-Bomig,
Federal Republic of Germany

Abstract

DOSIMETRY AND QUALITY CONTROL IN THE PROCESSING OF POLYMERS.
  For radiation processing of polymers very good reproducibility of the dose setting and dose distribution within the product is an important requisite because the physical and chemical properties of the polymers are strongly dose dependent, while processing steps following the radiation treatment require rather constant material properties. As some physical and chemical properties are not only strongly and reproducibly dose dependent but are also measurable with high accuracy, they can be used for dosimetric purposes. Accuracies of about ± 2% are easily obtainable in routine dose determinations. The methods for measuring solubility and swelling, elastic modulus and other mechanical properties, the melt flow index and viscosity changes of solutions of irradiated polymers are outlined by means of typical calibration curves. Normal variations of the environmental conditions (humidity, temperature, storage time, etc.) before, during and after irradiation have no measurable influence on property changes. The major drawback of these dosimetric methods, with the exception of the polymer degradation dosimeter, is the dose-rate dependence of the property changes in the presence of oxygen. As this dependence is primarily due to diffusion of oxygen into the polymeric material, the effect is negligible in the dose-rate range in which electron irradiations are carried out. The possibilities of standardizing and modifying these methods for general use as dosimeters and adjusting the dose range in which they are applicable are discussed.

1. REQUIREMENTS OF DOSE AND DOSE DISTRIBUTION CONTROL

Most discussions on high-dose dosimetry as a process and quality control measure are conducted with regard to biological applications of radiation processing such as the sterilization of medical supplies and other products and the irradiation of foodstuffs. In such applications, however, relatively large dose inhomogeneity ratios can usually be tolerated, as long as the treatment does not fall short of a specified minimum dose.

They must be tolerated because gamma radiation is used by preference in areas of applications taking advantage of biological radiation effects. In industrial gamma irradiation facilities ratios of the maximum to minimum dose below about 1.25 can be obtained (even for products with a low average density) only by reductions of efficiency in utilization of radiation energy which are economically unacceptable. For the same reason dose inhomogeneity ratios of up to about 2.0 have to be accepted in the gamma radiation of foodstuffs, usually with average densities above 0.4 g/cm^3 in the packaged form in which they are irradiated.

In the radiation processing of polymers for modification and improvement of the physical and chemical properties of these materials such large dose variations within a batch of product can often not be tolerated. The processing steps following radiation treatment require rather constant material properties within a batch as well as for batches irradiated at different times. Constant and reproducible properties can only be obtained if deviations of the actual absorbed dose from a specified optimum dose are sufficiently small. Therefore, established upper and lower limits of the radiation dose must be observed continuously and rigorously because the irradiated product will otherwise become completely useless scrap. An example of the very stringent requirements regarding dose variations is the radiation cross-linking of polymers for the manufacture of heat-shrinkable products and foams with a fine cell structure in the density range below about 0.1 g/cm^3.

Heat-shrinkable products with a large expansion ratio and therefore high shrink forces can only be manufactured from cross-linked materials with a so-called memory for the geometric form and dimensions at the time when the cross-linking of the polymer chains took place. Afterwards, cross-linked products such as tubes, fittings and fasteners are expanded at a suitable temperature and pressure difference between the inside and outside surfaces of the product.

A given set of process parameters for expansion requires a certain degree of cross-linking which may vary only by ± 3% around the ideal value. This corresponds typically to dose variations between ±8 and ±12% in the irradiated product as the still tolerable maximum. In completely automatic expansion facilities for heat-shrinkable tubes even smaller tolerances are required for the undisturbed continuous manufacture of high quality products. Larger deviations from the ideal dose make control of the expansion process and manufacture of heat-shrinkable products with a sufficiently constant diameter and wall thickness much more difficult and reduce the output. If the applied dose deviates in parts of the product by more than ±15% from the nominal value, expansion of the irradiated product becomes practically impossible.

A still more stringent control of the exact dose is required in cross-linking the matrix for foam production. Variations of the dose well below ±10% may already be visible in the foamed product by changes in the cell structure and so reduce the quality classification and the price of the product. Larger deviations make the matrix structure too rigid so that the foaming agent cannot exert its full action (if the dose was higher than the nominal value), or they result in a product with an inhomogeneous surface and cell structure which no customer will accept (if the applied dose was below the nominal value).

However, even for applications in the radiation processing of polymers in which only a specified minimum dose has to be observed, there is still a strong

incentive to have a narrow dose distribution within the product. Modification
and improvement of the properties of these materials usually require much higher
doses than the sterilization of medical supplies. The treatment cost per unit of
mass or volume is correspondingly higher. Hence, it is more desirable to irradiate
as close as possible to the specified lower dose limit in order to make the radiation
processing economically more attractive.

2. DOSIMETRY BY DOSE-DEPENDENT CHANGES OF POLYMER PROPERTIES

The preceding discussion of the requirements of dose and dose distribution
control indicates that rather accurate dosimetry systems are needed in the
processing of plastics. The commercially available systems for routine dosimetry,
which have been developed primarily for biological applications of radiation processing, are usually not accurate enough for the processing of plastics. They are
affected by rather large systematic and random errors, which are caused, among
other things, by variable environmental conditions during exposure in the
irradiation facility. As routine dosimetry usually does not allow control of all
environmental conditions such as humidity, temperature, dose rate and storage
time as carefully as in scientific dosimetry experiments, inaccuracies of dose
determination in the order of 10% may occur quite often [1, 2].

In the radiation processing of polymers such inaccuracies and uncertainties
in dose determination are unacceptable in quality control because they do not allow
investigation and laying down of processing parameters which give the kind of
narrow dose distributions in the product described in Section 1. Therefore,
established commercially available dosimetry systems such as red, amber and
colourless Perspex, dyed nylon, blue cellophane and polyvinylchloride, which are
based on changes of the optical density at a given wavelength (induced by radiation),
have very often only to be considered as orientating measuring devices for approximate dose setting in the processing of polymers.

As changes of the properties of polymer materials, produced by relatively
small variations of the dose, influence the quality of the irradiated products quite
noticeably, it is obvious that such changes can also be used for dose setting,
determination of dose distributions and quality control during production.
However, we use changes of properties of polymer materials in our service
irradiation centre [3] as dosimetric systems for all kinds of product, including
biological applications of radiation processing.

There are many dose-dependent changes of properties of polymeric materials.
For the fine tuning of dose setting and quality control only such properties can be
used that are sensitive to dose variations, at least in certain dose ranges, are well
reproducible under standardized test conditions and are not influenced by normal
variations of environmental conditions before, during and after irradiation. In the

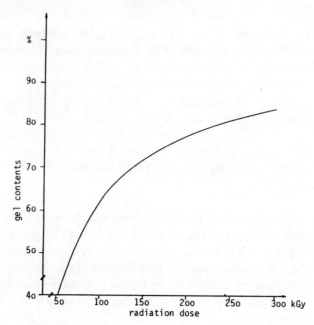

FIG.1. Gel contents of a polyethylene quality as a function of radiation dose.

following subsections a brief description will be given of those methods which, according to our experience, fulfil these requirements and thus are useful for dosimetric purposes and quality control. Their relative merits and drawbacks will also be discussed.

2.1. Gel contents and swelling of cross-linked polymers

Irradiation of cross-linkable polymers such as polyethylene produces a three-dimensional network of polymer chains, thereby reducing the solubility in organic solvents. The solubility can be determined by a solvent extraction test using, for example, xylene as a solvent for polyethylene. A typical curve for the dose dependence of the insoluble part, i.e. the gel fraction of a polyethylene sample, is shown in Fig.1.

The curves of different polyethylene qualities have somewhat different courses which are mainly due to differences in the average molecular weight, the width of the molecular weight distribution and the degree of branching of the polymer chains. However, if the same polyethylene quality is always used, the dose dependence of the gel content is perfectly reproducible. For more than 10 years we have observed the cross-linking behaviour of certain polyethylene qualities used for the insulation of electrical cables and wires without detecting any variation of the dose dependence of the gel content.

Without any special precautions, determinations of the gel contents are reproducible within ±0.5%. This allows dose determinations within an accuracy of ±2% in the range between about 50 and 200 kGy. As the gel contents do not increase very much at higher doses, the error of dose determination becomes greater.

However, the reliability and reproducibility of determinations of the gel content depend in a critical manner on the reduction of the sample to small pieces. As the extraction process is controlled by diffusion of the solvent into single pieces of the sample, the whole soluble part is not really dissolved during an extraction time of several hours if the single particles are too large. Therefore, rather careful control of the size of the single pieces is required for good reproducibility of the results of gel content determinations. A granulator is useful for the production of particles always of equal size. As polyethylene is usually delivered by polymer producers in granular form for further processing, these granules, with typical dimensions of a few 100 μm, can directly be used as dosimeters. Thereby, during the processing step any modification of the material properties can be excluded.

The greatest drawback of determinations of the gel contents is the relatively long time which elapses after the sample is taken until the result becomes available; for extraction of the soluble part of the sample 6 hours are needed in standardized testing procedures. After separation of the gel fraction from the solvent several more hours are needed for the evaporation of solvent residues under vacuum. Reductions of extraction and drying times are possible if larger inaccuracies of determination of the gel contents can be tolerated. However, a large number of determinations can be carried out simultaneously.

The formation of a three-dimensional network also reduces the swelling of polymers in solvents. It has been proposed that the swelling value be used, i.e. the weight ratio of the sample in the swollen and unswollen state as a measure of the degree of cross-linking [4]. Determination of the swelling has to be carried out in a suitable solvent at a temperature above the melting point of crystallites until equilibrium has been reached, i.e. when the weight of the swollen sample no longer increases. However, our experience has shown that the relative reproducibility of the swelling value is not as good as that of the gel contents because at temperatures above the melting point of the crystallites the still soluble fraction of the sample already starts to dissolve. Therefore, the swelling value depends much more than the gel contents on the geometry of the sample and on the swelling time.

2.2. Elastic modulus and other mechanical properties of cross-linked polymers

Polymers are cross-linked in order to make them usable at higher temperatures. A measure for the performance of cross-linked polymers is their elongation under a given load (typically 0.2 N/mm^2) at a suitable temperature, for example 150 or 200°C. The result of measurement is the elastic modulus, which can be deter-

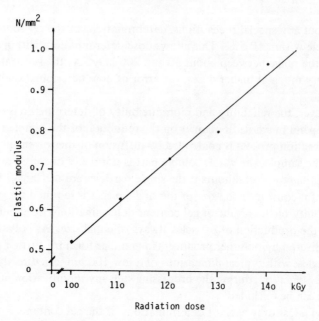

FIG.2. Elastic modulus of a polyethylene quality as a function of radiation dose.

mined routinely with an error of a single measurement up to about ±5%. As the elastic modulus of cross-linked polymers is strongly dependent on the degree of cross-linking and therefore on the amount of absorbed radiation energy, dose determinations which are reproducible within ±2% are easily possible.

This is illustrated in Fig.2, in which the variation of the elastic modulus within a small dose range is shown for a polyethylene quality used for the production of heat-shrinkable tubing. The points represent a set of single measurements carried out as part of the routine quality control of a batch of tubing in order to demonstrate the accuracy of dose determinations that can easily be obtained.

Figure 2 shows only a small dose range. Measurements of the elastic modulus can be performed, however, in a much wider dose range if the determinations are carried out with other specific loads and/or at other temperatures above the melting point of the crystallites. The best accuracy is obtained for elongations of the sample between 100 and 250% [5] and with test strips up to 50 mm long and 5 to 10 mm wide in accordance with existing standards for the determination of the elastic modulus. Measurement is performed rather rapidly; within about 30 min the results are available, taking into account the time needed for heating of the test strip in a suitable oven.

In a similar way the elongation or tensile strength of irradiated and cross-linked polymer samples may serve for dosimetric purposes. Also, the torsion modulus measured, for example, according to the Federal German standard

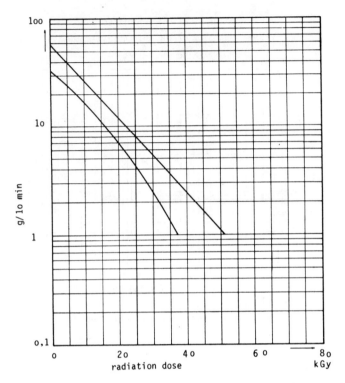

FIG.3. *Melt flow index (190/21.6) of different polyethylene types as a function of radiation dose.*

DIN 53 455 can be used on account of the relationship between this modulus and the density of cross-links, which is proportional to the radiation dose.

2.3. Changes of the molecular weight of irradiated polymers

Before a three-dimensional network springs up at the gel point dose, irradiation causes an additional polymerization. During this process the molecular weight rises sharply until it approaches practically infinite values when the gel point is reached. For a given polymer quality the melt flow index is a simple measure of molecular weight. It is determined by the mass of material which is pressed by a fixed load at a fixed temperature through a nozzle per unit of time.

While repeated measurements of the melt flow index of the same material may differ up to about ±10%, rather accurate dose determinations are nevertheless possible because such variations correspond only to much smaller dose differences, at least above 10 kGy. Figure 3 shows the measured dose dependence of the melt flow index for two polyethylene qualities.

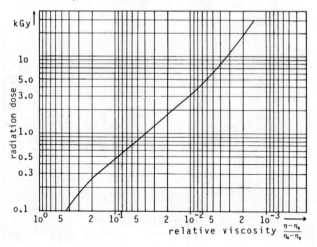

FIG.4. Relative viscosity of a polyisobutylene solution in heptane as a function of radiation dose.

As we have only recently started experiments using the melt flow index for dosimetric purposes in the lower 10 kGy range, still better accuracies can be expected in the future when the sources of errors have been identified and eliminated. Results of measurements of the melt flow index can be expected within less than 1 hour after sampling.

For the methods described polyethylenes (easily cross-linkable materials) were used. Other polymers are degraded by ionizing radiation. If solutions of such polymers in organic solvents are irradiated, their viscosity decreases rapidly. As viscosities can be measured very accurately, dose determinations with a mean statistical error of about ±1% are possible.

Such a polymer degradation dosimeter has been discussed in a previous publication [6]. Figure 4 shows the calibration curve for a solution of 20 g polyisobutylene with an average viscosity molecular weight of 219.000 in 1 L heptane for doses between 0.1 and more than 10 kGy. The same calibration curve can be used for polyisobutylene batches with other molecular weights. Only the radiation dose has to be added or subtracted; this is required to compensate for the differences in molecular weight between different batches. Doses of up to several 100 kGy can be measured by using a polyisobutylene with a substantially higher molecular weight.

Viscosity measurements require sample volumes of at least a few cubic centimetres. Therefore, polymer-degradation dosimeters are primarily suitable for dosimetry and quality control in gamma irradiation facilities. The degradation process of dissolved polyisobutylene is the same in air, pure oxygen and nitrogen. A dose-rate dependence was not observed between 0.01 and 2 kGy/h.

3. DISCUSSION AND CONCLUSIONS

The common feature of the property changes of polymers outlined in the preceding section is their direct relation to the changes of the molecular weight or the number of cross-links which are produced by a given absorbed radiation energy. For a specific type of polymer these effects are quantitatively reproducible. Even different batches of the same type of polymer do not show noticeable differences in the dose dependence of the molecular weight or the number of cross-links. For routine dosimetry it may be preferable to select and calibrate batches of polymers whereby a still better reproducibility of single dose measurements than ±2% may be achievable.

It is also obvious that usual changes of storage times and environmental conditions such as temperature and humidity before, during and after irradiation do not noticeably influence the changes in a polymer by a given dose. If such changes should occur, the successful radiation processing of polymer products, based on properties of the irradiated material which are constant within rather narrow limits, would in many cases be impossible because under industrial manufacturing conditions storage times and environmental factors cannot be held constant all the time. Thus, the handling of dosimeters, utilizing property changes of polymers induced by ionizing radiation, does not require special precautions in order to achieve high accuracy.

The major drawback of these methods, with the exception of the polymer degradation dosimeter, is the dependence of the G-value of cross-linking of most polymers on the dose rate if irradiation does not occur at sufficiently low partial pressures of oxygen. As diffusion of oxygen into the polymer during irradiation is preponderantly responsible for this dose-rate effect in most polymers, it is negligible for irradiations with electrons in air when the dose rates are always in the range of several 10 kGy/s. During the resulting short irradiation times no oxygen can diffuse into the polymer.

For gamma irradiations this dose-rate effect is a great disadvantage as the dose rates are typically in the order of 1 Gy/s. Moreover, they are usually not sufficiently constant in industrial facilities during the movement of the product through the irradiation chamber. One way of overcoming this disadvantage is irradiation of the dosimeter substance in small ampoules under a sufficiently reduced partial pressure of oxygen or in a heat-shrinkable envelop which prevents the penetration of oxygen into the dosimeter substance.

As the presence of oxygen does not influence further polymerization and cross-linking of other polymers as much as that of polyethylene (on which our experience is mainly based), it may well be that a systematic search will lead to a suitable dosimeter material for which the dose-rate effect in the presence of oxygen is sufficiently small, at least in the range of dose rates usually encountered in gamma irradiation facilities.

Another way of overcoming the dose-rate effect in the presence of oxygen is the addition of a sensitizer to the polymeric dosimeter substance. Such sensitizers promote the cross-linking reaction and reduce the dose required for a well measurable property change of the dosimeter substance. Thus, it is also possible to use measurements of gel contents, elastic modulus, etc. for accurate determination of doses in the range from a few kilograys upwards.

The property changes of polymers, which are caused by their exposure to ionizing radiation, offer many possibilities for relatively accurate routine dosimetry and quality control. However, as yet systematic efforts have not been made to make use of this potential for the benefit of radiation processing.

REFERENCES

[1] CHADWICK, K.H., Radiat. Phys. Chem. **14** (1979) 203.
[2] MILLER, A., CHADWICK, K.H., NAM, J.W., Radiat. Phys. Chem. **22** (1983) 31.
[3] WIESNER, L., Radiat. Phys. Chem. **22** (1983) 953.
[4] VOIGT, H.U., Kautschuk und Gummi, Kunststoffe **29** (1976) 17.
[5] BÄUERLEIN, R., Bundesministerium für Forschung und Technologie, Bonn, Rep. BMFT-FB-T 81-123 (1981) 24.
[6] WIESNER, L., "The use of polyisobutylene solutions for measuring doses from 10^3 rad up to about 10^{10} rad", Selected Topics in Radiation Dosimetry (Proc. Symp. Vienna, 1960), IAEA, Vienna (1961) 361.

IAEA-SM-272/23

LiF THERMOLUMINESCENCE DOSIMETRY FOR MAPPING ABSORBED DOSE DISTRIBUTIONS IN THE GAMMA RAY DISINFECTION OF MACHINE-BALED SHEEP WOOL

Dexi JIANG
Institute of Atomic Energy,
Bejing, China

Presented by Lishu Chen

Abstract

LiF THERMOLUMINESCENCE DOSIMETRY FOR MAPPING ABSORBED DOSE DISTRIBUTIONS IN THE GAMMA RAY DISINFECTION OF MACHINE-BALED SHEEP WOOL.

The measurement of absorbed dose distributions of ^{60}Co γ-rays in machine-baled sheep wool, which is disinfected of certain parasitic bacteria (e.g. *Brucella* bacilli) by γ-ray treatment, is summarized. The preparation and main physical properties of the LiF-TLD are described, as well as the shape, structure and the activity of the ^{60}Co source and typical dose distributions measured around the source in free air. The results of dose distributions measured by the LiF-TLD agreed within $\pm$ 5% with those given by a calibrated ionization chamber. The exposure rates (units R/min) at three typical measurement points inside a bale of sheep's wool were found to be quite uniform: centre 3.8×10^3 ($\pm$ 2.1%); upper region 3.9×10^3 ($\pm$ 2.4%); lower region 3.9×10^3 ($\pm$ 1.9%).

1. AIMS OF THE DOSIMETRY

For the protection of wool spinning workers from the *Brucella* bacilli, which colonize on sheep wool, disinfection by use of ^{60}Co irradiation has been widely used. It is important to measure the dose distribution due to γ-ray irradiation inside machine-formed bales of the wool. The irradiation is designed to kill efficiently the bacteria, but without exceeding doses that would cause deterioration of the wool. By accurate dosimetry it is possible to improve the efficiency and provide uniform dose distribution when using radionuclide sources for such disinfection procedures.

2. SELECTION OF THE DOSIMETRY METHOD

To measure satisfactorily the dose distribution of ^{60}Co γ-rays in machine-baled sheep wool, the detector should be inserted in the bale in such a way that the material and the geometry at the measurement points in the bale are not appreciably distorted by the detector. It is required that the size of the detector be negligible compared with the size of the bale of wool, and that the material of the detector should have similar radiation absorption properties to wool. LiF-TLD was found to be satisfactory for this purpose.

3. MAIN PHYSICAL PROPERTIES AND PREPARATION OF THE LiF-TLD

Li_2CO_3 was converted chemically into LiF, and specified trace concentrations of activator elements (Mg, Ti) were added. The doped LiF was then drawn to form a monocrystal of LiF (Mg, Ti) consisting of a homogeneous mixture. Because of the intrinsic inhomogeneity of the distribution of trace elements (Mg, Ti) during the process of drawing monocrystal, it had to be smashed, and the powder (about 100 ~ 180 mesh in size) was then sieved. This monocrystalline powder was heated and finally pressed into LiF (Mg, Ti) tablets after cooling to room temperature. The tablets were machined into 1 × 1 × 6 mm elements.

Since LiF (Mg, Ti) [1] may be considered approximately equivalent to wool in terms of radiation absorption properties, and since the size of these elements (1 × 1 × 6 mm) meets the requirements of the Bragg-Gray principle, it is feasible to use this material directly to measure absorbed doses of γ-rays in the wool bale.

To extend the linear range of the dose measurement, a sensitization processing of the LiF-TLD was made. The linear range used in the measurement could thus be extended to the range 2×10^{-3} to 5×10^4 R after such processing.[1]

4. DESCRIPTION OF ^{60}Co SOURCE AND RESULTS OF DOSE FIELD MAPPING

The ^{60}Co source is a rectangular plate source containing 27 source capsules (94 bars of radioactive cobalt altogether). The source plate activity is vertically symmetrical, with dimensions of 1.0 m in length and 0.75 m in width. The total activity at the time of the measurements was 1×10^5 Ci (3.7×10^6 GBq).

To measure the dose distribution of ^{60}Co γ-rays in free air, a large wooden frame was positioned perpendicular to the vertical source plate. A horizontal thread was strung along the central normal line of the vertical plate, and along this thread were placed at 10-cm intervals the LiF-TLD elements, each of which was held in a paper pouch having a wall thickness of 1.08 mg/cm². There were 11 measurement points along the thread 10 cm to 110 cm from the outside surface of the source plate. The exposure rates measured in this geometry were in the range 2200 to 280 R/min. The values obtained by the LiF-TLD and the data

---

[1] 1 R = 2.58×10^{-4} C/kg.

measured by a calibrated ionization chamber[2] agreed within 5%. The PTW ionization chamber had been compared with a 2502/3 Farmer-NE-English meter, with an agreement to within 1.5%.

5. EXPOSURE RATES AT THREE LOCATIONS ALONG THE DIAGONAL LINE ACROSS THE CENTRE OF THE WOOL BALE

A bale of sheep's wool was placed very close to the source plate for irradiation, and three typical measurement locations were selected along the diagonal line across the centre of the machine bale side closest to the source plate: at the centre of the wool bale, near one of the upper corners, and near one of the lower corners. LiF-TLD elements were placed at these measurement points along a small hole (ϕ 1.6 mm) through the bale. The mean values (in R/min) of 12 measured exposure rates for each of the three locations were respectively 3.9×10^3 ($\pm 2.4\%$), 3.8×10^3 ($\pm 2.1\%$) and 3.9×10^3 ($\pm 1.9\%$).

6. CONCLUSIONS

It has been shown that small calibrated LiF-TLDs are useful to map dose rate distributions or exposure rate distributions, both in air and in a material being processed or disinfected at high dose rates. In the present study, the exposure rates in sheep wool baled by machine and irradiated by gamma radiation close to a ^{60}Co source were found to be quite uniform, at least along a plane close to the source plate.

REFERENCE

[1] CAMERON, J., "Lithium fluoride thermoluminescent dosimetry," Manual on Radiation Dosimetry (HOLM, N.W., BERRY, R.J., Eds), Marcel Dekker, New York (1970) 405.

---

[2] The ionization chamber used was supplied by the Physikalisch-Technische Werkstätte (PTW) Freiburg, Federal Republic of Germany.

MEASUREMENT OF HIGH DOSES NEAR METAL AND CERAMIC INTERFACES*

W.L. McLAUGHLIN, J.C. HUMPHREYS,
M. FARAHANI
Center for Radiation Research,
National Bureau of Standards,
Gaithersburg, Maryland,
United States of America

A. MILLER
Accelerator Department,
Risø National Laboratory,
Roskilde, Denmark

Abstract

MEASUREMENT OF HIGH DOSES NEAR METAL AND CERAMIC INTERFACES.
 Radiochromic dosimeters consisting of leuco dyes dissolved and cast in very thin (5 to 100 μm) plastic films have been shown to be accurate and reproducible dosimeters for measuring absorbed doses in the range 10^3 to 10^6 Gy. There are also thin, optical-quality ceramic crystals (e.g. LiF, NaCl and CaF_2) having thicknesses ~0.1 to 2 mm that can provide precise absorbed dose readings in the range 10^2 to 10^9 Gy by spectrophotometric readings of a series of radiation-induced colour-centre absorption bands. Besides their relatively broad response ranges, these dosimeters have the advantages of being useful in both photon and electron radiation fields, without great losses in accuracy due to rate or temperature dependence. The plastic films are particularly useful for mapping high-resolution dose distributions, such as depth-dose or isodose contours in thin layers, tubing and wire insulation. It has been shown that, by suitable selection of these plastic and crystalline systems, a fairly wide assortment of materials can be simulated in terms of radiation absorption properties over wide photon and electron spectral ranges (0.01 to 10 MeV). To minimize the energy dependence of response of many dosimetric materials relative to that of silicon or silicon dioxide layers in solid-state devices, for example, certain radiochromic dosimeters can be selected to simulate the ionizing electron and photon absorption properties of Si and SiO_2 over the above energy range. These dosimeters are then used to measure high-resolution dose distributions in 'silicon' and 'silicon dioxide' close to metal interfaces. For degraded ^{60}Co gamma ray spectra, polychlorostyrene having 25% chlorine simulates SiO_2 and 71% NaCl plus 29% LiF simulates Si; for electrons, polybromostyrene having 43% bromine simulates SiO_2 and 71% CaF_2 plus 29% LiF simulates Si in terms of radiation absorption cross-sections.

---

* Work carried out under NBS-IAEA Research Agreement 3061/CF.

1. INTRODUCTION

When multi-layer materials are irradiated with ionizing photons or electrons the absorbed dose varies sharply near interfaces, particularly when there is a large difference in electron densities and radiation absorption properties of the materials making up each interface. This effect is sometimes called a 'transition zone' dose excursion [1] or, in the case of semiconductor assemblies, it is referred to as 'dose enhancement' in the sensitive layer (Si, SiO_2) near a high-Z thin layer (Au) or other metal [2–4]. Discontinuities of dose distributions near plastic-metal interfaces are also of concern in electron or gamma ray radiation processing of materials [5–8] and in dosimetry itself, as when aqueous dosimeter solutions are held in glass-walled containers [9–11].

Determination of such dose distributions is accomplished either by making radiation transport calculations [12–14] or by using high-resolution measurement systems, such as thin calorimeters [15, 16], extrapolation ionization chambers [17–21], thermoluminescence dosimeters [22], electron beam resists [23], or by thin radiochromic film dosimeters [5, 8, 10, 11, 24, 25].

In the present work, thin radiochromic and ceramic dosimeters are designed to simulate the electron or photon absorption properties of sensitive layers in electronic devices (Si or SiO_2 typically). These thin dosimeters are then calibrated and are used to measure absorbed doses close to metal interfaces when irradiated by electron beams or ^{60}Co or ^{137}Cs gamma radiation. The measured dose distributions close to such interfaces may be compared with the Monte Carlo calculations of Seltzer [12].

2. DOSIMETER CHARACTERISTICS

Thin radiochromic films are supplied commercially[1] in a number of formulations [26]. The present work deals with polyhalostyrene thin-film dosimeters [27, 28] and contributes particular designs and combinations intended to simulate the electron and photon absorption properties of silicon and silicon oxide layers in semiconductors. These special radiochromic films are compared with conventional nylon-base films [26] for use close to interfaces of metals, ceramics, plastics and air irradiated with high-energy photons and electrons.

Thin, polished, optical-quality, alkali halide crystalline materials are also commercially available[2] and, using colour-centre (F and M-centre) analysis

---

[1] Far West Technology, Inc., Goleta, California, United States of America.
[2] Harshaw/Filtrol, Solon, Ohio, United States of America.

The listing of these commercial suppliers is only for acknowledgement of the source of material and does not imply endorsement or suggestion that these materials are superior to similar materials from other sources.

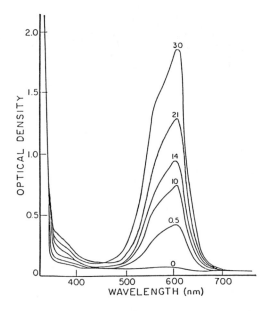

FIG.1. *Absorption spectra of radiochromic nylon film containing hexa(hydroxyethyl) pararosaniline cyanide. Values of absorbed dose are indicated in kGy.*

by spectrophotometry, serve for high-dose dosimetry [29, 30]. Instead of using the more common dosimeter, namely LiF [29, 31], F and M-centres in NaCl plus CaF_2 [32] and pure NaCl [33, 34] as well as combinations of NaCl and LiF are considered here, since these higher atomic-number crystalline materials simulate more closely the radiation absorption properties of Si and SiO_2. The optical-quality crystalline materials have radiation-induced colour centres (F and M) that are stable enough at room temperature to serve for high-dose dosimetry.

Tables I and II list the radiochromic thin-film and alkali halide thin-chip dosimeter materials, thicknesses, densities and elemental compositions, as well as the typical wavelengths of radiation-induced absorption band maxima for spectrophotometric analysis. The thinner dosimeter pieces are designed for measurement of absorbed dose distributions close to metal and ceramic interfaces. Radiation-induced absorption spectra for different gamma-ray absorbed doses in five of the materials of interest are shown in Figs 1 to 5. In Figs 4 and 5, the vertical lines are at the spectral wavelengths of analysis for dosimetry in the absorbed dose range of interest (~1 to 10 kGy). Figures 6 and 7 show typical response curves for these dosimeters.

The absorption bands formed in the radiochromic dye films are due to radiolytic scission of the −CN or −OCH_3 group from the colourless leuco dye

TABLE I. RADIOCHROMIC THIN-FILM DOSIMETERS

Film type	Radiochromic leuco dye (wavelengths for analysis)	Thickness (mm)	Density (g/cm³)	Area density (g/cm²)	Fractional elemental composition by weight	
Nylon	Hexa(hydroxyethyl) pararosaniline CN (605 nm)	0.0095 0.056	1.08	1.03 6.05	H: C: N: O:	0.104 0.648 0.099 0.149
Polychlorostyrene (25% Cl)	Malachite green methoxide (628 nm)	0.011 0.056	1.29	1.42 7.22	H: C: Cl:	0.058 0.692 0.250
Polybromostyrene (43% Br)	Malachite green methoxide (632 nm)	0.0075 0.081	1.495	1.12 12.1	H: C: Br:	0.044 0.526 0.430
Polybromostyrene (10% Br)	Malachite green methoxide (630 nm)	0.080	1.26	10.1	H: C: Br:	0.070 0.830 0.100
Polybromostyrene (2% Br)	Malachite green methoxide (630 nm)	0.072	1.065	7.67	H: C: Br:	0.075 0.905 0.020

TABLE II. ALKALI HALIDE THIN CRYSTAL DOSIMETERS

Crystal type	Wavelength for spectro-photometry (nm)	Thickness (mm)	Density (g/cm³)	Area density (mg/cm²)	Fractional elemental composition by weight	
NaCl	460, 727 or 380	0.12	2.17	26.0	Na:	0.393
					Cl:	0.607
71% NaCl 29% LiF	460, 727 or 380	0.082	2.24	18.4	Na:	0.333
					Cl:	0.513
					Li:	0.041
					F:	0.113
LiF	247 or 444	1.0	2.63	263	Li:	0.268
					F:	0.732
71% CaF₂ 29% LiF	257, 595 or 385	0.073	2.83	20.7	Ca:	0.452
					F:	0.516
					Li:	0.032

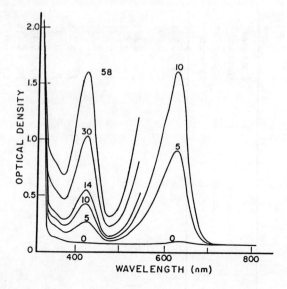

FIG.2. Absorption spectra of radiochromic polychlorostyrene (25% Cl) film containing malachite green methoxide. Values of absorbed dose are indicated in kGy.

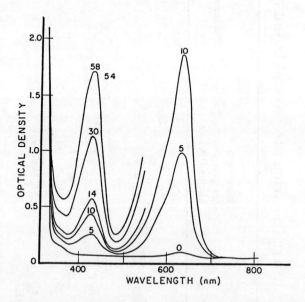

FIG.3. Absorption spectra of radiochromic polybromostyrene (43% Br) film containing malachite green methoxide. Values of absorbed dose are indicated in kGy.

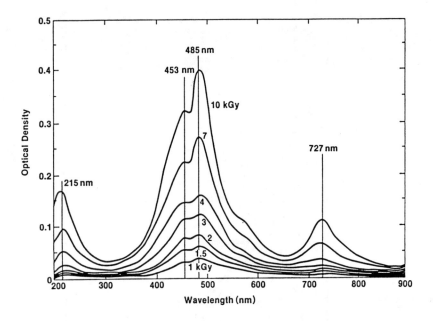

FIG.4. Absorption spectra of optical-quality thin crystal, 71% NaCl plus 29% LiF. Values of absorbed dose are indicated. Vertical lines indicate wavelengths for analysis of F- and M-centre absorption band amplitudes. These spectra were measured 24 hours after irradiation.

molecule, resulting in isomerization to the highly coloured polar carbonium cation of the dye [35]. These bands increase slightly in intensity during the first few hours after irradiation.

The bands induced in the alkali halide materials, which are due to F- and M-centres (electron-filled negative ion vacancies), are less stable than those in the radiochromic films and require about a 24-hour delay period after irradiation before spectrophotometric analysis should begin. The F-centre band generally decreases in intensity whereas the M-centre band increases in intensity during this period. For these materials, slight instabilities continue for some weeks (except in the case of nylon film and LiF crystals), but suitable corrections for fading or growth of colour can be made.

3. ENERGY DEPENDENCE RELATIVE TO Si AND SiO_2

The main purpose of using these thin dosimeter systems is to simulate the ionizing photons and electron absorption properties of silicon and silicon dioxide, so that meaningful absorbed dose distributions close to interfaces with

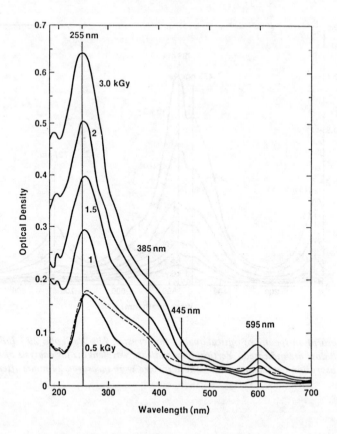

FIG.5. Absorption spectra of optical-quality thin crystal, 71% CaF_2 plus 29% LiF. Values of absorbed dose are indicated. Vertical lines indicate wavelengths for analysis of F- and M-centre absorption band amplitudes. These absorption spectra were measured 24 hours after irradiation, except for the dashed-line spectrum, which was made for a 0.5 kGy absorbed dose 10 minutes after the end of irradiation.

other materials can be measured. Tables III and IV list, respectively, the computed photon mass energy-absorption coefficients versus photon energy [36] and the electron mass collision stopping powers versus electron energy [37] for Si and SiO_2, as well as for the selected dosimeter systems tabulated in Tables I and II. From these data it is evident that, for the broad photon and electron spectral regions, the following comparison may be made. For ionizing photons, 71% NaCl plus 29% LiF or polybromostyrene with 10% bromine (Br-10) are similar to silicon, and polychlorostyrene with 25% chlorine (Cl-25) simulates silicon dioxide. For electrons, 71% CaF_2 plus 29% LiF resembles silicon, and LiF or polybromostyrene with 43% Br (Br-43) may be compared with silicon dioxide.

The comparisons are illustrated graphically by comparing absorption coefficients and their ratios versus radiation spectral energy for the different

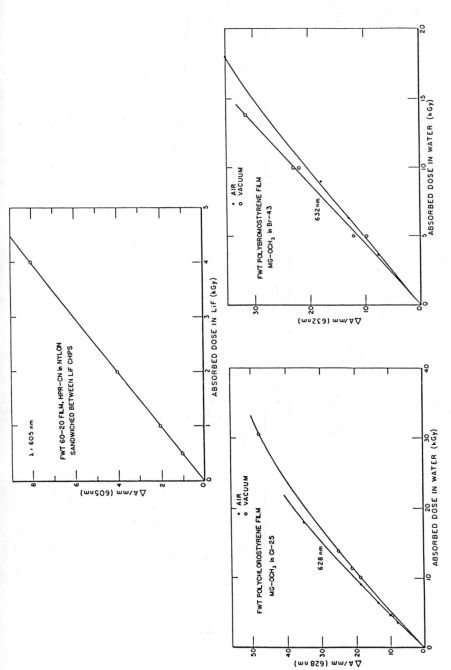

FIG.6. Response curves in terms of increase in optical density per unit film thickness ($\Delta A/mm$) as a function of absorbed dose for three types of radiochromic film dosimeters. For polyhalostyrene films, differences in the response functions are seen between irradiations in air and in vacuum. There is a negligible difference in response between these conditions for nylon film.

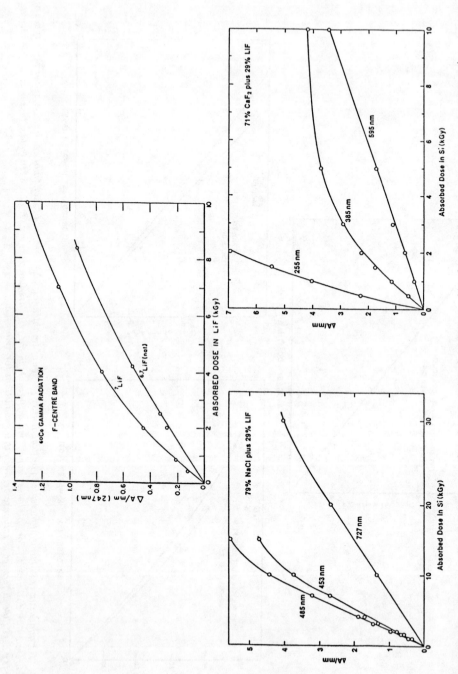

FIG.7. Response curves of three kinds of optical-quality thin crystal dosimeters. Wavelengths of spectrophotometric analysis are indicated. The LiF dosimeter is represented in two forms, natural pure LiF and ^{7}Li-enriched pure LiF.

TABLE III. PHOTON MASS ENERGY-ABSORPTION COEFFICIENTS $\left(\dfrac{\mu}{\rho}\right)^{en}$ (cm^2/g) [35]

Energy (MeV)	Si	SiO$_2$	NaCl	71% NaCl 29% LiF	71% CaF$_2$ 29% LiF	CaF$_2$	Nylon	Cl-25	Polyhalostyrenes[a] Br-2	Br-10	Br-43
0.03	1.15	0.627	1.46	1.09	1.46	1.98	0.0856	0.568	0.310	1.39	5.48
0.06	0.142	0.0831	0.180	0.137	0.186	0.249	0.0243	0.0796	0.0617	0.227	0.875
0.10	0.0449	0.0335	0.0522	0.0435	0.0540	0.0671	0.0237	0.0339	0.0320	0.0697	0.218
0.15	0.0308	0.0277	0.0321	0.0296	0.0329	0.0367	0.0268	0.0288	0.0289	0.0401	0.0840
0.20	0.0291	0.0278	0.0290	0.0278	0.0293	0.0311	0.0290	0.0291	0.0296	0.0342	0.0524
0.60	0.0295	0.0296	0.0284	0.0285	0.285	0.0289	0.0322	0.0310	0.0319	0.0315	0.0307
1.00	0.0278	0.0278	0.0267	0.0265	0.0271	0.0271	0.0304	0.0292	0.0301	0.0295	0.0282
4.00	0.0196	0.0192	0.0189	0.0185	0.0186	0.0192	0.0201	0.0196	0.0199	0.0198	0.0202
10.00	0.0175	0.0161	0.0175	0.0160	0.0164	0.0174	0.0148	0.0154	0.0150	0.0153	0.0188

[a] The percentages by weight of the halogens in polystyrene are indicated (e.g. Cl-25 indicates 25% by weight of chlorine in polystyrene).

TABLE IV. ELECTRON MASS COLLISION STOPPING POWERS $\left(\frac{1}{\rho} \cdot \frac{dE}{dx}\right)^{coll}$ (MeV·cm²·g⁻¹) [36]

Energy (MeV)	Si	SiO$_2$	NaCl	LiF	71% CaF$_2$ 29% LiF	CaF$_2$	Nylon	Cl-25	Polyhalostyrenes[a] Br-2	Br-10	Br-43
0.03	7.48	7.80	7.17	7.75	7.49	7.39	9.78	8.8	9.0	8.9	7.8
0.06	4.56	4.74	4.38	4.67	4.56	4.50	5.86	5.33	5.69	5.48	4.76
0.10	3.27	3.39	3.14	3.32	3.26	3.22	4.15	3.79	4.03	3.91	3.41
0.15	2.59	2.68	2.48	2.62	3.57	3.55	3.26	2.98	3.18	3.08	2.69
0.20	2.24	2.32	2.15	2.26	2.23	2.21	2.81	2.57	2.74	2.66	2.32
0.60	1.60	1.64	1.53	1.58	1.58	1.57	1.96	1.81	1.91	1.86	1.64
1.00	1.51	1.54	1.46	1.49	1.49	1.48	1.84	1.72	1.81	1.75	1.56
4.00	1.58	1.60	1.53	1.51	1.54	1.53	1.87	1.78	1.84	1.81	1.64
10.00	1.69	1.70	1.63	1.59	1.64	1.63	1.96	1.91	1.96	1.92	1.78

[a] The percentages by weight of the halogens in polystyrene are indicated (e.g. Cl-25 indicates 25% by weight of chlorine in polystyrene).

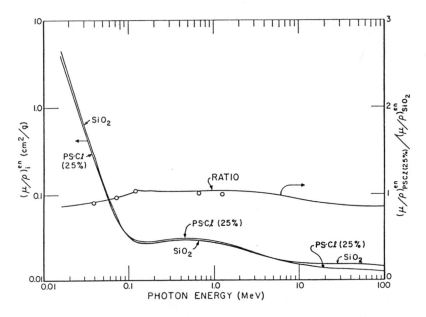

FIG.8. Left ordinate: Calculated mass energy-absorption coefficients of polychlorostyrene (25% Cl) and of SiO_2 versus photon energy. Right ordinate: Ratio of the mass energy-absorption coefficients versus photon energy. Experimental points are plotted against the right ordinate and represent the response of the polychlorostyrene dosimeter to X- and γ-rays of various photon energies relative to the response to 1.25 MeV photons when the absorbed dose is expressed in kGy in SiO_2.

materials [11]. In the case of ionizing photons, Figs 8 and 9 show plots of mass energy-absorption coefficients (left ordinate) versus photon energy, which predict a flat energy dependence of response of the two dosimeters, polychlorostyrene Cl-25 radiochromic film relative to SiO_2 and 29% LiF plus 71% NaCl relative to Si. The ratios of absorption coefficients (right ordinate) are very close to unity over the entire photon energy range. The points represent experimental data, where the responses of the dosimeters (increase of optical absorbance versus dose) relative to that at 1.25 MeV (gamma radiation from ^{60}Co) are plotted on the right ordinate, and the absorbed doses are expressed in terms of energy imparted to SiO_2 (Fig.7) or Si (Fig.9). The experimental arrangement and photon energy values are described in Ref.[38].

Figures 10 to 12 show, for electrons, plots of computed [11] mass collision stopping powers and their ratios as a function of electron energy for three of the dosimeter materials relative to Si or SiO_2. These plots thus predict a relatively energy-independent response of the dosimeters over a broad spectral energy range.

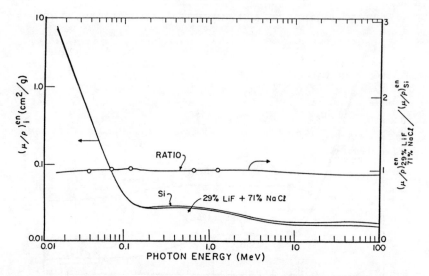

FIG. 9. Left ordinate: Calculated mass energy-absorption coefficients of 29% LiF plus 71% NaCl and of Si versus photon energy. Right ordinate: Ratio of the mass energy-absorption coefficients versus photon energy. Experimental points are plotted against the right ordinate and represent the response of the 29% LiF plus 71% NaCl dosimeter to X- and γ-rays of various photon energies relative to the response to 1.25 MeV photons when the absorbed dose is expressed in kGy in SiO_2.

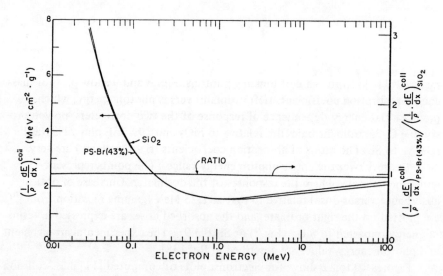

FIG. 10. Left ordinate: Calculated mass collision stopping powers of polybromostyrene (43% Br) and of SiO_2 versus electron energy. Right ordinate: Ratio of the mass collision stopping powers versus electron energy.

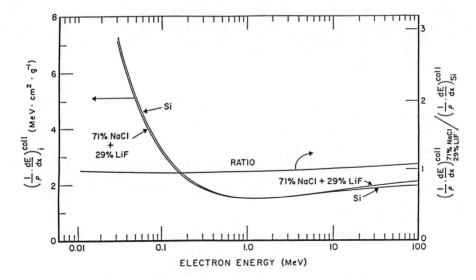

FIG.11. *Left ordinate: Calculated mass collision stopping powers of 71% NaCl plus 29% LiF and of Si versus electron energy. Right ordinate: Ratio of mass collision stopping powers versus electron energy.*

4. DOSE MEASUREMENTS CLOSE TO INTERFACES

Four radiation geometries and spectral distributions were used: (1) ^{60}Co gamma radiation of approximately isotropic incidence in air, as provided in a cylindrical stainless-steel canister surrounded closely by 12 ^{60}Co source rods and a large pool of water (described in Refs [39, 40]); (2) ^{60}Co gamma radiation from a single source giving a Pb-collimated beam (5 × 5 cm field at 80 cm from the source) typical of cancer therapy sources; (3) a scanned 10 MeV electron beam having a nearly monoenergetic electron spectral distribution [41]; (4) a scanned 0.4 MeV electron beam having a broad low-energy spectral distribution [42]. The calculated photon spectrum for case (1) is shown in Fig.13, having a ratio of low-energy scattered photons to total photons of 0.40 [40]. The photon spectrum for case (2) was measured by Ehrlich et al. [43] and found to have a 0.13 ratio of scattered low-energy to total photons, also peaking at ~200 keV.

(a) Geometry (1), cylindrically isotropic gamma ray field

The irradiation geometry is illustrated in Fig.14, where the axis of the interface and dosimeter cylindrical stack is perpendicular to the axis of the annular ^{60}Co source geometry, being supported on a piece of styrofoam at a position 6.5 cm from the bottom of the can, where the isodose pattern is quite uniform [44].

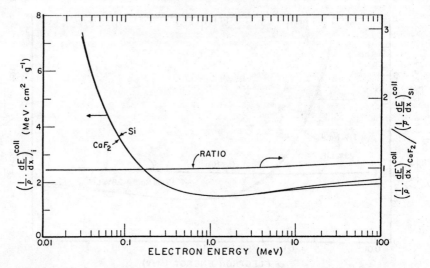

FIG.12. Left ordinate: Calculated mass collision stopping powers of CaF_2 and of Si versus electron energy. Right ordinate: Ratio of the mass collision stopping powers versus electron energy.

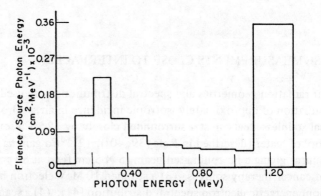

FIG.13. Calculated photon spectrum [40] in units of number of photons per $cm^{-2} \cdot MeV^{-1}$ for the central position in air at a height of 6.5 cm from the bottom of the stainless-steel can surrounded by 12 ^{60}Co source rods and water, as illustrated in Fig.14. The ratio $(\frac{S}{T})$ of scattered (<1.1 MeV) to total fluence is 0.404.

The interface materials in all cases are thin layers of either gold (0.3 mm), silicon (2 mm), Kovar[3] (0.3 mm) or air. The spacers are either nylon or silicon dioxide to match the film materials used in the measurements. Figures 15 and 16 show the measured dose distributions in nylon and SiO_2 (or polychlorostyrene-25% Cl) as a function of thickness of material to ~0.5 mg/cm² from the interface. The

---

[3] 'Kovar' is an alloy consisting of 54% Fe, 28% Ni and 18% Co by weight with a thin Au coating and is the lid material for many metal oxide semiconductors.

IAEA-SM-272/10

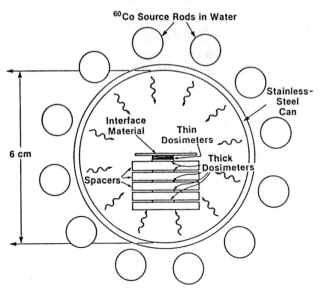

FIG.14. *Diagram of the annular source geometry for cylindrically isotropic irradiation in air of the dosimeter stack against a thin interface material. When air is the interface material, the thin layer is omitted. The thicknesses of the dosimeter layers are not in scale and are actually much thinner than represented here.*

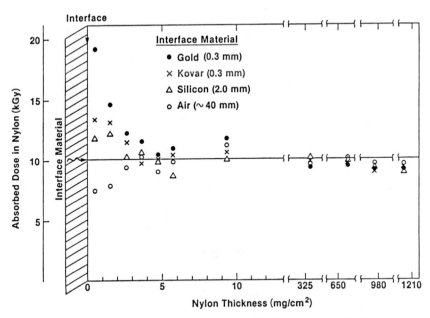

FIG.15. *For the irradiation geometry of Fig.14 (nominal absorbed dose 10 kGy in nylon), the readings of dose by calibrated films in the stack placed against the four different interface materials are plotted as a function of area density (thickness) away from the interface. The spacer materials and the dosimeters are both nylon. Note that the interface material is actually much thicker than represented by the scale of the abscissa.*

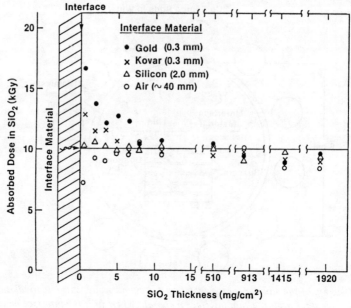

FIG.16. *For the irradiation geometry of Fig.14 (nominal absorbed dose 10 kGy in SiO_2), the readings of dose by calibrated films in the stack placed against the four different interface materials are plotted as a function of area density (thickness) away from the interface. The spacer materials are quartz (SiO_2), and the dosimeters are thin polychlorostyrene (25% Cl) radiochromic films.*

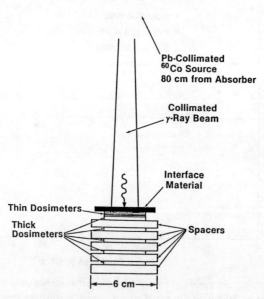

FIG.17. *Diagram of the collimated gamma ray beam geometry for irradiation in air of the dosimeter stack against a thin interface material. When air is the interface material, the thin layer is omitted. The thicknesses of the dosimeter layers are not in scale and are actually much thinner than represented here.*

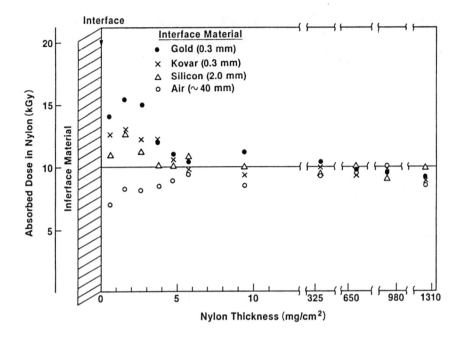

FIG.18. *For the irradiation geometry of Fig.17 (nominal absorbed dose 10 kGy in nylon), the readings of dose by the calibrated films in the stack placed against the four different interface materials are plotted as a function of area density (thickness) away from the interface. The spacer materials and the dosimeters are both nylon.*

results show that, at least over dimensions of up to 1 or 2 g/cm² thickness from the interface, the dose distribution is fairly uniform, except within ~10 mg/cm² thickness where there is a marked dose 'excursion'.

(b) Geometry (2), collimated gamma ray beam

The irradiation geometry is shown in Fig.17, where the beam axis coincides with the axis of the interface and dosimeter cylindrical stack. The uniform beam cross-section at the location of the interface is 2 × 2 cm². The interface and spacer materials are the same as those used with geometry (1). Figures 18 and 19 show the measured dose distributions in nylon and SiO_2 (or polychlorostyrene-25% Cl) as a function of thickness of material to ~0.6 mg/cm² from the interface. The results are similar to those for 'isotropic' radiation incidence, except that there is a dose build-up region within about 2 or 3 mg/cm² thickness from the interface.

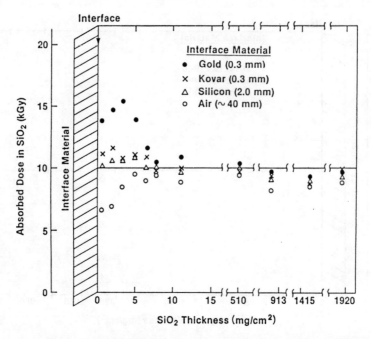

FIG.19. For the irradiation geometry of Fig.17 (nominal absorbed dose 10 kGy in SiO$_2$), the readings of dose by the calibrated films in the stack placed against the four different interface materials are plotted as a function of area density (thickness) away from the interface. The spacer materials are quartz (SiO$_2$), and the dosimeters are thin polychlorostyrene (25% Cl) radiochromic films.

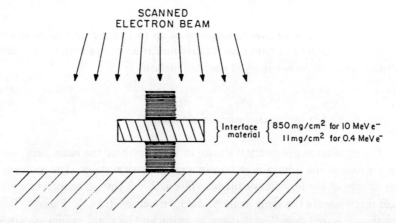

FIG.20. Diagram of scanned electron irradiation geometry for both 10 and 0.4 MeV beams incident on radiochromic film stacks on both sides of an interface material. The stack assembly is backed by near unit density semi-infinite material (polystyrene or polystyrene on wood). In some cases thinner films are on the outsides of both top and bottom film stacks. For 10 MeV beam irradiation, the interface material is much thicker relative to the film stack thicknesses than represented here.

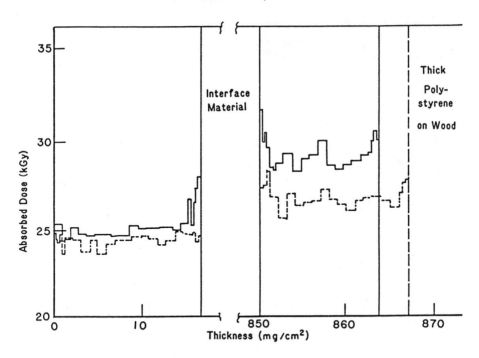

FIG.21. *The histograms are readings of absorbed dose (in water) by stacked nylon radiochromic films on both sides of a thick interface material (——— aluminium; ———— nylon) irradiated by scanned 10 MeV electron beam incident from the left side. The total thickness of the back film stack differed somewhat. Before striking the film stacks, the beam passes through two 0.5 mm Al windows and a 60 cm layer of air.*

(c and d) Geometries (3) and (4), scanned 10 MeV and 0.4 MeV electron beams

The irradiation geometry is shown in Fig.20. The electron beams are scanned across the interface and are of broad-beam incidence rather than collimated [42, 43]. The interface materials are aluminium and nylon. Figures 21 and 22 show the measured dose distributions in nylon and polychlorostyrene (25% Cl) as a function of thickness of material to ~0.6 mg/cm² on both sides of the interface. The results show that for diffuse incidence of broad-beam 10 MeV electrons on the double stack there is a thin region within about 3 mg/cm² on both sides of the interface where dose is enhanced, but only when the interface material has a higher electron density than its surroundings. In the case of 0.4 MeV electrons, this effect is more pronounced due to backscattering than to forward scattering.

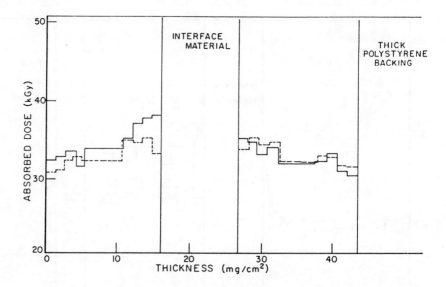

FIG.22. The histograms are readings of absorbed dose (in water) by stacked nylon radiochromic films on both sides of an interface material (———— aluminium; ———— nylon) irradiated by scanned 0.4 MeV electron beam incident from the left side. Before striking the film stacks, the beam passes through a 25 μm Ti window and an 8 cm layer of air.

5. SUMMARY

5.1. Material simulation

By selecting suitable thin radiochromic film or alkali halide dosimeter materials capable of reasonably accurate and precise dosimetry by spectrophotometric analysis, it is possible to simulate either the ionizing photon or electron response characteristics (mass energy-absorption coefficients or collision stopping powers) of sensitive semiconductor materials, namely silicon or silicon dioxide. To simulate silicon for photon irradiation over the energy range 0.01 to 10 MeV, a 71% NaCl plus 29% LiF dosimeter may be used; for electrons over the same energy range, 71% CaF_2 plus 29% LiF is recommended. To match the radiation absorption properties of silicon dioxide, the radiochromic film polychlorostyrene containing 25% chlorine by weight is recommended for photons and polybromostyrene containing 43% Br for electrons over the above energy range.

5.2. Interface dosimetry

By stacking thin-film radiochromic dosimeters close to interfaces of different materials and irradiating these assemblies with ionizing photons or electrons, it is possible to determine fairly high-resolution dose distributions in silicon dioxide and differences in dose enhancement close to high-Z interfaces, such as gold, Kovar or silicon, and dose build-up near an air interface.

ACKNOWLEDGEMENTS

The authors wish to thank K.C. Humphreys of Far West Technology, Inc. for preparing the very thin polyhalostyrene radiochromic dosimeters. They are also grateful to A.S. Yue of the Los Angeles School of Engineering and Applied Science, University of California, for supplying the optical-quality mixed alkali halide materials and to H.E. Fischer, Jr., of Harshaw/Filtrol for supplying the pure alkali halide materials.

REFERENCES

[1] SPIERS, F.W., "Transition-zone dosimetry", Radiation Dosimetry, Vol.III (ATTIX, F.H., TOCHILIN, E., Eds), Academic Press, New York (1969) 809–867.
[2] LONG, D.M., MILLWARD, D.G., WALLACE, J., Dose-enhancement effects in semiconductor devices, IEEE Trans. Nucl. Sci. **NS-29** 6 (1982) 1980–1984.
[3] KELLY, J.G., LUERA, T.F., POSEY, L.D., VEHARD, D.M., Dose enhancement effects in MOSFET ICs exposed in typical ^{60}Co facilities, IEEE Trans. Nucl. Sci. **NS-30** 6 (1983) 4388–4393.
[4] GARTH, J.C., Present status of the simple model for dose enhancement, IEEE Trans. Nucl. Sci. **NS-31** 6 (1984) (in press).
[5] EISEN, H., ROSENSTEIN, M., SILVERMAN, J., "Electron dosimetry using Chalkley-McLaughlin dye cyanide thin films", Dosimetry in Agriculture, Industry, Biology and Medicine (Proc. Symp. Vienna, 1972), IAEA, Vienna (1973) 615–625.
[6] McLAUGHLIN, W.L., HJORTENBERG, P.E., BATSBERG PEDERSEN, W., Low energy scanned electron-beam dose distributions in thin layers, Int. J. Appl. Radiat. Isot. **26** (1975) 95–106.
[7] McLAUGHLIN, W.L., MILLER, A., PEJTERSEN, K., BATSBERG PEDERSEN, W., Distribution of energy deposited in plastic tubing and copper-wire insulation by electron beam irradiation, Radiat. Phys. Chem. **11** (1978) 39–52.
[8] HUSSMANN, E.K., McLAUGHLIN, W.L., "Dye films and gels for megarad dosimetry", Absolute and Radiative Measurements and High Doses and Dose Rates (Proc. UK Panel on Gamma and Electron Irradiation) (DEALER, J., ELLIS, S.C., Eds), National Physical Laboratory, Teddington, UK (1971) 35–41.
[9] JARRETT, R.D., BRYNJOLFSSON, A., WANG, C.P., "Wall effects in gamma-ray dosimetry", National and International Standardization of Radiation Dosimetry, (Proc. Symp. Atlanta, 1977), Vol. II, IAEA, Vienna (1978) 91–102.
[10] McLAUGHLIN, W.L., "A national standardization programme for high-dose measurements", High-Dose Measurements in Industrial Radiation Processing, Technical Reports Series No.205, IAEA, Vienna (1981) 17–32.
[11] MILLER, A., McLAUGHLIN, W.L., "Calculation of the energy dependence of dosimeter response to ionizing photons", Trends in Radiation Dosimetry (McLAUGHLIN, W.L., Ed.), Pergamon Press, Oxford (1982); Int. J. Appl. Radiat. Isot. **33** (1982) 1299–1310.

[12] SELTZER, S.M., "Calculated depth-dose distributions in multi-layer media irradiated by electrons", Application of Accelerators in Research and Industry (Proc. 8th Conf. Denton, 1984), United States Department of Energy, Washington, DC (in press).

[13] TABATA, T., ITO, R., An algorithm for electron depth-dose distributions in multilayer slab absorbers, Jpn. J. Appl. Phys. **20** (1981) 249–258.

[14] BERGER, M.J., SELTZER, S.M., EISEN, H., SILVERMAN, J., Absorption of electron energy in multi-layer targets, Trans. Am. Nucl. Soc. **14** (1971) 887–888.

[15] CHAPPELL, S.E., HUMPHREYS, J.C., The dose rate response of a dye-polychlorostyrene film dosimeter, IEEE Trans. Nucl. Sci. **NS-19** 6 (1972) 175–180.

[16] MILLER, G.H., LOCKWOOD, G.J., HALBLEIB, J.A., Improved calorimetric method for energy deposition measurement, IEEE Trans. Nucl. Sci. **NS-21** 6 (1974) 359–365.

[17] DUTREIX, J., DUTREIX, M., Dosimetry at interfaces for high energy X and gamma rays, Br. J. Radiol. **39** (1966) 205–210.

[18] WINGATE, C.L., GROSS, W., FAILLA, C., Experimental determination of absorbed dose from X rays near the interface of soft tissue and other materials, Radiology **79** (1962) 984–999.

[19] LOWE, L.F., CAPPELLI BURKE, E.A., Dosimetry errors in Co-60 gamma cells due to transition zone phenomena, IEEE Trans. Nucl. Sci. **NS-29** 6 (1982) 1992–1995.

[20] KLEVENHAGEN, S.C., LAMBERT, G.D., ARBABI, A., Backscattering in electron beam therapy for energies between 3 and 35 MeV, Phys. Med. Biol. **27** (1982) 363–373.

[21] LAMBERT, G.D., KLEVENHAGEN, S.C., Penetration of backscattered electrons in polystyrene for energies between 1 and 25 MeV, Phys. Med. Biol. **27** (1982) 721–725.

[22] ALM-CARLSSON, G., Dosimetry at interfaces, Acta Radiol. Suppl. **332** (1973) 1–64.

[23] GARTH, J.C., MURRAY, B.W., DOLAN, R.P., Soft X-ray-induced energy deposition in a three-layered system: Au/C/PBS, IEEE Trans. Nucl. Sci. **NS-29** 6 (1982) 1985–1991.

[24] McLAUGHLIN, W.L., HUSSMANN, E.K., "The measurement of electron and gamma-ray dose distribution in various media", Large Radiation Sources for Industrial Processing (Proc. Symp. Munich, 1969), IAEA, Vienna (1969) 579–590.

[25] EISEN, H., ROSENSTEIN, M., SILVERMAN, J., Electron depth-dose distribution measurements in two-layer slab absorbers, Radiat. Res. **52** (1972) 429–447.

[26] KANTZ, A.D., HUMPHREYS, K.C., "Radiochromics: A radiation monitoring system", Radiation Processing (Trans. 1st Int. Meeting Puerto Rico, 1976) (SILVERMAN, J., VAN DYKEN, A., Eds) (1977); Radiat. Phys. Chem. **9** (1977) 737–748.

[27] BISHOP, W.P., HUMPHREYS, K.C., RANDTKE, P.T., Poly(halo)styrene thin-film dosimeters for high doses, Rev. Sci. Instrum. **44** (1973) 443–452.

[28] HARRAH, L.A., Chemical dosimetry with doped poly(halostyrene) film, Radiat. Res. **41** (1970) 229–246.

[29] McLAUGHLIN, W.L., MILLER, A., ELLIS, S.C., LUCAS, A.C., KAPSAR, B.M., "Radiation-induced color centers in LiF for dosimetry at high absorbed dose rates", Nucl. Instrum. Methods **75** (1980) 17–18.

[30] VAUGHN, W.J., MILLER, L.O., Dosimetry using optical density changes in LiF, Health Phys. **18** (1970) 578–579.

[31] McLAUGHLIN, W.L., LUCAS, A.C., KAPSAR, B.M., MILLER, A., "Electron and gamma-ray dosimetry using radiation-induced color centers in LiF", Advances in Radiation Processing (Trans. Int. Meeting Miami, 1978), Vol. II (SILVERMAN, J., Ed.) (1978); Radiat. Phys. Chem. **14** (1979) 467–480.

[32] PATTERSON, D.A., FULLER, R.G., F band in X- and electron-irradiated CaF_2, Phys. Rev. Lett. **18** (1967) 1123–1124.

[33] RABIN, H., KLICK, C.C., Formation of F centers at low and room temperature, Phys. Rev. **117** (1960) 1005–1010.

[34] FREIDMAN, H., GLOVER, C.P., Radiosensitivity of alkali-halide crystals, Nucleonics **10** 6 (1952) 24–29.

[35] McLAUGHLIN, W.L., Microscopic visualization of dose distributions, Int. J. Appl. Radiat. Isot. **17** (1966) 85–96.

[36] HUBBELL, J.H., "Photon mass attenuation and energy-absorption coefficients from 1 keV to 20 MeV", Trends in Radiation Dosimetry (McLAUGHLIN, W.L., Ed.), Pergamon Press, Oxford (1982); Int. J. Appl. Radiat. Isot. **33** (1982) 1269–1290.

[37] SELTZER, S.M., BERGER, M.J., Improved procedure for calculating the collision stopping power of elements and compounds for electrons and positions, Int. J. Appl. Radiat. Isot. **35** (1984) 665–680.

[38] McLAUGHLIN, W.L., MILLER, A., URIBE, R.M., KRONENBERG, S., SIEBENTRITT, C.R., Paper IAEA-SM-272/9, these Proceedings.

[39] McLAUGHLIN, W.L., HUMPHREYS, J.C., MILLER, A., "Dosimetry for industrial radiation processing", Proceedings of a Meeting on Traceability for Ionizing Radiation Measurements, NBS Special Publication 609 (HEATON, H.T., Ed.), National Bureau of Standards, Gaithersburg (1982) 171–178.

[40] WOOLF, S., BURKE, E.A., Monte Carlo calculations of irradiation test photon sources, IEEE Trans. Nucl. Sci. **NS-31** 6 (in press).

[41] MILLER, A., Dosimetry for Electron Beam Application, Risø National Laboratory, Roskilde, Rep. M-2401 (1983).

[42] MILLER, A., Beam spot measurements on a 400 keV electron accelerator, Radiat. Phys. Chem. **13** (1979) 1–4.

[43] EHRLICH, M., SELTZER, S.M., BIELEFELD, M.J., TROMBKA, J.I., Spectrometry of a ^{60}Co gamma-ray beam used for instrument calibration, Metrologia **12** (1976) 169–179.

[44] HUMPHREYS, J.C., McLAUGHLIN, W.L., Dye film dosimetry for radiation processing, IEEE Trans. Nucl. Sci. **NS-28** (1981) 1797–1801.

IAEA-SM-272/33

EVALUATION OF IRRADIATED ETHANOL-MONOCHLOROBENZENE DOSIMETERS BY THE CONDUCTIVITY METHOD*

A. KOVÁCS, V. STENGER, G. FÖLDIÁK
Institute of Isotopes,
Hungarian Academy of Sciences,
Budapest

L. LEGEZA
L. Petrik Technical School of Chemistry,
Budapest

Hungary

Abstract

EVALUATION OF IRRADIATED ETHANOL-MONOCHLOROBENZENE DOSIMETERS BY THE CONDUCTIVITY METHOD.
    Evaluation of irradiated ethanol-monochlorobenzene dosimeters can be carried out in different dose ranges with various analytical methods such as spectrophotometry, oscillometry and titration. The conductivity evaluation method, which is applicable in a wide dose range, has recently been worked out. The conductivity of unirradiated and irradiated dosimeter samples was measured by an OK-102/1 conductivity meter (Radelkis Electrochemical Instruments, Budapest) using a bell electrode. The conductivity of the solutions versus the absorbed dose can be described by the function $y = ax^b$. The temperature dependence of the conductivity of the solutions was also investigated and it was found to follow the Nernst equation. By fitting the experimental data, the activation energy was found to be 11.4 kJ·mol^{-1}. A check was also carried out on the difference between the conductivity values of the irradiated dosimeter solutions and those containing the same concentration of HCl, prepared by dilution of the ethanol-monochlorobenzene solution containing 10^{-1} M·L^{-1} HCl. The agreement was satisfactory. This evaluation method is simple, cheap and applicable in a wide dose range both for gamma and electron irradiations.

1. INTRODUCTION

Application of ionizing radiation is of increasing importance and, for example, for sterilization of medical products it has become a routine, large-scale method. As radiation processing requires accurate and reproducible measurements of

---

    * Research carried out with the support of the IAEA under Research Contract No. 2389/RB.

absorbed dose, the introduction and development of new dosimetric systems and methods is of basic importance.

1.1. The ethanol-monochlorobenzene dosimeter

This dosimeter system was introduced by Dvornik and co-workers [1]. It consists of an alcoholic solution of monochlorobenzene. Chloride ions are produced in the radiation-induced reaction by dissociative electron attachment to chlorobenzene and then stabilized in the system by solvation. The measured radiolytic product is hydrogen chloride. The solution has a high HCl yield, which is independent of the dose rate up to 10^5 $Gy \cdot h^{-1}$, temperatures in the range of 293 to 360 K and the dose at a given monochlorobenzene concentration [2]. The system can be utilized in the dose range of 10 Gy to 1 MGy using various evaluation methods. The solution can be stored for long periods and the only limitation is that the effect of UV light should be avoided during storage. By choosing a suitable dosimeter solution tissue-equivalent, electron density can be set. The dosimeter system is also applicable for electron beam dosimetry [3].

For the gamma- or electron-irradiated ethanol-monochlorobenzene dosimeter system the dose absorbed by the dosimeter solution can be evaluated by a number of different methods.

1.1.1. Titration method

The original evaluation method is titration, i.e. when the absorbed dose is evaluated by measuring the hydrogen ion concentration using alkalimetric titration according to the reaction

$$KHCO_3 + HCl = KCl + H_2O + CO_2$$

and/or by determining the chloride ion concentration using mercurimetric titration based on the process

$$Hg^{2+} + 2Cl^- = HgCl_2$$

Alkalimetric titration should be used only for doses above 2 kGy, whereas the mercurimetric method is valid over the entire dose range, i.e. 0.4 kGy to 400 kGy.

1.1.2. Evaluation by spectrophotometry

The spectrophotometric readout method was introduced by Ražem et al. [4]. This method is based on the reaction of radiolytically generated Cl^- ions with mercury(II) thiocyanate and the subsequent reaction of the liberated thiocyanate ions with iron(III) to give the familiar red colour of the ferric thiocyanate complex. The method is applicable in the 10 to 10^4 Gy dose range.

1.1.3. Oscillometric evaluation method

For quick routine analysis of irradiated monochlorobenzene dosimeter samples the oscillometric evaluation method was introduced [5]. Since the conductivity of a solution is changed by altering the amount of ions in the solution, the conductivity can be followed by high frequency oscillometry. Because there is no galvanic contact between the solution and the electrodes, the measurements can be carried out in a closed system. The method, which requires calibration, is applicable in the 0.1 to 100 kGy dose range. The system, along with the ceric-cerous dosimeter, is recommended by the IAEA as a back-up system for absorbed dose control in radiation processing in the medium and high dose range, i.e. 5 to 100 kGy [6].

Since evaluation can be carried out using sealed ampoules, the measurements can be repeated later because, in the absence of UV light, the irradiated dosimeters can be stored for years. However, if oscillometric evaluation is used the size of the dosimeter ampoule and the temperature of the solution to be measured should be similar to those of the calibration ampoules. International intercomparison programmes have proved the reliability of the oscillometric ethanol-monochlorobenzene dosimeter system for quality control of radiation processing.

2. EXPERIMENTAL

2.1. Chemicals

Our dosimeter solution contained 24 vol.% monochlorobenzene, 4 vol.% water, 0.04 vol.% acetone and benzene and 71.92 vol.% ethanol. The monochlorobenzene was 'Merck-Schuchardt' puriss., used as received. Ethanol and other chemicals were of an analytical grade, also used as received; distilled water was added to the system.

2.2. Irradiation and dosimetry

Irradiations were carried out using the 3 PBq (80 kCi) nominal activity ^{60}Co gamma irradiation facility at the Institute. The dose rates were 10 Gy·s^{-1} (3.6 Mrad·h^{-1}) and 10^{-1} Gy·s^{-1} (36 krad·h^{-1}); the doses varied between 50 Gy and 1 MGy (5 krad and 100 Mrad). Irradiations were performed at room temperature (20 to 25°C) in flame-sealed glass ampoules.

The dose delivered to the samples was measured partly by using Fricke dosimeters with spectrophotometric evaluation and partly by ethanol-monochlorobenzene dosimeters using oscillometric evaluation.

2.3. Conductivity measurements

The conductivity of the irradiated ethanol-monochlorobenzene dosimeter samples was measured using an OK-102/1 conductivity meter (Radelkis Electrochemical Instruments, Budapest). This instrument has a direct reading Siemens scale (ohm^{-1}) with a measuring range of 0.1 μS to 0.5 S. Its main operating principle is that a geometrically well-defined pair of indifferent electrodes, i.e. the measuring cell, is immersed into the solution, followed by measurement of the change of voltage across the cell. In our case the cell is a bell electrode, which includes platinum rings coated with platinum black in order to provide a larger surface area. It is well known that the conductivity of electrolytes can be measured only by means of AC current because of polarization processes. The higher the conductivity of the solution, the higher is the frequency required. We used an instrument that incorporates a special oscillator, generating 80 Hz and 3 kHz. At or above 500 μS, automatic changeover is carried out to the higher frequency, together with extension of the measuring range. A personal computer (Sinclair ZX81) was used for the calculations. The basic program required 16 kbyte RAM.

3. RESULTS AND DISCUSSION

In solutions the electric current is transferred by the ions, which start to migrate under the influence of the electric field strength between the indifferent electrodes. The conductivity of a solution is the sum of the conductivities of the electrolyte and that of the solvent. Measurement of the conductivity is carried out by measuring the resistance (r).

The conductivity of a solution is usually defined in terms of specific conductivity (κ), which is given by the conductivity of a solution between electrodes of 1 cm^2 surface area, placed 1 cm from each other

$$\kappa = \frac{1}{r} \frac{L}{q} \text{ ohm}^{-1} \cdot \text{cm}^{-1}$$

where $\frac{1}{r}$ is the conductivity, V (ohm^{-1}) (Siemens reading, obtained from meter scale)

L is the distance between the electrodes (cm)

q is the electrode surface area (cm^2)

$\frac{L}{q}$ is the cell constant, which can be determined by calibrating the measuring cell against a solution of known κ.

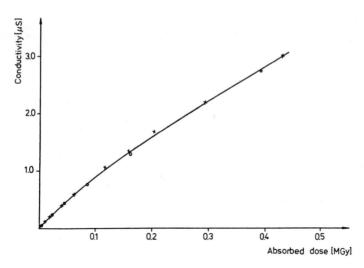

FIG.1. *Conductivity results of the irradiated and prepared dosimeter solutions (o = irradiated solutions; x = prepared solutions; continuous line = calculated data).*

In our case determination of the cell constant is not necessary because the 1/r conductivity values are used directly to determine the absorbed dose after calibration.

3.1. Conductivity of the irradiated dosimeter samples

The conductivity of the dosimeter solutions was measured first at room temperature. Before measuring the conductivity of the unirradiated and irradiated ethanol-monochlorobenzene dosimeter samples the conductivity of the components of the solution was determined. Thus, the conductivity of the distilled water was 1.90 μS, while that of the ethanol and monochlorobenzene was 0.48 and 0.10 μS, respectively. The value for the unirradiated dosimeter solution was found to be 0.30 μS. The dosimeter samples were irradiated in the range of 50 Gy to 1 MGy and the conductivity of the solutions was determined several times. The results are shown in Fig.1.

One of our aims was to determine that function which gives the best approximation of the measured points in the dose range examined. The measured points were approximated by an unknown function having a single variable. Various mathematical methods are available for determining this unknown function. Approximation is based on the results of the conductivity measurements, which include a specific uncertainty. In such cases, the least squares method is usually applied for approximation. Using this method we attempted to fit different functions to the measured points. In the dose range under consideration we found that the function $y = ax^b$ gave the best approximation; that is,

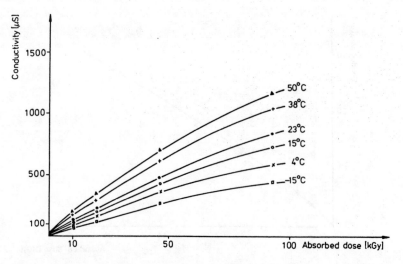

FIG.2. Conductivity of irradiated dosimeter solutions at different temperatures.

the conductivity of the solutions as a function of the absorbed dose can be described by that function. In our case a = 0.3404 and b = 0.8529 were found. On comparing the measured results and the calculated data, the differences were within the experimental error limit (±5%).

3.2. Temperature dependence of the conductivity of solutions

The temperature dependence of the conductivity of the irradiated solution was also investigated in the 258 to 323 K range (Fig.2). It was found that it follows the Nernst equation, i.e.

$$V = A \exp\left(-\frac{E_A}{RT}\right)$$

To calculate the activation energy, approximation of the measured points was again carried out using the least squares method. However, in this case the shape of the function was well known, i.e. after an ln V versus (1/T) transformation it has to be linear. Therefore, only the linear regression (y = mx + b) was applied using the following transformation

$$y = \ln V$$

$$x = \frac{1}{(273 + t)}$$

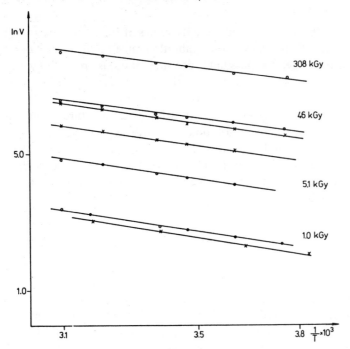

FIG.3. *Temperature dependence of dosimeter solutions by linear regression (o = irradiated solutions; x = prepared solutions; continuous line = calculated data).*

where V is the conductivity of the solutions measured at different temperatures t (°C). The results are shown in Fig.3.

The average of the slope values (E_A/R) of the regression lines obtained by fitting the experimental data by the above-mentioned transformation is 1.378×10^3, with a standard deviation of 9.32×10^1. Thus, the activation energy was found to be 11.4 kJ·mol^{-1}, which is within the range of diffusion-controlled reactions.

3.3. Calibration for dose determination

As was shown previously, by measuring the conductivity of irradiated ethanol-monochlorobenzene dosimeter solutions there is a possibility to determine the dose absorbed by the solutions. However, the method needs calibration, i.e. a calibration curve and table have to be taken previously. This can be done using solutions irradiated with known doses (e.g. at Fricke calibrated places). Also, by means of the Nernst equation

$$V_1 = V_2 \exp\left[\frac{E_A}{R}\left(\frac{1}{T_2} - \frac{1}{T_1}\right)\right]$$

correction of the measured conductivity values of the solution with regard to the temperature of the data of the calibration table can be carried out.

We also checked the possibility of using solutions prepared by dilution of stock ethanol-monochlorobenzene solution containing 10^{-1} M·L^{-1} HCl (instead of irradiation). A series of these solutions was investigated and their conductivity measured several times at different temperatures. The agreement, compared with the irradiated solutions, was satisfactory concerning both the conductivity-absorbed dose relationship (see Fig.1) and the slope value calculated from the ln V versus (1/T) plot (see Fig.3). Thus, the calibration procedure can be carried out by using prepared solutions if no calibrated gamma source is available.

4. CONCLUSIONS

Determination of absorbed dose is possible by directly measuring the conductivity of irradiated ethanol-monochlorobenzene dosimeter solutions. The method requires calibration, but it can be carried out using not only irradiated but also prepared ethanol-monochlorobenzene solutions. Thus, the method can also be applied where calibrated gamma sources are not available.

This evaluation method is very simple, applicable in a wide dose range (50 Gy to 1 MGy), can be used for gamma- and electron-irradiated samples and also shows prospects for the preservation and sterilization of food.

REFERENCES

[1] DVORNIK, I., et al., Manual on Radiation Dosimetry (HOLM, N.W., BERRY, R.J., Eds), Marcel Dekker, New York (1970) 345.
[2] RAŽEM, D., DVORNIK, I., "Application of the ethanol-chlorobenzene dosimeter to electron-beam and gamma-radiation dosimetry. II. Cobalt-60 gamma rays", Dosimetry in Agriculture, Industry, Biology and Medicine (Proc. Symp. Vienna, 1972), IAEA, Vienna (1973) 405.
[3] DVORNIK, I., RAŽEM, D., BARIĆ, M., "Application of the ethanol-chlorobenzene dosimeter to electron beam dosimetry: pulsed 10 MeV electrons", Large Radiation Sources for Industrial Processes (Proc. Symp. Munich, 1969), IAEA, Vienna (1969) 613.
[4] RAŽEM, D., OČIĆ, G., JAMIČIĆ, J., DVORNIK, I., Int. J. Appl. Radiat. Isot. 32 (1981) 705.
[5] HORVÁTH, Zs., BÁNYAI, É., FÖLDIÁK, G., Radiochim. Acta 13 (1970) 150.
[6] NAM, J.W., Food Irradiat. Newsl. 5 (1981) 16.

CONSISTENCY OF ETHANOL-CHLOROBENZENE DOSIMETRY*

D. RAŽEM, L. ANDELIĆ, I. DVORNIK
'Ruder Bošković' Institute,
Zagreb, Yugoslavia

Abstract

CONSISTENCY OF ETHANOL-CHLOROBENZENE DOSIMETRY.
 The consistency of the ethanol-chlorobenzene (ECB) dosimetry system was investigated with respect to both radiation-chemical response ($G(Cl^-)$ value) and analytical methods of evaluation. The $G(Cl^-)$ values of several characteristic dosimeter formulations to ^{60}Co gamma radiation were redetermined against a Fricke dosimeter. Molar absorptivities of ferric ions were determined for all three spectrophotometers used. The wavelength and absorbance linearity calibration of the spectrophotometers were also checked. The influence of the composition of dosimetric solution on the oscillometric method of evaluation was examined. Since the ECB dosimetry system represents a family of dosimetric solutions, whereby energy absorption characteristics can be continuously varied by changing chlorobenzene concentration, use of the applicable $G(Cl^-)$ value at any concentration is essential for the correct procedure. Earlier established $G(Cl^-)$ values for several characteristic chlorobenzene concentrations were reconfirmed, as well as the molar absorptivity of secondary complexes used in spectrophotometric methods of readout. A linear response of the oscillotitrator with Cl^- concentration was found to be independent of the composition of the dosimetric solution.

1. INTRODUCTION

This work was undertaken in order to explain the systematically high readings obtained with the ethanol-chlorobenzene (ECB) dosimetry system in the course of the High-Dose Standardization and Intercomparison for Industrial Radiation Processing Programme of the International Atomic Energy Agency (1977 to 1982) [1–14]. Evaluation of irradiated dosimeters in this programme was carried out by oscillometry, which did not require the opening of dosimeter ampoules for analysis [15]. This is an advantage if the information must be saved for future reference. On the other hand, this prevents a direct relationship between the actual amount of chloride ions formed by irradiation and the deflection of the instrument scale to be established for each dosimeter.

Readings higher than nominal might have been caused either by improper calibration of the dosimetric response (the $G(Cl^-)$ value) or by improper calibration of the analytical methods of evaluation. This paper deals with both, i.e. reconsideration of the radiation-chemical response of the ECB system against the ferrous sulphate (Fricke) dosimetry system, and examination of the oscillometric method.

---

* Research carried out with the support of the IAEA under Research Contract No.3505/RB.

2. EXPERIMENTAL PROCEDURES

2.1. Chemicals, containers and systems

Chlorobenzene (CB) ('Fluka', puriss. and 'Merck', for synthesis) and absolute ethanol ('Merck') were used as supplied. Distilled water was additionally distilled from alkaline permanganate, acidic dichromate and without additives. Triply distilled water was used for the preparation of 96 vol.% ethanol, which was subsequently used for preparation of ECB dosimetric solutions. Fresh triply distilled water was also used for preparation of Fricke dosimetric solution. ECB dosimetric solutions containing five characteristic concentrations (4, 10, 20, 25 and 40 vol.% of chlorobenzene in 96 vol.% ethanol) were prepared. All other chemicals necessary for the preparation, evaluation and standardization of the two dosimetric systems were of analytical reagent grade purity and were used as received.

Pyrex test tubes (16 mm outer diameter × 75 mm) with ground glass stoppers were boiled in acid, rinsed with water and steam and baked at 300°C for several hours before use. For oscillometry, commercial 5 mL pharmaceutical ampoules (16 mm outer diameter) were used as received. Each dosimeter contained 5 mL of dosimetric solution. ECB systems were deoxygenated by bubbling with nitrogen for 5 minutes immediately before use or before flame sealing. Sealed dosimeters were kept in the dark.

2.2. Standardization of the ferric solution

The procedure for standardization of the ferric solution as described in Ref.[16] was followed.

2.2.1. Volumetric measurements

The calibration of volumetric glassware was checked by weighing water or mercury contained in a specified volume. The actual volume of ordinary glassware was found to differ from the specified one by ±15% on average. Taking into account actual volumes and the weight of $K_2Cr_2O_7$ in air, the actual concentration of the titrant solution of $K_2Cr_2O_7$ was calculated. The concentration of the stock solution of ferric ions was calculated taking into account corrections for both the volume of the burette and the consumption of the titrant due to the indicator diphenylamine. The latter correction is represented by an intercept of the straight line: consumption of $K_2Cr_2O_7$ versus millilitre of ferric solution. The stock solution of ferric ions was used for determination of the molar absorptivity of Fe^{3+}.

2.2.2. Spectrophotometric measurements

Three available spectrophotometers belonging to different generations of instruments were used in order to examine the effects of design and possible performance degradation on the quality of data: Beckman DU-2, Cary 17 and Perkin Elmer 55B. The oldest model, DU-2, and the most recent, PE 55B, are single-beam instruments. There is no automatic control of the bandwidth, and wavelength setting is done manually. For these two instruments wavelength calibration was checked by a didymium glass filter in the visible region [17], and a potassium dichromate solution in the UV region [18]. Small positive (additive) corrections were found in the visible (0.3 and 0.2 nm for DU-2 and PE 55B, respectively) and UV regions (0.2 nm for both instruments). The position of the maximum absorption of ferric ion was found to be 302 nm with all three instruments. This is slightly lower than the recommended 304 nm and recently reported 303 nm [19]. However, the consistency of the three instruments was decisive in selecting this wavelength for determination of the molar absorptivity of Fe^{3+} ions, as well as for the subsequent evaluation of irradiated dosimeters.

Cary 17 was checked for linearity at 350 nm and PE 235, 257, 313, 350 and 440 nm with $K_2Cr_2O_7$ solution. Correlation coefficients, on average better than 0.9995, were found for both instruments.

As suggested in Ref.[19], molar absorptivity of ferric ions was determined for each spectrophotometer. The concentration of Fe^{3+} was adjusted in the range where a linear dependence of absorbance versus concentration holds. The possible existence of the instrument's dark current was checked by fitting the data by the method of least squares to the straight line: absorbance versus concentration. Dark current is represented by the intercept, and molar absorptivity by the slope of the straight line.

2.3. Irradiation

Low doses of ^{60}Co gamma radiation in the range 0.1 to 0.4 kGy (10 to 40 krad) and medium doses in the range 1 to 15 kGy (0.1 to 1.5 Mrad) were applied. The systems were irradiated in the low-dose range by the well-type source and in the medium-dose range by the panoramic source. Depending on the geometry of irradiation, the dose rate varied between 2 and 100 Gy/min.

The useful volume of the well-type source has the shape of a cylindrical cavity formed by arranging 12 ^{60}Co rods in a circle. The samples to be irradiated were introduced into this cavity in a cylindrical container which is manually operated. The reproducibility of irradiation is provided by a holder which can accommodate four dosimeter test tubes at any desired height within the volume of the container. Four dosimeters containing the same composition of the Fricke dosimetric system were irradiated simultaneously. Three different doses were applied at any selected height. Four characteristic formulations of the ECB dosi-

TABLE I. MOLAR ABSORPTIVITY OF FERRIC ION AT 302 nm AND 20°C ($L \cdot mol^{-1} \cdot cm^{-1}$)

Instrument	Light source	λ nm	Linearity check		Ferric solution		
			Dark current[a]	r^2	Dark current[a]	ϵ	r^2
PE 55B	W				0.007	2032 ± 30	0.99993
	D_2	313	0.009	0.99995	0.008	2070 ± 23	0.99983
Cary 17	W	350	0.002	0.99942	−0.003	2170 ± 19	0.99990
Beckman DU-2	Hg				0.007	2056 ± 72	0.99960

[a] Absorbance units.

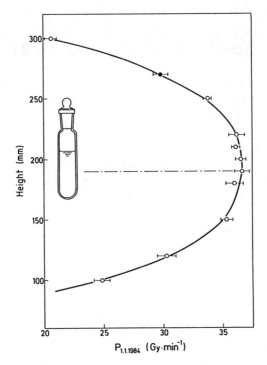

FIG.1. Dose rate distribution with height inside the well-type source measured by Fricke dosimetry. Filled circle: no NaCl added. Dose range 0.1 to 0.4 kGy (10 to 40 krad).

metry system (4, 10, 20 and 40 vol.% CB) were irradiated together as a group. The irradiations at every selected height were performed with four different doses.

For each irradiation using the panoramic source about 100 ECB dosimeters in sealed ampoules were prepared, about 20 ampoules each of the five selected concentrations. The ampoules were arranged so that the concentrations were regularly alternating along a wooden lath which was used to support them, with the axes of the ampoules 2 cm apart. The lath could be put in a horizontal or a vertical position at a desired distance from the source.

2.4. Dosimetry

The dosimetry of the well-type source was performed by Fricke dosimetry. Doses absorbed in Fricke dosimeters were expressed as the average of four values obtained with four identical dosimeters irradiated simultaneously. The mean values were fitted by the method of least squares to the straight line: absorbed dose versus irradiation time. The slope of this straight line represents the dose rate,

while the intercept represents the transit dose. All data were reduced to
1 January 1984 applying a half-life period of 5.26 years for ^{60}Co. An average
transit dose of 3.5 Gy (350 rad) was used for all positions of the holder.

Some Fricke dosimeters were prepared without the usual addition of
10^{-3} M NaCl to check for the presence of impurities. The procedure was scrutinized until no significant difference between the readings of these dosimeters and the ones containing NaCl was observed.

Dose mapping of the radiation field of the panoramic source was made by the ECB system itself. Four dosimeter concentrations (4, 10, 20 and 40 vol.% CB), which have been thoroughly characterized in earlier work [20–22], were used. The dose rates in positions occupied by dosimeters containing 25 vol.% CB were obtained by interpolation to smoothed data.

2.5. Evaluation of irradiated ECB dosimeters

Irradiated ECB dosimeters were analysed for chloride ion concentration by several methods. The systems in stoppered test tubes, which were irradiated in the low-dose range, were analysed by spectrophotometry [22]. The systems in sealed ampoules, which were irradiated in the medium-dose range by the panoramic source, were first read out by oscillometry before opening; after opening, aliquots were taken for mercurimetric titration [23] and spectrophotometry [22]. Oscillometry was performed by oscillotitrators made by Radelkis Electrochemical Instruments, Budapest. Two types of instruments were used: 140 MHz (the older) and 48 MHz (the more recent). Four readings of each ampoule were taken, the ampoule being rotated by 90° between readings. Before reading irradiated dosimeters the instrument was zeroed by an unirradiated solution of the same composition.

3. RESULTS

3.1. Preliminary measurements

The results of determination of molar absorptivity for ferric ion at 302 nm and 20°C with three generations of spectrophotometers, dating from the 1980s, 1970s and 1960s, respectively, are given in Table I, along with the results of the linearity check. The quality of data, as judged from both the dark current and the correlation coefficient for linearity r^2, was best for Cary 17. This is not surprising as it was the most sophisticated of the three instruments used, although not the most recent.

The results of dose mapping by Fricke dosimetry of the well-type source are shown in Fig. 1. This was used as a standard irradiation geometry for calibrating the response of the ECB dosimetry system in the low-dose range. The results of

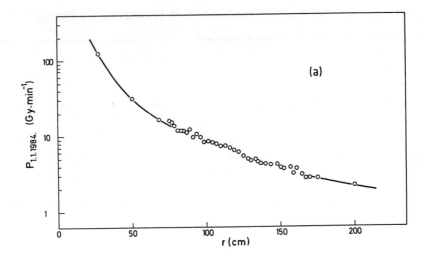
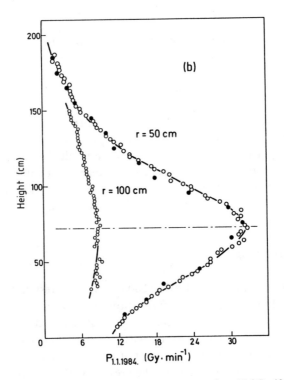

FIG.2. *Dose rate data for the panoramic source. Dose range 1 to 15 kGy (0.1 to 1.5 Mrad). (a) Dependence on the distance from the centre in the horizontal plane of symmetry 72 cm above the chamber floor. (b) Dependence on the distance from the chamber floor in the vertical planes 50 and 100 cm from the axis of the source. Filled circles: system containing 4 vol.% CB, based on $G(Cl^-) = 4.00$.*

TABLE II. G(Cl⁻) VALUES FOR SOME ECB DOSIMETRY SYSTEMS

CB concentration (vol.%)	G(Cl⁻) values	
	Present work 0.1 – 0.4 kGy	Ref.[22] 0.1 – 1 kGy
4	4.25 ± 0.25	4.06 ± 0.19
10	5.32 ± 0.25	5.07 ± 0.33
20	5.97 ± 0.27	5.66 ± 0.28
40	6.35 ± 0.30	6.16 ± 0.32

dose mapping of the panoramic source by the ECB system are shown in Fig.2(a) and (b).

3.2. Determination of G(Cl⁻) values

For practical reasons, only several characteristic concentrations have been well characterized, particularly with respect to the presence of air and dose. It was found [21] that incomplete deoxygenation results in G(Cl⁻) values which are constant to within ± 1 to 3% over a dose range of 1 to 100 kGy (0.1 to 10 Mrad). In the low-dose range, 0.1 to 1 kGy (10 to 100 krad), the systems containing lower concentrations of CB exhibited a stronger dose dependence than those with higher CB concentrations.

3.2.1. Low-dose range

A low-dose range, below 1 kGy (100 krad), is interesting for some food irradiation applications, particularly for sprout inhibition and disinfestation of agricultural products. This dose range is not easily covered by titration, spectrophotometry being more sensitive. The G(Cl⁻) values obtained by this method are given in Table II. In previous work, G(Cl⁻) values were averaged over a broader dose range, 0.1 to 1 kGy [22], while the present work was concentrated on a lower range, 0.1 to 0.4 kGy. The G(Cl⁻) values obtained are in accord with the general tendency noted, namely that higher G(Cl⁻) values apply at lower doses, and that dose dependence is more pronounced at the lower CB concentration. On average, however, the difference is smaller than 5%.

IAEA-SM-272/13

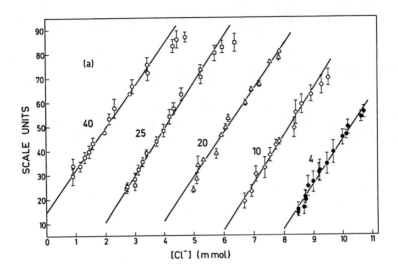

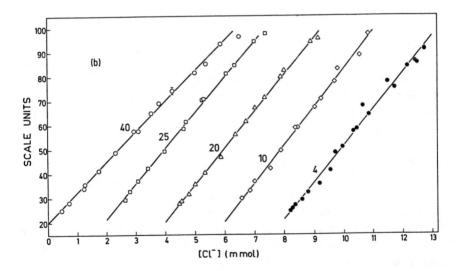

FIG.3. Dependence of the oscillotitrator reading on the concentration of Cl^- ions in various formulations of the ECB system. The data for each subsequent concentration translated by 2 mmol units to the right of the preceding data for improved clarity. (a) Oscillotitrator model OK-302; (b) Oscillotitrator model OK-302/1.

3.2.2. Medium-dose range

Doses in the medium range, 1 to 10 kGy (0.1 to 1 Mrad), are required by many applications of radiation. The four characteristic formulations of the ECB system have already been adequately characterized in this dose range. However, it seems that the system containing 25 vol.% CB, and having electron density of muscle, was used throughout the IAEA Intercomparison Programme [24]; in view of this fact it is appropriate that a pertaining radiation chemical response of that system be determined by an independent measurement. Relative calibration of this system against four other members of the series gave the value 5.77 ± 0.08.

3.3. Oscillometric measurements

We were concerned with the possibility that oscillometric measurements depend on the composition of dosimetric solution. Some variations of composition could be expected, based on the following considerations:

(a) A single concentration 10 vol.% CB system was suggested by the original formulation [23], also containing 0.04 vol.% acetone.
(b) The system containing 24 vol.% CB, 0.04 vol.% acetone and 0.04 vol.% benzene was used in the first oscillometric measurements [15]. The addition of acetone and benzene was originally proposed [25] in order to saturate the effect of eventual impurities in ethanol. Since our practice has shown that ethanol of sufficient purity is generally available, these additives were subsequently omitted, which resulted in a small increase of sensitivity (the G-values) [26].
(c) Our subsequent papers [20, 21] were restricted to characterization of the systems containing 4, 10, 20 and 40 vol.% CB only. Since then all the $G(Cl^-)$ values we have published refer to impurity-free systems.
(d) The issuing laboratory oscillated between 25 vol.% CB [27] and 24 vol.% CB systems [28].

The results of oscillometric measurements (Fig.3(a) and (b)) are shown in terms of concentrations of Cl^- ions rather than absorbed doses. They show a linear relationship between the concentration and meter readings for both instrument models. The more recent model has an appreciably smaller scattering. It is also evident that no significant differences exist in the response of the instrument to the same concentration of Cl^- ions in systems with different composition. The minor as well as major variations mentioned in items (a) to (d) above could not cause overestimation of the dose. However, the lower response of the model OK–302/1 to Cl^- ions in the system containing 40 vol.% CB, which was also observed at other higher and lower sensitivity settings of the instrument, cannot be explained at this point.

4. DISCUSSION

The ethanol-chlorobenzene dosimetry system is actually a family of dosimetric solutions which feature continuously tunable energy absorption characteristics in the range from water to bone [21]. Adjustment of the absorption characteristics is done by changing the concentration of CB. The radiation-chemical response of the system (the $G(Cl^-)$ value) depends, besides other factors, primarily on the concentration of CB. The choice of the applicable $G(Cl^-)$ value is critical for correct use of ECB dosimetry. This has been recognized by the authors of reviews and manuals [29–31] who list published, well-defined $G(Cl^-)$ values for several ECB formulations.

Another unique property of this dosimetry system — the possibility of checking the internal consistency of one's measurements by using several CB concentrations — seems not to have been exploited (based on the evidence published in Refs [1–15, 32–37]), with one exception [38]. This possibility is illustrated in Fig.2(b): filled circles show dose rate data calculated for the system containing 4 vol.% CB on the basis of $G(Cl^-) = 4.00$. This $G(Cl^-)$ value is not applicable in the dose range used (1 to 15 kGy) [21, 22], which is easily seen, although the discrepancy is relatively small, about 3% on average.

A direct relationship between Fricke and ECB dosimetry was sought in the low-dose range, which is the range of applicability of the Fricke system. This was done in order to have both systems exposed to the same sources and kinds of error, originating mostly from manual operation of the sample holder. For example, an average transit dose had to be used, although it is expected to vary with the position in the holder. However, omitting the data for the shortest irradiation times, of which transit dose may represent up to 5%, did not change the values of $G(Cl^-)$.

In the low-dose range the spectrophotometric method of readout had to be applied. It should be recognized that the product $G\epsilon$ is actually determined by this method, and that both quantities, G and ϵ, are potentially variable. Using previously determined absorptivity of coloured secondary complex, $\epsilon = 3990 \text{ L} \cdot \text{mol}^{-1} \cdot \text{cm}^{-1}$ [22], the $G(Cl^-)$ values listed in Table II were obtained. Considering a probable dose dependence in the low-dose range, possible discrepancy of less than 5% with respect to the previously determined values is not alarming.

5. CONCLUSIONS

Current re-examination of the ECB dosimetry has confirmed the internal consistency of the system. No discrepancies with previously established $G(Cl^-)$ values were found. The most plausible explanation for the systematically high readings performed by this system throughout the IAEA Intercomparison

Programme is the use of an underestimated $G(Cl^-)$ value. The issuing laboratory has also recognized this independently [24]. Their corrected $G(Cl^-)$ value (5.64) is lower than the value reported herein for the system containing 25 vol.% CB (5.77), which can be accounted for by the presence of acetone and oxygen in the system causing lower $G(Cl^-)$ values. Application of these $G(Cl^-)$ values (5.64 and 5.77) to the Intercomparison Programme results averaged over the period 1977 to 1981 [12] would give the average ratio (estimated dose)/(nominal dose) = 0.993 and 0.971, respectively.

Future investigations should include the relationship between $G(Cl^-)$ values at high doses (up to 100 kGy) and Fricke dosimetry, and the combined effects of acetone, oxygen and dose on the $G(Cl^-)$ values.

ACKNOWLEDGEMENTS

The contribution of the IAEA through Research Contract No.3505/RB is gratefully acknowledged. The authors thank V. Stenger and A. Kovács for making available Refs [24, 27, 28, 33, 34] and L. Fistrić for technical assistance.

REFERENCES

[1] Anon., Food Irradiat. Newsl. **1** 3 (1977) 15.
[2] Anon., ibid. **2** 1 (1978) 15.
[3] Anon., ibid. **2** 2 (1978) 24.
[4] NAM, J.W., ibid. **3** 1 (1979) 18.
[5] Anon., ibid. **4** 1 (1980) 24.
[6] Anon., ibid. **4** 1 (1980) 29.
[7] NAM, J.W., ibid. **5** 1 (1981) 3.
[8] NAM, J.W., ibid. **5** 3 (1981) 16.
[9] INTERNATIONAL ATOMIC ENERGY AGENCY, High-Dose Measurements in Industrial Radiation Processing, Technical Reports Series No.205, IAEA, Vienna (1981).
[10] Advisory Group on High-Dose Pilot Intercomparison, Vienna, November 1981, IAEA Internal Report.
[11] NAM, J.W., "The High-Dose Standardization and Intercomparison for Industrial Radiation Processing Programme of the International Atomic Energy Agency (Lecture Note, IAEA Seminar on High-Dose Dosimetry in Industrial Radiation Processing, 1982, Roskilde, Denmark).
[12] CHADWICK, K.H., IAEA Bull. **24** 3 (1982) 21.
[13] MILLER, A., CHADWICK, K.H., NAM, J.W., Radiat. Phys. Chem. **22** 1-2 (1983) 31.
[14] Co-ordinated Research Programme on High-Dose Standardization and Intercomparison for Industrial Radiation Processing, Munich, 1983, Report of The Final Research Co-ordination Meeting.
[15] HORVÁTH, Zs., BÁNYAI, E., FÖLDIÁK, G., Radiochim. Acta **13** 3 (1970) 150.
[16] INTERNATIONAL ATOMIC ENERGY AGENCY, Training Manual on Food Irradiation Technology and Techniques, Technical Reports Series No.114, IAEA, Vienna (1970) 110.

[17] VENABLE, W.H., Jr., ECKERLE, K.L., Didymium Glass Filters for Calibrating the Wavelength Scale of Spectrophotometers SRM 2009, 2010, 2013 and 2014, NBS Special Publication 260–66, National Bureau of Standards, Washington, DC (1979).
[18] BURKE, R.W., MAVRODINEANU, R., "Acidic potassium dichromate solutions as ultraviolet absorbance standards", Standardization in Spectrophotometry and Luminescence Measurements (MIELENZ, K.D., VELAPOLDI, R.A., MAVRODINEANU, R., Eds), NBS Special Publication 466, National Bureau of Standards, Washington, DC (1977) 121.
[19] EGGERMONT, G., BUYSSE, J., JANSENS, A., THIELENS, G., JACOBS, R., "Discrepancies in molar extinction coefficients of Fe^{3+} in Fricke Dosimetry", National and International Standardization of Radiation Dosimetry (Proc. Symp. Atlanta, 1977), Vol.2, IAEA, Vienna (1978) 317.
[20] RAŽEM, D., DVORNIK, I., "Application of the ethanol-chlorobenzene dosimeter to electron beam and gamma radiation dosimetry. II. Cobalt-60 gamma rays", Dosimetry in Agriculture, Industry, Biology and Medicine (Proc. Symp. Vienna, 1972), IAEA, Vienna (1973) 405.
[21] RAŽEM, D., DVORNIK, I., "Application of the ethanol-chlorobenzene dosimeter to electron beam and gamma radiation dosimetry. III. Tissue-equivalent dosimetry", Radiation Preservation of Food (Proc. Symp. Bombay, 1972), IAEA, Vienna (1973) 537.
[22] RAŽEM, D., OČIĆ, G., JAMIČIĆ, J., DVORNIK, I., Int. J. Appl. Radiat. Isot. 32 (1981) 705.
[23] DVORNIK, I., "The ethanol-chlorobenzene dosimeter", Manual on Radiation Dosimetry (HOLM, N.W., BERRY, R.J., Eds), Marcel Dekker, New York (1970) 345.
[24] KOVÁCS, A., STENGER, V., "Environmental effects on the ethanol-monochlorobenzene dosimeter system before, during and after irradiation", Final Report 1979–1983, Institute of Isotopes, Hungarian Academy of Sciences, Budapest (1983).
[25] DVORNIK, I., ZEC, U., RANOGAJEC, F., "The ethanol-chlorobenzene aerated system as a new high-level dosimeter for routine measurements", Food Irradiation (Proc. Symp. Karlsruhe, 1966), IAEA, Vienna (1966) 81.
[26] DVORNIK, I., RAŽEM, D., BARIĆ, M., "Application of the ethanol-chlorobenzene dosimeter to electron beam dosimetry: Pulsed 10 MeV electrons", Large Radiation Sources for Industrial Processes (Proc. Symp. Munich, 1969), IAEA, Vienna (1969) 613.
[27] Chemical Dosimetry Course, A Laboratory Aid, Institute of Isotopes, Hungarian Academy of Sciences, Budapest (1971).
[28] HORVÁTH, Zs., STENGER, V., FÖLDIÁK, G., Dosimetry for Large Gamma-Irradiation Facilities, Institute of Isotopes, Hungarian Academy of Sciences, Budapest (1976) (in Russian).
[29] DRAGANIĆ, I., "Liquid chemical dosimeters", Sterilization by Ionizing Radiation (Proc. Conf. Vienna, 1974) (GAUGHRAN, E.R.L., GOUDIE, A., Eds), Multiscience Publications, Montreal (1974) 253.
[30] PIKAEV, A.K., Dosimetry in Radiation Chemistry, Nauka, Moscow (1975) (in Russian).
[31] SARAEVA, V.V. (Ed.), Training Manual on Radiation Chemistry, Moscow University, Moscow (1982) (in Russian).
[32] FÖLDIÁK, G., HORVÁTH, Zs., STENGER, V., "Routine dosimetry for high-activity gamma irradiation facilities", Dosimetry in Agriculture, Industry, Biology and Medicine (Proc. Symp. Vienna, 1972), IAEA, Vienna (1972) 367.
[33] PENKALA, V., PETSHAK, M., "The application of chlorobenzene dosimetry in radiation sterilization of catgut", Introduction of Radiation Facilities and Radiation Technology (Proc. COMECON Conf. Budapest, 1972), Institute of Isotopes, Hungarian Academy of Sciences, Rep. KFKI-73-8885 (1973) 383.

[34] HORVÁTH, Zs., FÖLDIÁK, G., "The control of sterilization dose by the ethanol-chlorobenzene dosimetry", ibid., p.395 (in Russian).
[35] HORVÁTH, Zs., "Dosimetric inspection system of gamma-irradiation facilities", Sterilization by Ionizing Radiation (Proc. Conf. Vienna, 1974) (GAUGHRAN, E.R.L., GOUDIE, A., Eds), Multiscience Publications, Montreal (1974) 253.
[36] SAID, F.I.A., FÖLDIÁK, G., WOJNÁROVITS, L., Radiochem. Radioanal. Lett. **32** 3–4 (1978) 223.
[37] BLAHA, J., VANCL, V., Int. J. Appl. Radiat. Isot. **29** (1978) 217.
[38] FISCHER, I., Isotopenpraxis **10** 7 (1974) 269.

RADIATION DOSIMETRY APPLICATIONS OF GLASS OPTICAL FIBRES

P.P. PANTA
Institute of Nuclear Chemistry and Technology*,
Warsaw

R. ROMANIUK, K. JĘDRZEJEWSKI
Technical University of Warsaw,
Warsaw

Poland

Abstract

RADIATION DOSIMETRY APPLICATIONS OF GLASS OPTICAL FIBRES.
Permanent radiation-induced coloration of low-loss glass optical waveguides was investigated to determine the parameters affecting the usefulness of these materials in gamma ray and electron dosimetry. As a convenient dosimetric effect, optical absorption measurements were chosen at wavelengths from 400 to 1700 nm after gamma and electron irradiations. Preliminary experimental results seem promising as this new dosimeter gives reasonably good responses over a wide dynamic range of electron and gamma doses (about 10^{-3} Gy to 10^5 Gy or 0.1 rad to 10 Mrad). The average error of the suggested method is 2 to 10% for a sufficiently reproducible procedure, which is slightly better than routine bulk glass dosimetry.

1. INTRODUCTION

Radiation-induced transmission losses in optical fibres are detrimental to telecommunication systems [1, 2]. On the other hand, it is possible to use such fibres for wide-range radiation dosimetry, based on the darkening of glass due to the absorption of ionizing radiation. A combination of ultra-high purity (the pp 10^9 level of contamination and low internal losses) and the high ratio of length to cross-section inside optical fibres contributes towards designing a dosimeter with a range of sensitivity that is several orders of magnitude greater than conventional glass or plastic bulk dosimeters. The suggested optical fibre dosimeter can be applied to monitoring radiation doses inside large volumes, and also to measuring the average dose at great distances. An alternative version is also possible, i.e. measurements inside small cavities, such as the glass capillaries for irradiation associated with ESR measurements, and also inside radiolysis cells.

---

* Formerly the Institute of Nuclear Research.

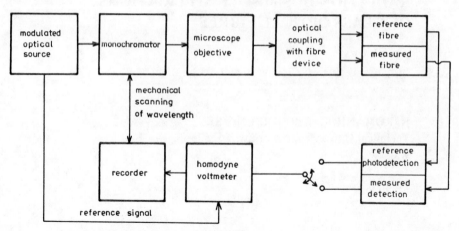

FIG.1. *Schematic diagram of apparatus used for measuring the optical absorption change in glass fibre.*

Moreover, telemetering of radiation of very intense fields is possible, as performed by connection of a conventional telecommunication waveguide system to a specified segment of measuring fibre. In such cases, part of the connection link would be made of radiation-resistant fibre.

The radiation-induced increase of optical absorbance of the fibre was the basis of the dosimetric effect as related to a given wavelength of analytical light.

Other effects such as luminescence and Cerenkov radiation are omitted here, but they will be investigated in the near future.

2. EXPERIMENTAL PROCEDURES

Gamma ray irradiations of optical fibres have been performed using ^{60}Co sources (Issledovatel, Mineyola 1000 and Mineza) and the electron linac (LAE-13/9) [3–6].

Permanent radiation-induced absorption of the fibres was measured after gamma ray and electron irradiations at optical wavelengths from 400 to 1700 nm.

The basic measuring system of optical fibre dosimetry (Fig.1) consists of:
(1) An optical converter with an electroluminescence diode
(2) An optical coupler
(3) A suitable length of optical fibre (optical absorbance changed as a result of irradiation)
(4) A photodetector such as a photodiode or a phototransistor.

In certain cases, instead of the electroluminescence diode, a He-Ne laser ($\lambda = 632.8$ nm) was used. For measurements at a constant wavelength of 880 to 900 nm the measuring set-up can be simpler than that shown in Fig.1 (i.e. a fully

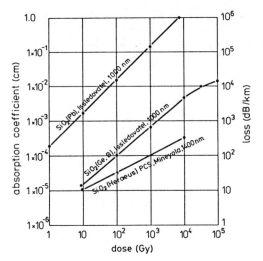

FIG.2. *Absorption coefficient increase in three gamma-irradiated groups of optical fibres as a function of dose.*

portable variant). The radiation response of several conventional and prototype optical fibres of Polish manufacture has been tested as follows:

(a) Telecommunication fibres made by a chemical vapour deposition (CVD) of Ge-doped silica
(b) Telecommunication fibres made by a modified CVD of Ge-doped silica
(c) Polymer-clad silica (PCS) fibres
(d) Multicomponent glass fibres made by a double crucible technique of Pb-doped silica
(e) Some prototype glass fibres made of suitable preform (on a laboratory-scale only)
(f) Preliminary small samples of polymethyl methacrylate (PMMA) fibre
(g) Strips of polyvinyl chloride (PVC) measured along lengths up to 10 cm (in contrast to common perpendicular measurement of foils).

These optical fibres were tested in the irradiation range of 10 to 10^5 Gy incident electron beams (10 and 13 MeV) and ^{60}Co gamma rays.

A linear relationship was obtained between the logarithm of the radiation-induced absorbance and the logarithm of the total absorbed dose.

3. RESULTS AND DISCUSSION

CVD Suprasil fibres doped with Ge, Pb and B display the best linear characteristics from the standpoint of possible dosimetric applications. From Fig.2 it can

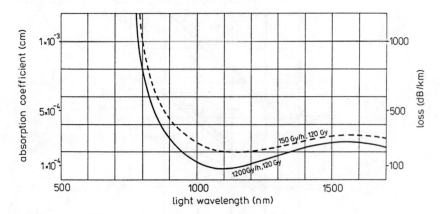

FIG.3. Optical absorption spectra of glass fibres at different gamma irradiations (gamma source, Mineyola 1000).

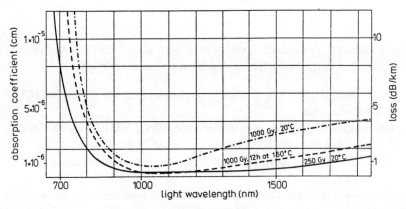

FIG.4. Optical absorption spectra of glass fibres at different electron irradiations (LAE-13/9).

be seen that CVD UMCS L30181 optical fibres (of Polish manufacture) doped with Ge or B show a suitable linearity of readout in the dose range from 10 to 10^4 Gy (10^3 to 10^6 rad) for gamma radiation and that the saturation effect is not yet apparent. It is noteworthy that silica glass doped with Pb shows strict linearity up to doses above 10^5 Gy (10^7 rad) (see Fig.2, upper line).

Some spectral characteristics of optical losses of CVD SiO_2 (Ge, B) 182 fibre and the effects of dose rates are shown in Figs 3 and 4. Figure 3 also reveals that at wavelengths below 900 nm the attenuation light loss is relatively strong and for this reason optical fibre dosimeters provide high readout sensitivity. For detection

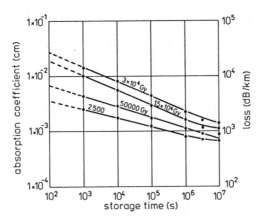

FIG.5. *Fading characteristics of CVD SiO_2 (Ge, B) 182 glass fibres at various doses and storage temperatures (dose rate, 5000 Gy/h; gamma source, Issledovatel).*

of megarad doses, longer values of the analytical light wavelength (from 900 to 1700 nm) may be used as intrinsic attenuation is much lower. The influence of dose rate when varying within one order of magnitude is negligible for wavelengths below 900 nm; this is also true from 1400 to 1700 nm. It should be noted that a wide assortment of commercially available optical fibres offers a broad selection of fibre materials for various ranges of doses and dose rates of ionizing radiation.

For example, PCS ultra-high purity silica fibres display very intensive absorption in the range from 600 to 1100 nm even for doses of several grays and lower sensitivity between 1400 and 1700 nm [7, 8]. A bismuth lead borate glass, suggested by Bishay [9] for conventional glass dosimetry, showed considerable promise as an optical fibre material for both low- and high-level electron beam and gamma ray measurements. For instance, in the case of typical telecommunication fibres a dose of 0.01 Gy is easily detected at a distance of approximately 1 km. On the other hand, using a special glass such as bismuth lead borate the sensitivity increases to a level such that 0.01 Gy can be detected by a fibre several metres in length.

Figure 5 shows the fading characteristics of radiation-induced changes of absorbance (or its equivalent in terms of transmission loss) as a function of storage time after irradiation. The data of Fig.5 imply that for storage times from minutes to one month the initial absorbance value decreases to nearly within one order of magnitude. A selection of suitable materials with negligible changes of absorbance is also possible from the standpoint of dosimetry, particularly when the chemical composition of typical dosimetric glass for the production of fibres is feasible.

A repeatable radiation-induced effect on the same specimen of optical fibre is possible after thermal annealing of radiation-induced changes. For the majority of known glasses, the optimal temperature for complete annealing of optical radiation-induced darkening is ca. 400°C, and an annealing time of one hour is

TABLE I. COMPARISON OF THE TECHNICAL PARAMETERS OF OPTICAL FIBRE DOSIMETERS WITH CONVENTIONAL BULK DOSIMETERS MADE OF THE SAME MATERIAL

Dosimetric material	Bulk dosimeter	Fibre dosimeter	Remarks
	Typical thickness or length, wavelength and range		
Inorganic glass	1.5 to 5 mm 255 to 800 nm 10 to 1000 Gy	20 cm to 2 km 400 to 1700 nm 10^{-5} to 10^8 Gy	
PMMA or Perspex	1 to 3 mm 250 to 305 nm 10^3 to 10^5 Gy	1 m 400 to 1700 nm 0.1 to 10^6 Gy	
Polystyrene	up to 2 mm 250 to 500 nm 10^4 to 10^6 Gy	0.5 to 1 mm 400 to 1700 nm 10^2 to 10^4 Gy	
Polyvinyl chloride	0.2 mm 396 nm 10^3 to 10^5 Gy	10 cm 400 to 1700 nm 10 to 1000 Gy	In the shape of strips $0.2 \times 8 \times 100$ mm

sufficient. For fibres protected by silicon rubber clad, it is better to use lower temperatures for annealing and accordingly longer times. For glass fibres irradiated inside liquid samples the effect of degraded gamma radiation spectra on the dosimetric response is negligible [10, 11]. Nevertheless, there is a significant systematic error when glass fibres are irradiated directly in air, particularly in the Issledovatel facility. This is caused by the influence of thick (30 mm) steel shielding around the radiation chamber. In this case, the degraded spectrum of energy of ^{60}Co gamma rays contains a soft 'tail' of low-energy photons, which are strongly absorbed in the layer of thickness comparable to the diameter of the glass fibre (100 to 125 μm). However, the overall error should not be greater than ±20%.

The activation energy of glass annealing depends strongly on its chemical composition and can fluctuate within wide limits. For example, in the case of borosilicate glass (used among others in the dosimetric study) the activation energy of annealing was found to be between 1 and 1.5 eV [12].

The optical transmission of irradiated glass is usually a power function of storage time, as is clearly visible in Fig.5 (this relationship is linearized on the logarithmic scale).

Friebele et al. [13] report that radiation-induced bands of optical absorption are associated with silicon centres E' (i.e. with electrons trapped by a Si atom on the sp^3 orbital). A silicon atom fills the place usually occupied by an oxygen atom (oxygen vacancy).

In highly irradiated silicon glass Evans and Sigel [12] observed some hyperfine spectral lines associated with the atomic nuclei of ^{29}Si.

However, the influence of irradiation temperature is negligible in the case of radionuclide gamma ray sources for electron irradiation at very high dose rates; even the self-annealing of defects can cause optical changes in the glass. For irradiations using the LAE-13/9 linac the temperature rise induced by a dose of 10^4 Gy is equal to about 10°C for the fibre in air. This temperature rise decays in milliseconds, and it is advisable to avoid irradiations with higher pulse repetition rates, for example close to 900 Hz.

The average error of the suggested method is estimated to be 2 to 10% for a sufficiently reproducible procedure, which is slightly better than routine bulk glass dosimetry. Becker [14] and Attix and Roesch [15] report useful data on the range and accuracy of routine bulk glass dosimetry.

A quantitative comparison of conventional and optical fibre methods of dosimetry is made in Table I.

4. CONCLUSIONS

Low intrinsic loss and extended optical path length, coupled with small size and convenient real-time monitoring, make carefully designed optical fibres an attractive dosimetry system.

As fibre glass dosimetry is a relatively new technology, much research still needs to be done; however, potentially it covers a wide range of doses from 10^{-3} Gy (personnel dosimetry) up to maximal radiation processing doses (10^5 Gy).

ACKNOWLEDGEMENTS

The authors wish to thank W.L. McLaughlin for his valuable assistance and critical remarks. Discussions held with Z.P. Zagórski are also much appreciated.

REFERENCES

[1] KARANY, N.S., Fiber Optics, Principles and Applications, Academic Press, New York (1967).
[2] SUEMATSU, Y., Introduction to Optic Fiber Communication, Wiley, New York (1982).
[3] KULISH, E.E. (Ed.), Raschet i konstruirovanie radioizotopnykh radiatsionno-khimicheskikh ustanovok, Atomizdat, Moscow (1975) 155.
[4] NEY, W., Nukleonika 11 (1966) 415.
[5] MINC, S., ZAGÓRSKI, Z.P., Nature (London) 193 (1962) 1290.
[6] ZIMEK, Z., KOŁYGA, S., LEVIN, V., NIKOLAEV, V., RUMIANCEV, V., FOMIN, L., Nukleonika 17 (1972) 75.
[7] ROMANIUK, R., JĘDRZEJEWSKI, K., PANTA, P.P., Elektronika 24 (1983) 18.
[8] ROMANIUK, R., JĘDRZEJEWSKI, K., PANTA, P.P., Elektronika 24 (1983) 38.
[9] BISHAY, A.M., Phys. Chem. Glasses 2 (1961) 33.
[10] KRONENBERG, S., SIEBENTRITT, C., Nucl. Instrum. Methods 175 (1980) 109.
[11] KRONENBERG, S., McLAUGHLIN, W.L., SIEBENTRITT, C., Nucl. Instrum. Methods 190 (1981) 365.
[12] EVANS, B.D., SIGEL, G., Appl. Phys. Lett. 24 (1974) 410.
[13] FRIEBELE, E., GINTHER, R., SIGEL, G., Appl. Phys. Lett. 24 (1974) 412.
[14] BECKER, K., Solid State Dosimetry, CRC Press, Cleveland (1973).
[15] ATTIX, F., ROESCH, W.C. (Eds), Radiation Dosimetry, 2nd edn, Vol.I, Academic Press, New York (1966) 248.

GEIGER-MÜLLER GAMMA DETECTORS OPERATED AT UP TO 1000 G/h

P.L. LECUYER, P.M. CHAISE
LLC/Thomson-CSF,
Bollène, France

Abstract

GEIGER-MÜLLER GAMMA DETECTORS OPERATED AT UP TO 1000 G/h.
   Unique halogen-quenched Geiger-Müller (GM) gamma detectors are discussed. These devices can be operated up to 1000 G/h, which is 30 times more than common GM detectors. Thus, very rugged instruments can be designed to make reliable and accurate measurements of very high gamma exposure rates. Experimental data are given and commented upon.

1. INTRODUCTION

Geiger-Müller (GM) detectors are widely used because they are very rugged devices and require quite simple associated electronics.

The upper exposure rate limit of common GM detectors is typically 3000 R/h when counting losses become very large at high exposure rate levels.[1]

Unique halogen-quenched GM detectors that can be operated at exposure rate levels as high as 100 000 R/h are discussed. The counting life under these stringent conditions is also very long.

2. MECHANICAL DATA

The dimensions of the detectors are given in Fig.1.

The weight is 1.5 g; the wall comprises envelope material: lead glass and cathode material: chrome iron; the total areal density is 380 mg/cm^2; the electrical connections are tinned dumet leads with a diameter of 0.5 mm; and the gas fill is neon-argon-halogens.

3. MOST IMPORTANT CHARACTERISTICS OF THE DEVICES

The main nuclear characteristics of 4 G 300 (and similar devices), 3 G 70 and 3 G 10 are given in Table I.

---

[1] 1 R = 2.58 × 10^{-4} C/kg.

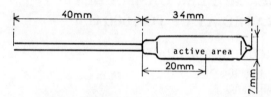

FIG.1. Dimensions of 4 G 300, 3 G 70 and 3 G 10.

TABLE I. MAIN NUCLEAR CHARACTERISTICS

Type	4 G 300 (and similar devices)	3 G 70	3 G 10
Mean sensitivity ^{60}Co			
(counts/s per R/h)	400	70	12
(counts/min per mR/h)	24	4.2	0.7
Minimum plateau (V)	500–600	420–500	420–500
Maximum slope (%/V)	0.30	0.35	0.40
Typical slope (%/V)	0.15	0.20	0.25
Typical background (counts/min)	1.2	0.6	1.0
Operating temperature (°C)	−40/+70	−40/+70	−40/+70
Maximum high voltage (V)	600	500	500
Minimum load resistor (MΩ)	2.2	1	1
Maximum capacitance (pF)	10	10	10

The energy response is within ∓20% from 80 keV to 3 MeV with an energy compensation filter (0.45 mm lead, +0.25 mm tin).

4. COUNT RATE VERSUS EXPOSURE RATE

4.1. Test circuit

The test circuit that was used is described in Fig.2. The tubes were operated at the mid-plateau: 4 G 300 at 550 V, 3 G 70 at 460 V, and 3 G 10 at 460 V.

IAEA-SM-272/34

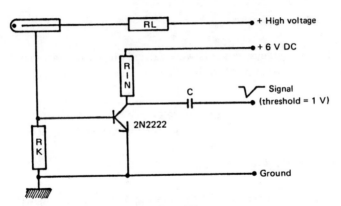

RL (load resistor) = 2.2 MΩ
RIN (input resistance) = 10 kΩ
C (capacitor) = 10000 pF
RK (cathode resistor) = 2.2 MΩ

FIG.2.   Test circuit.

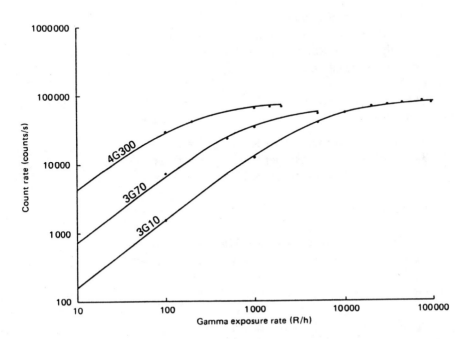

FIG.3.   Count rate versus gamma exposure rate (● experimental data; ▬▬ calculated curves).

4.2. Gamma irradiations

The following measurements were carried out in the National Bureau of Metrology (Laboratoire de Métrologie des Rayonnements Ionisants) at Saclay:

(1) 2 to 100 R/h : primary reference beam ^{60}Co No.1
(2) 200 to 10 000 R/h : secondary reference beam ^{60}Co No.2
(3) 20 000 to 100 000 R/h : calibrated industrial irradiator PAGURE.

4.3. Experimental results

The experimental results are given in Fig.3. Measurements were made up to 100 000 R/h.

4.4. Calculated saturation curves

Theoretical saturation curves can be calculated according to the following formula

$$N = \frac{N_0}{1 + N_0 \tau}$$

where N_0 is $\Sigma \cdot D$
  N is the calculated count rate in counts/s
  Σ is the detector sensitivity in counts/s per R/h
  D is the gamma exposure rate in R/h
  τ is the resolving time in s.

The saturation curves of 4 G 300, 3 G 70 and 3 G 10, which are plotted on Fig.3, were calculated with the following numerical values:

  4 G 300: Σ = 450 counts/s per R/h τ = 13 μs
  3 G 70: Σ = 72 counts/s per R/h τ = 15 μs
  3 G 10: Σ = 15 counts/s per R/h τ = 12 μs.

4.5. Discussion

Accurate measurements (less than 10% counting losses) can be made without any electronic adjustment up to the following exposure rate levels: 4 G 300 at 20 R/h, 3 G 70 at 100 R/h, and 3 G 10 at 600 R/h.

Figure 3 shows very good agreement between the experimental results and the calculated saturation curves. Thus, a very efficient electronic adjustment can be designed on the basis of the formula given in subsection 4.4. With such a correction, which is well known to instrument designers, accurate measurements can be made at much higher exposure rate levels, typically: 4 G 300 at 1000 R/h, 3 G 70 at 5000 R/h, and 3 G 10 at 30 000 R/h.

The count rate of the 3 G 10 detector was also found to be stable and reproducible up to 100 000 R/h, which was the upper limit of the measurements.

The resolving time, τ, depends on the detector and the test circuit. A short resolving time requires:

(1) A low value load resistor, R_L (measurements were made with 2.2 MΩ); use of 1MΩ would shorten the resolving time
(2) A low value signal resistor, R_k (2.2 kΩ was used).

5. LIFE TEST

A life test of one 3 G 10 detector was performed at 10 000 R/h. The detector operated satisfactorily for 140 hours, that is 2.8×10^{10} counts or 1.4×10^6 R. If necessary, the load resistor value could even be decreased to 1 MΩ in order to shorten the resolving time.

6. CONCLUSIONS

The halogen-quenched GM gamma detectors 3 G 70 and 3 G 10 appear to be perfectly suited for measurement of the very high gamma exposure rates that can be encountered in industrial irradiations because they are rugged devices and require quite simple associated electronics, they give accurate measurements of exposure rate levels as high as 30 000 R/h, they operate up to at least 100 000 R/h, and they have good counting stability and an excellent life.

SILVER DICHROMATE AS A ROUTINE DOSIMETER IN THE RANGE 1 TO 12 kGy

J. THOMASSEN
Institute for Energy Technology,
Kjeller, Norway

Abstract

SILVER DICHROMATE AS A ROUTINE DOSIMETER IN THE RANGE 1 TO 12 kGy.
 A dosimeter system based on the radiation-induced reduction of silver dichromate in perchloric acid was investigated. Measurement of radiation yield is done by a spectrophotometric method. The molar extinction coefficient of the dosimeter solution was determined. The effect of dichromate concentration on absorbance measurements and reduction yield was investigated. A formula for calculating the dose based on absorbance measurements is presented. The G-value for dichromate reduction when this dosimeter is exposed to gamma irradiation was determined. The value found, 0.397, is close to some of the values found by other investigators for potassium dichromate in sulphuric acid. The system was investigated in the medium dose range, 1 to 10 kGy, but it may easily be adapted to higher doses. In routine use no serious disadvantages or systematic errors have so far been discovered. The stability, accuracy and reproducibility of response are such that this system should be seriously considered as a possible new secondary standard dosimeter.

1. INTRODUCTION

The radiation-induced reduction of dichromate ions to chromic ions in dilute acid has been known for many years [1, 2]. Several investigators have studied this reaction with the aim of using it as a dosimeter system.

In the work of McLaren [3] potassium dichromate in dilute sulphuric acid was used as the dosimeter system and the measurement of dichromate reduction yield was done by a potentiometric method.

Matthews [4] used spectrophotometric measurement of dichromate reduction yield. He found that addition of silver ions to the dichromate solution had a remarkable positive effect on the linearity and precision of the response of dichromate to radiation. This effect was ascribed to inhibition of impurity effects and protection of the hydrogen by the silver ions.

In a recent publication by Sharpe et al. [5] the sulphuric acid was replaced by perchloric acid. This gives a more stable dichromate solution. They also investigated the effect of silver ions in the dosimeter solution.

The aim of the present work has been to find the details necessary for using this system as a dosimeter on a routine basis. During this work it was found that

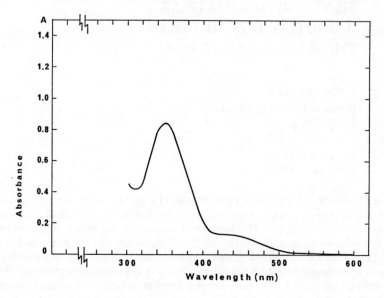

FIG.1. Spectrum of silver dichromate in 0.1M $HClO_4$.

a dosimeter system based on silver dichromate in perchloric acid exhibited several attractive qualities and it was decided to investigate the system as a possible secondary standard dosimetry system.

2. EXPERIMENTAL METHODS

All dichromate dosimeter solutions were made up of analytical grade perchloric acid diluted to 0.1M with water double-distilled in a quartz still. Silver dichromate was of analytical reagent grade and was used without any further purification.

Irradiation of samples of the dosimeter solution was done in our ^{60}Co irradiation plant at an air temperature of 22 ± 2°C. For comparison, a set of ampoules with dosimeter solution was irradiated at the Risø National Laboratory, Denmark, in their 10 kCi ^{60}Co calibration facility.

To establish electron equilibrium conditions the samples of dosimeter solution in glass ampoules were surrounded by a polymethyl metacrylate cylinder with a wall thickness of 1 cm. The space in the ampoules above the dosimeter solution was filled with ordinary air.

The concentration of dichromate was measured by spectrophotometry at a wavelength of 350 nm. As can be seen from Fig. 1 this corresponds to the maximum in the light absorption spectrum of silver dichromate in perchloric acid solutions. Because of the high light absorption of dichromate ions it is necessary to work with low concentrations, in the range 0.1 to 1.0mM. On the other hand, this gives a system with a high sensitivity to radiation-induced reduction of dichromate ions to chromic ions, Cr(III).

To find any temperature dependence on absorbance readings a series of samples was measured in the temperature range 20 to 32°C. The temperature coefficient found was −0.06% per °C.

In an irradiated dichromate solution there will be chromic ions present in an amount corresponding to the reduced amount of dichromate ions. Experiments were carried out to find the contribution of chromic ions to the measured absorbance. Pure chromic oxide was dissolved in 0.1M perchloric acid and the absorbance measured. It was found that the extinction coefficient of chromic ions in this solution measured at 350 nm is 7 L·mol^{-1}·cm^{-1}. This is a small value compared with the value for the extinction coefficient of dichromate. The contribution of absorbance due to the reduced form of dichromate, Cr(III), can therefore be ignored except for the most demanding measurements.

The Fricke solution used to calibrate the dose rate at the irradiation point in the ^{60}Co plant had the following composition

392 mg Fe(NH$_4$)$_2$(SO$_4$)$_2$·6H$_2$O
58 mg NaCl
1 L 0.4M H$_2$SO$_4$

This gives a solution which is 1mM in Fe(II) and Cl$^-$. The absorbance of this solution was measured against the unirradiated solution at a wavelength of 304 nm and the absorbed dose, D, in gray calculated using the formula

$$D = A \cdot 275$$

where A is the measured absorbance. The G-value of oxidation of Fe(II) to Fe(III) was taken as 15.6, the extinction coefficient at 304 nm as 2195 L·mol^{-1}·cm^{-1}, and the solution density 1.024 g·cm^{-3} [6]. The temperature at the absorbance measurements was 23 ± 1°C.

3. ABSORBANCE VERSUS SILVER DICHROMATE CONCENTRATION

To find the connection between silver dichromate concentration and measured absorbance a series of samples was prepared. The concentration range was from 0.05 to 0.55mM. Absorbance measurements were done against 0.1M perchloric acid.

The results are given in Table I and shown graphically in Fig. 2. As can be seen from this figure the relation between concentration and the measured absorbance at 350 nm is strictly linear. This means that the dosimeter solution obeys the Beer-Lamberts law in the concentration range investigated. Table I shows that the absorbance of solutions with a concentration higher than about 0.5mM is in the

TABLE I. RELATION BETWEEN SILVER DICHROMATE CONCENTRATION AND ABSORBANCE AT 350 nm

Silver dichromate concentration (mM)	Absorbance
0.055	0.173
0.138	0.432
0.208	0.646
0.277	0.862
0.415	1.302
0.553	1.751

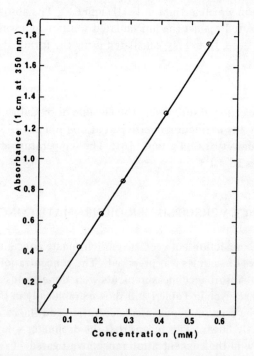

FIG.2. *Absorbance of silver dichromate in 0.1M $HClO_4$.*

TABLE II. DEPENDENCE OF REDUCTION YIELD ON SILVER DICHROMATE CONCENTRATION (dose 3.5 kGy)

Silver dichromate concentration (mM)	Decrease in absorbance
0.25	0.403
0.30	0.405
0.40	0.413
0.50	0.411
1.0	0.414

range 1.5 and upwards. This range is not suitable for accurate absorbance measurements. Thus it is recommended that in cases where the initial concentration must be higher than about 0.5mM the solution should be diluted before absorbance measurements.

A higher silver dichromate concentration will be required if one is going to use the system in the range 10 to 20 kGy. In this case concentrations between 0.5 and 1.0mM will be necessary. In general, one should choose a silver dichromate concentration which has an optimum response in the desired dose range.

4. EFFECT OF DICHROMATE CONCENTRATION ON REDUCTION YIELD

When using the dosimeter solution presented here as a routine dosimeter it may be convenient to choose a dichromate concentration which matches the most accurate range of absorbance reading of the spectrophotometer at hand and the actual dose to be measured.

For these reasons and from a theoretical point of view it was of interest to see if the decrease in absorbance or reduction yield showed any dependence on the initial dichromate concentration.

A series of dosimeter samples ranging in silver dichromate concentrations from 0.25 to 1.0mM was irradiated at the same time on a rotating disc so as to receive the same radiation dose. The results are given in Table II. It can be seen that no systematic difference is found in the reduction yield, measured as decrease in the absorbance.

5. DETERMINATION OF EXTINCTION COEFFICIENT

To derive a formula for calculating the radiation dose absorbed by a silver dichromate solution the following theoretical considerations may be made.

The G-value for the reaction

$$Cr_2O_7^{--} \rightarrow 2\, Cr^{+++}$$

is by definition

$$G = \frac{\text{Number of dichromate ions reduced}}{\text{Radiation energy absorbed (eV)}} \times 100 \tag{1}$$

Assume the dosimeter solution is given a radiation dose equal to D kGy. This will give rise to a reduction in the dichromate concentration equal to C mol per litre. Then the number of dichromate ions reduced per cubic centimetre of the dosimeter solution is

$$\frac{C \times N}{1000} \tag{2}$$

where N = Avogadro's number = 6.023×10^{23}.

A radiation dose of D kGy to the dosimeter solution corresponds to

$$D \times 6.242 \times 10^{18} \text{ eV per gram dosimeter solution} \tag{3}$$

If the density of the dosimeter solution is d gram per cubic centimetre, then Eq. (3) can be written

$$D \times d \times 6.242 \times 10^{18} \text{ eV per cubic centimetre dosimeter solution} \tag{4}$$

Substitution in Eq. (1) with (2) and (4) gives

$$G = \frac{C \times N \times 100}{D \times d \times 6.242 \times 10^{18} \times 1000} = \frac{C \times 9649}{D \times d} \tag{5}$$

If the reduction in dichromate concentration, C, is determined by spectrophotometry, then C may be substituted in Eq. (5) by the well known relationship

$$A = C \times E \times L \tag{6}$$

where A is the decrease in absorbance of the irradiated solution compared with the unirradiated, E is the extinction coefficient of the dichromate ions and L is the light path in the spectrophotometer cell.

Substitution and rearranging gives the final form

$$D = \frac{A \times 9649}{G \times E \times L \times d} \text{ kGy} \tag{7}$$

To use formula (7) for the purpose of calculating the dose, D, received by the dosimeter solution it is necessary to know the G-value and the extinction coefficient of dichromate ions in the actual solution.

According to European Pharmacopoeia [7] the molar extinction coefficient of a solution of potassium dichromate in 0.01N sulphuric acid measured at a wavelength of 350 nm is 3136 ± 48 $L \cdot mol^{-1} \cdot cm^{-1}$. It is to be expected that the extinction coefficient of silver dichromate in 0.1M perchloric acid measured at the same wavelength will have a value not far from this.

It was found in our experiments that the absorbance of potassium dichromate in 0.1M perchloric acid did not change when adding an equivalent amount of silver ions in the form of silver nitrate. Since the exact stoichiometric composition of the silver dichromate may be doubtful, resulting in uncertain dichromate concentrations by weighing, and since potassium dichromate solutions may be considered as well known and accepted calibration standards in spectrophotometry, it was decided to use potassium dichromate instead of silver dichromate in our experiments to determine the extinction coefficient.

Potassium dichromate of analytical grade was dried at 130°C for one hour and a series of known concentrations were made in the range 0.10 to 0.25mM by weighing and diluting.

The absorbance, A, was read out against 0.1M perchloric acid at a wavelength of 350 nm and a temperature of $22 \pm 2°C$. The extinction coefficient, E, was calculated from the relation

$$A = E \times C \times L$$

where C is the dichromate concentration and L is the light path in the spectrophotometer cell.

The results are given in Table III. Each value is the mean of three parallel measurements.

The mean value found, 3122 $L \cdot mol^{-1} \cdot cm^{-1}$, is very close to the value given in Ref. [7] for potassium dichromate in 0.01N sulphuric acid, 3136 $L \cdot mol^{-1} \cdot cm^{-1}$.

6. DETERMINATION OF G-VALUE

Matthews [4] found G-values ranging from 0.297 to 0.410 for solutions containing potassium dichromate and silver nitrate in sulphuric acid.

TABLE III. MEASURED VALUES OF EXTINCTION COEFFICIENT FOR DICHROMATE IN 0.1M PERCHLORIC ACID

Concentration (mM)	Absorbance	Extinction coefficient
0.1010	0.315	3119
0.2020	0.631	3125
0.2510	0.783	3120
0.2525	0.789	3125

Sharpe et al. [5] found by extrapolation to zero silver concentration a value of 0.405. They state that at higher silver ion concentrations the G-value is considerably lower.

Because of the need for a reliable value of the reduction yield necessary when using formula [7] to calculate the absorbed dose, it was decided to determine the G-value as exactly as possible.

Determination of the G-value for the reduction of dichromate is dependent on the measurement of the decrease in dichromate concentration. If this measurement is done by spectrophotometry, which is the most common method, one should take care to attain sufficient accuracy of the absorbance measured with the instrument at hand. A procedure for the control of absorbance values of dichromate measured at 350 nm is given in Ref. [7]. Another important factor is the extinction coefficient used to calculate the concentration. Because of this we have done considerable work in order to determine the extinction coefficient as accurately as possible (see Section 5).

The G-value determinations were done with 0.4mM silver dichromate solutions in 0.1M perchloric acid. Since Sharpe et al. [5] found no dose-rate effect on the reduction yield all our radiations were done at approximately the same dose rate, 0.7 kGy per hour. The dose rate at the irradiation point in the ^{60}Co plant was accurately determined by ordinary Fricke dosimeters.

Samples of silver dichromate containing 10 cm^3 solution were irradiated to accurately known doses, D. The temperature during the irradiations was 22 ± 2°C. The decrease in absorbance, A, was measured at a wavelength of 350 nm. The G-value was calculated according to formula (7) using a value for the extinction coefficient, E, of 3122 L·mol^{-1}·cm^{-1} (see Section 5). The value for the density of the dosimeter solution, d, was determined by weighing accurately known volumes of the solution and was found to be 1.004 g·cm^{-3} at 22°C.

The results are given in Table IV. According to these experiments the mean value is

$$G = 0.397 \pm 0.008$$

TABLE IV. MEASURED G-VALUES FOR RADIATION-INDUCED REDUCTION OF SILVER DICHROMATE IN 0.1M PERCHLORIC ACID

Dose (kGy)	Decrease in absorbance	G-value
2.89	0.367	0.391
3.46	0.455	0.405
5.13	0.666	0.399
6.44	0.821	0.392
7.75	1.001	0.397

7. CALIBRATION CURVE

A reliable and accurate calibration curve can only be made if several conditions are fulfilled. One of the most important conditions is to have a well defined point in the radiation field with an accurately known dose rate.

If this condition is difficult to fulfil at your own irradiation plant it should be possible to send a series of dosimetry samples in sealed ampoules to a reference calibration facility. Here they may be given accurately known doses and then returned for absorbance measurements with your own spectrophotometer. As we have shown, the absorbance reading will not change even many days after irradiation. The unirradiated dosimeter solution may be stored in closed glass flasks for several months without a change in absorbance reading.

We have found it convenient to hold the dosimeter solution in heat-sealed polyethylene tubing during irradiation. The tubing should have an outer diameter of at least 10 mm. In routine use one should be careful to use the same tubing material and size as in calibration measurements. In this way trouble caused by differences in electron equilibrium and radiation absorption in the calibration samples and routine dosimeters will be kept to a minimum.

Figure 3 shows an example of a calibration curve in the dose range 1 to 10 kGy.

8. CONCLUSIONS

It has been found and verified that silver dichromate dissolved in dilute perchloric acid has several attractive qualities as a dosimeter system in the medium dose range.

The response of the dosimeter to radiation is easily measured by ordinary spectrophotometry. It is seen from the experiments that measurements of change in concentration may be done at the wavelength of maximum light absorption for

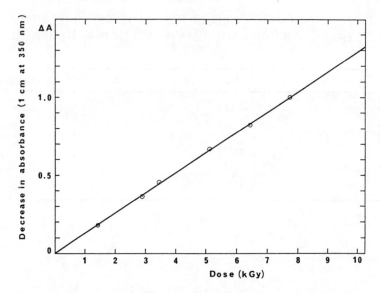

FIG. 3. *Calibration curve for silver dichromate dosimeter.*

the dichromate ions. This means high sensitivity in radiation dose measurements. Absorbance measurements of dichromate at this wavelength have a good reproducibility and the spectrophotometer response may be controlled by internationally accepted standard procedures. The absorbance values in the low concentration range used in these experiments show strict adherence to the Lambert-Beer law.

A formula for calculating the dose based on absorbance measurements only has been derived. The necessary constants in this formula have been established by experiments.

Determination of the molar extinction coefficient of dichromate in the actual solution has been made and the value found is very close to that of potassium dichromate in dilute sulphuric acid.

The G-value of radiation-induced reduction of dichromate in this solution is found to be close to some of the values published for dichromate in other solutions.

So far we have found few drawbacks with this dosimeter system. Used as a routine dosimeter it may in some cases be inconvenient with a system in the form of a liquid. Breakage of the ampoules followed by spillage of the solution on to expensive goods should be considered a potential risk.

We have found that commercially available silver dichromate does not have a reliable stoichiometric composition. This means that one cannot make up solutions with an exactly known concentration by simply weighing out the corresponding amount of the salt. The absorbance response of the dosimeter solution has a high, long stability, both for unirradiated and irradiated samples.

It is the opinion of the author that silver dichromate in dilute perchloric acid is a candidate to be considered as a new secondary standard radiation dosimeter system.

REFERENCES

[1] FRICKE, H., BROWNSCOMBE, E.R., J. Am. Chem. Soc. **55** (1933) 2358.
[2] ANDERSON, A.R., FARHATAZIZ, Trans. Faraday Soc. **59** (1963) 1299.
[3] McLAREN, K.G., Int. J. Appl. Radiat. Isot. **32** (1981) 803.
[4] MATTHEWS, R.W., Int. J. Appl. Radiat. Isot. **32** (1981) 861.
[5] SHARPE, P.H.G., BARRET, J.H., BERKLEY, A.M., Dichromate Solution As a Reference Dosemeter For Use In Industrial Irradiation Plants, National Physical Laboratory, Rep. RS(EXT)60 (1982).
[6] SEHESTED, K., Manual On Radiation Dosimetry (HOLM, N.W., BERRY, R.J., Eds), Marcel Dekker, New York (1970) 313.
[7] European Pharmacopoeia, Part I, 2nd edn, Vol. 6.19, Maisonneuve, S.A., France (1980).

РАБОЧИЕ ХИМИЧЕСКИЕ ДЕТЕКТОРЫ И ИНДИКАТОРЫ ПОГЛОЩЕННОЙ ДОЗЫ И ИХ МЕТРОЛОГИЧЕСКОЕ ОБЕСПЕЧЕНИЕ

С.В. КЛИМОВ, Б.М. ВАНЮШКИН, Н.Г. КОНЬКОВ, С.М. НИКОЛАЕВ
Всесоюзный научно-исследовательский институт радиационной техники,
Москва

В.В. ГЕНЕРАЛОВА, М.Н. ГУРСКИЙ
Госстандарт СССР,
Москва

В.К. АМБРОСИМОВ, Б.В. ТОЛКАЧЕВ
Московское научно-производственное объединение "НИОПИК",
Москва

М.П. ГРИНЕВ
Институт биофизики Министерства здравоохранения СССР,
Москва

Союз Советских Социалистических Республик

Abstract—Аннотация

OPERATIONAL CHEMICAL DETECTORS AND INDICATORS OF ABSORBED DOSE AND THE ASSOCIATED METROLOGY.
The main characteristics of operational chemical detectors and indicators used in radiation chemistry processes involving photon and electron radiations are presented. Questions of the calibration of chemical detectors on a set of standards designed for the calibration, verification and certification of detectors in the dose range from 10^2 to 10^6 Gy are discussed. Descriptions are given of systems of standards for the calibration of chemical detectors in electron radiation fields ranging in energy from 0.3 to 10 MeV.

РАБОЧИЕ ХИМИЧЕСКИЕ ДЕТЕКТОРЫ И ИНДИКАТОРЫ ПОГЛОЩЕННОЙ ДОЗЫ И ИХ МЕТРОЛОГИЧЕСКОЕ ОБЕСПЕЧЕНИЕ.
Приведены основные характеристики рабочих химических детекторов и индикаторов, используемых в радиационно-химических процессах с применением фотонного и электронного излучений. Рассмотрены вопросы градуировки химических детекторов на эталонном комплексе, предназначенном для градуировки, поверки и аттестации детекторов в интервале доз от 10^2 до 10^6 Гр. Описаны образцовые установки для градуировки химических детекторов в полях электронного излучения с энергией от 0,3 до 10 МэВ.

Одним из основных условий, определяющих качество радиационно-химического эксперимента или процесса, является точное знание дозы и мощности поглощенной дозы ионизирующего излучения. Поглощенная доза как и мощность поглощенной дозы являются одними из основных физических величин, от которых зависит эффект взаимодействия излучения с веществом. Знание этих величин во многом способствует качественному проведению научных исследований и получению с наименьшими затратами продукции с заданными эксплуатационными свойствами [1].

Специфические особенности проведения радиационно-химических процессов и необходимость измерения поглощенной дозы в различных по химическому составу и агрегатному состоянию веществах, облучаемых как в статистических, так и в динамических режимах, накладывает определенные трудности на выбор средств и методов измерения поглощенной дозы ионизирующего излучения и делает практически непригодным использование таких классических методов, как ионизационный и сцинтиляционный.

Перспективными в дозиметрии больших доз следует считать химические методы, основанные на использовании стандартных композиций веществ, свойства которых изменяются определенным образом в зависимости от дозы излучения. Химические дозиметрические системы способны моделировать любой состав объектов и условия их облучения (стационарные, динамические), давая возможность определять поглощенную дозу либо в конкретных точках объекта, либо усредненную по облучаемому объему. Поскольку химический состав и эффективный атомный номер дозиметра может быть подобран близким к облучаемому объекту, химический дозиметр практически не вносит искажения в поле поглощенных доз и обеспечивает возможность корректного определения поглощенной дозы в облучаемом объекте без дополнительного пересчета, приводящего к потере точности.

Наиболее перспективными для указанной цели оказались химические радиохромные пленочные детекторы и цветовые индикаторы поглощенных доз, представляющие собой полимерную матрицу с введенными в нее радиационно-чувствительными соединениями. Изготавливают детекторы и индикаторы дозы путем полива из раствора, экструзией или литьем под давлением.

Под действием ионизирующего излучения в матрице возникают активные частицы, которые, взаимодействуя с радиационно-чувствительным соединением, образуют продукты, изменяющие спектр поглощения детектора или индикатора дозы. Индикаторы позволяют определять поглощенную дозу путем визуального сравнения их цвета с калиброванной цветовой шкалой. Поглощенную дозу определяют по изменению оптической плотности в максимуме радиационно-индуцированной полосы оптического поглощения. Изменение оптической плотности детектора в максимуме радиационно-индуцированной полосы поглощения пропорционально поглощенной дозе.

Погрешность определения поглощенной дозы с помощью рабочих детекторов не превышает 10-15% в диапазоне мощности дозы от 0,1 до $3 \cdot 10^3$ Гр/с.

Индикаторы позволяют определять поглощенную дозу с погрешностью 25-30%.

В таблице I приведены основные характеристики рабочих химических детекторов и индикаторов, разработанных и внедренных в производство в СССР [2-4].

Детектор СОПДА-0,1/1,0 – средство измерения поглощенной дозы фотонного и электронного излучений в диапазоне доз от 0,1 до 1,0 кГр.

Детектор изготавливают в виде пленки толщиной 200 мкм. Цвет необлученного детектора – оранжевый. При облучении детектор приобретает фиолетовый цвет за счет появления новой полосы поглощения с максимумом на длине волны 530 нм.

ТАБЛИЦА I. ОСНОВНЫЕ ХАРАКТЕРИСТИКИ РАБОЧИХ ХИМИЧЕСКИХ ДЕТЕКТОРОВ И ИНДИКАТОРОВ

Тип детектора или индикатора	Диапазон поглощенной дозы кГр	Длина волны при измерении нм	Толщина детектора мкм
Детекторы			
СОПДА-0,1-1,0	0,1-1,0	530	200
СОПДФ-3/30	3,0-30	520	60
ДРД-0,4/4,0	4,0-40	314	2000
СОПДС-10/300	10,0-300	510	60; 100
Индикаторы			
ЦВИД-0,1/1,0	1,0-10	–	70
ЦИД-1/5	10,0-50	–	200
ЦИД-2/20	20,0-200	–	200

Рабочий диапазон температур детектора — 20-40°C.

Показания детектора стабильны в течение двух суток. В дальнейшем следует вносить поправку на постэффект. Детектор нашел применение в области радиобиологии.

Детектор СОПДФ-3/30 — средство измерения поглощенной дозы фотонного и электронного излучений в диапазоне доз от 3,0 до 30 кГр. Детектор изготавливают в виде пленки толщиной 60 мкм. Цвет необлученного детектора — желтый. При облучении детектор приобретает красный цвет за счет образования новой полосы поглощения с максимумом на длине волны 520 нм. Показания детектора стабильны в течение 20 суток в рабочем диапазоне температур 20-50°C. Детектор нашел применение при контроле процессов радиационного модифицирования тканей.

Детектор СОПДС-10/300 — средство измерения поглощенной дозы фотонного и электронного излучений в диапазоне доз от 10 до 300 кГр. Детектор изготавливают в виде пленки толщиной 60 и 100 мкм. Цвет необлученного детектора — светло-зеленый. При облучении детектор приобретает коричневый цвет за счет образования дополнительной полосы поглощения с максимумом на длине волны 510 нм. Рабочий диапазон температур — 20-50°C. Детектор нашел применение при контроле процессов радиационного модифицирования полимерных материалов и изделий из них.

Детектор ДРД-0,4/4 — средство измерения поглощенной дозы фотонного и электронного излучений в диапазоне доз от 4,0 до 40 кГр. Детектор предназначен для измерения поглощенной дозы фотонного излучения с энергией 0,66 и 1,25 МэВ и ускоренных электронов с энергией выше 3 МэВ при мощности дозы от 0,5 до 10^4 Гр/с. Показания детектора в процессе облучения не зависят от температуры в диапазоне от +10°C до 50°C при относительной влажности от 0 до 100%.

Указанные характеристики детектора обеспечиваются применением соответствующих материалов детектора (низкомолекулярный полиметилметакрилат (ПММА) и технологией изготовления. Размеры детектора – 35×15×2 мм. Рабочая длина волны – 314 нм.

При использовании детектора ДРД-0,4/4 отсутствует необходимость контролировать толщину и начальную оптическую плотность каждого детектора, что выгодно отличает его от аналогов, известных в мировой практике под названием Red Perspex, Perspex HX. Детекторы не нуждаются в дополнительной проверке градуировочной характеристики после выпуска в течение трех лет.

Детектор используется в качестве основного средства измерения дозы на промышленных установках для радиационной стерилизации.

Цветовые индикаторы дозы ЦИД-1/5 и ЦИД-2/20 имеют близкие цветовые переходы: синий-фиолетовый-красный-желтый, лежащие в интервале доз от 10 до 50 кГр и от 20 до 200 кГр, соответственно. Принцип действия этих индикаторов – конкурентное обесцвечивание в полимерной матрице двух красителей с различным радиационно-химическим выходом разложения.

Цветовой индикатор ЦВИД-0,1/1,0 представляет собой пленку желтого цвета толщиной 70 мкм. При облучении в интервале доз от 0,1 до 1,0 кГр цвет индикатора постепенно изменяется от желтого к красному (5 кГр – оранжевый, 10 кГр – красный).

Области применения цветовых индикаторов дозы в соответствии с рабочим диапазоном доз. Основное назначение – разделение облученной и необлученной продукции, визуальная оценка поглощенной дозы электронного излучения пучков.

Для градуировки химических детекторов разработан и метрологически исследован эталонный комплекс, предназначенный для градуировки, поверки и аттестации детекторов в интервале доз от $1 \cdot 10^2$ до $1 \cdot 10^6$ Гр (рис. 1).

В рабочем эталоне в качестве источников излучения использовались радиоизотопные самозащищенные гамма-установки серийного производства.

В этих установках источники излучения кобальт-60 и цезий-137 расположены по образующим цилиндрических камер для облучения и создают близкое к изотропному поле излучения. В качестве стандартных материалов для поглотителей калориметров были выбраны графит, полистирол, вода, как наиболее близкие по химическому составу многим типам дозиметров [5].

Поглотитель имел форму полого цилиндра диаметром и высотой 22 мм с толщиной стенки 3 мм. При выбранной форме и размерах поглотитель обеспечивал условия электронного равновесия для фотонного излучения кобальта-60 и цезия-137 при незначительном ослаблении излучения в самом калориметре. Градуировку калориметра производили по тепловому действию электрического тока с помощью нагревателя из манганиновой проволоки диаметром 30 мкм, вмонтированного в стенки поглотителя. Выходящий из поглотителя калориметра тепловой поток регистрировался термобатареей, равномерно размещенной по боковой поверхности и торцам поглотителя.

Снижение влияния окружающей среды на температуру поглотителя достигалось термостатированием его. Схема автоматического регулирования температуры обеспечивала термостатирование с точностью $10^{-4}°C$ в диапазоне температур от 20 до 50°C. Чувствительность калориметра по мощности тепловыделения составляла 50 мВ/Вт, по

IAEA-SM-272/46

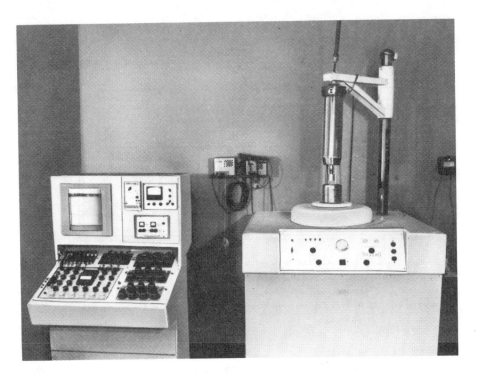

Рис. 1. Общий вид рабочего эталона.

мощности поглощенной дозы 500 мкВ · Гр$^{-1}$ · с и обеспечивала возможность измерения мощности дозы начиная с 0,01 Гр/с и выше. Постоянная времени калориметра τ составляла 70-100 с. Оптимальное количество термопар определялось экспериментальным путем [6].

В результате метрологического исследования было установлено, что среднее квадратичное отклонение результатов измерения мощности поглощенной дозы фотонного излучения в диапазоне от 0,01 до 10 Гр/с не превышает 0,1% при неисключенных систематических погрешностях не более 1% [3].

Измерения спектра излучения в радиоизотопных установках, входящих в состав рабочего эталона, проводились с помощью сцинтиляционного (на модельной установке) и калориметрического (на действующих установках) методов и подтвердили наличие в спектре низкоэнергетической компоненты фотонного излучения. По полученным гистограммам спектра был рассчитан вклад в поглощенную дозу рассеянного гамма-излучения с энергией от 0,05 до 0,2 МэВ для материалов, наиболее часто используемых в радиационной технологии. Он составил для графита, полистирола, воды и стекла (Пирекс) 15, 15, 19, 26%, соответственно. Полученные данные подтверждают, что определение поглощенной дозы в материале произвольного состава путем пересчета, полученного с использованием массовых коэффициентов поглощения энергии для начальных значений энергии излучения кобальта-60 и цезия-137, может быть в ряде случаев выполнено с большой

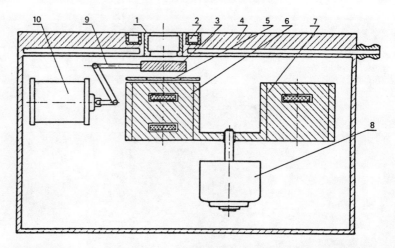

Рис. 2. Схема образцовой установки для градуировки дозиметров в полях электронного излучения с энергией электронов от 3 до 10 МэВ:

1 – коллиматор, 2 – коллектор электронов, 3 – затвор, 4 – алюминиевый экран, 5 – монитор, 6 – фантом-калориметр, 7 – фантом с градуируемым калориметром, 8 – электродвигатель, 9 – рычаг, 10 – электромагнит.

погрешностью, уменьшение которой возможно лишь при корректном учете истинного спектра излучения [7].

Принципы градуировки дозиметров в полях электронного излучения с энергией электронов от 3 до 10 МэВ имеют много общего с определением поглощенной дозы в различных материалах при облучении их фотонным излучением. Доминирующей составляющей погрешности градуировки по-прежнему следует считать погрешность передачи размера единицы поглощенной дозы от калориметра к дозиметрам.

Градуировка дозиметров производится методом замещения. В фантоме, на глубине приблизительно равной 1/3 длины экстраполированного пробега электронов, с помощью калориметра измеряют поглощенную дозу в поглотителе, изготовленном из того же материала, что и фантом (рис. 2). Затем в эту область среды помещают дозиметр и поглощенную дозу в нем определяют путем пересчета, с учетом возмущения потока электронов полостью, т.е. градуируемым дозиметром:

$$D_g = D_k \cdot S_k^g \cdot P_k^g,$$

где S_k^g – отношение тормозных способностей материала дозиметра и поглотителя калориметра;

P_k^g – коэффициент, учитывающий возмущение потока электронов полостью.

Чтобы свести к минимуму погрешность, связанную с определением тормозных способностей и коэффициента замещения, поглотители калориметра были изготовлены из графита, полистирола и стекла, идентичными по форме и размерам градуируемым дозиметрам.

Рис. 3. Образцовая установка для градуировки дозиметров в полях электронного излучения с энергией электронов от 3 до 10 МэВ.

На основе результатов метрологического исследования в качестве образцовой аттестована калориметрическая установка для градуировки дозиметров на ускорителях электронов с энергией электронов от 3,0 до 10 МэВ при мощностях доз от 10 до $2 \cdot 10^4$ Гр/с (рис.3).

Установка обеспечивает возможность градуировки дозиметров в диапазоне от $1 \cdot 10^2$ до $1 \cdot 10^6$ Гр. Погрешность измерения мощности поглощенной дозы калориметрами не превышает 1,5% при доверительной вероятности 0,95; погрешность градуировки разных типов дозиметров, зависящая от материала дозиметра и его размеров, а также от нестабильности работы ускорителя, достигает 2,5-3,5% при доверительной вероятности 0,95 [8].

Для градуировки химических дозиметров электронного излучения с энергией электронов от 0,3 до 3,5 МэВ на основе калориметра полного поглощения была разработана образцовая установка такой конструкции (рис.4), чтобы обеспечить измерение плотности потока энергии электронного излучения в диапазоне от 10^{-3} до 10 Вт/см^2 (условия облучения на промышленных ускорителях электронов) с погрешностью, не превышающей 1% при доверительной вероятности 0,95. Учитывая, что градуировочные характеристики почти всех пленочных дозиметров, используемых для определения поглощенных доз низкоэнергетического электронного излучения, как правило, нелинейны, необходимо было также разработать соответствующую методику градуировки таких дозиметров с помощью калориметра полного поглощения и исследовать влияние факторов, определяющих погрешность градуировки. Основой разработанной установки

Рис. 4. *Образцовая установка для градуировки дозиметров в полях электронного излучения с энергией электронов от 0,3 до 3,5 МэВ.*

является калориметр полного поглощения, выполненный из графита и имеющий форму полого цилиндра. Использование материала с низким атомным номером и выбранная конструкция поглотителя калориметра обеспечивали малые потери энергии на тормозное излучение и на обратное отражение электронов. Для снижения флуктуации температуры окружающей среды в установке использовались два идентичных калориметра (дифференциальная схема), помещенных в термостат.

При работе ускорителя один из калориметров облучался потоком электронов, а второй был защищен алюминиевым экраном. Чувствительность калориметра составляет 40 мВ/Вт при работе в стационарном режиме и 4 мВ/Вт — в динамическом, а время измерения — 1,5-3 и 4-5 мин, соответственно [3].

Предельная погрешность, возникающая при определении поглощенной пакетом пленочных дозиметров энергии излучения, не превышает 3% при доверительной вероятности 0,95 во всем диапазоне значений энергии электронов. Разработанная методика градуировки позволяет на основе измерения плотности потока энергии излучения определить поглощенную дозу отдельным пленочным дозиметром. Погрешность градуировки пленочного дозиметра зависит от воспроизводимости показаний химического дозиметра.

Создание системы образцовых мер и методов передачи с минимальной потерей точности размера единицы поглощенной дозы от исходных средств измерений к образцовым и рабочим химическим дозиметрам обеспечивает единство и достоверность измерений этой величины в радиационно-химической технологии.

Градуируемые на эталонных установках химические дозиметры используются для метрологической аттестации радиационных установок и текущего дозиметрического контроля различных радиационно-технологических процессов. Результаты исследований показали, что наиболее надежными для использования в дозиметрии являются растворы глюкозы; разложение глюкозы не зависит в широких пределах $(0{,}01\text{-}1\cdot10^2\text{ Гр/с})$ от мощности дозы, примесей, чистоты реактивов, кислорода и от времени приготовления исходного раствора. Детальное метрологическое исследование водных 5 и 20% растворов глюкозы позволило рекомендовать их для использования в качестве образцового средства измерения в интервале поглощенных доз от 0,05 до 2,0 МГр с погрешностью 5%.

С помощью образцовых глюкозных дозиметров в период с 1970 г. по 1983 г. аттестовано около 150 установок, из них 32 опытно-промышленных и промышленных.

Для организации дозиметрического контроля разработаны методические рекомендации проведения измерений поглощенной дозы с помощью глюкозного и пленочных дозиметров, а также общие положения по организации дозиметрического контроля на всех стадиях проведения радиационных процессов с радиоизотопными источниками и ускорителями электронов [9, 10].

ЛИТЕРАТУРА

[1] КОДЮКОВ, В.М., и др., В сб. докладов конференции по вопросам внедрения радиационных установок и радиационной технологии, Постоянная комиссия СЭВ по использованию атомной энергии в мирных целях, Будапешт, 1973.
[2] ГРИНЕВ, М.П., и др., В кн. Радиационная Дозиметрия, Ташкент, Фан (1982) 59.
[3] АМБРОСИМОВ, В.К., и др., Вопросы атомной науки и техники, сер. Радиационная техника, **2** 24 (1982) 9.
[4] АМБРОСИМОВ, В.К., и др., В кн. Радиационная Дозиметрия, Ташкент, Фан (1982) 101.
[5] ГЕНЕРАЛОВА, В.В., ГУРСКИЙ, М.Н., Дозиметрия в Радиационной Технологии, М., Изд-во стандартов, 1981.
[6] БЕРЛЯНД, В.А., ГЕНЕРАЛОВА, В.В., ГУРСКИЙ, М.Н., Ат. Энерг. **38** 4 (1975) 253.
[7] АЛЕЙКИН, В.В., и др., Измер. Тех. 8 (1980) 65.
[8] БЕРЛЯНД, В.А., ГЕНЕРАЛОВА, В.В., ГУРСКИЙ, М.Н., Измер. Тех. 8 (1980) 52.
[9] Методические Рекомендации по Дозиметрии Радиационно-технологических Установок с Радиоизотопными Источниками Излучения, М., СЭВ, 1976.
[10] Методические Материалы по Дозиметрии на Радиационно-технологических Установках с Ускорителями Электронов, М., СЭВ, 1980.

DOSIMETRO DE NITRATO/NITRITO DE POTASIO PARA ALTAS DOSIS

E.M. DORDA, S.S MUÑOZ
Comisión Nacional de Energía Atómica,
Buenos Aires, Argentina

Abstract—Resumen

POTASSIUM NITRATE/NITRITE DOSIMETER FOR HIGH DOSES.
 A dosimetric method for monitoring doses of the order of 10 kGy (1 Mrad) is presented. The nitrate undergoes radiolytic reduction which can be related to the dose delivered. The analytical technique employed when using potassium nitrate for dosimeters, and the effects of the irradiation temperature, the pH, the nitrate concentration and post-irradiation stability are described. For the range 1 kGy to 150 kGy (0.10 to 15 Mrad), over which the dosimeter was tested, the error did not exceed 3%. The experiment was carried out in the pilot irradiation plant of the Ezeiza Atomic Centre, using a 220-Gammacell irradiator.

DOSIMETRO DE NITRATO/NITRITO DE POTASIO PARA ALTAS DOSIS.
 La puesta a punto del método dosimétrico objeto de este trabajo permite controlar dosis del orden de 10 kGy (1 Mrad). El nitrato sufre un fenómeno de reducción radiolítica, el cual se puede relacionar con la dosis entregada. Se describen la técnica analítica empleada para la utilización del nitrato de potasio como dosímetro, así como también los efectos de la temperatura de irradiación, del pH, de la concentración de nitrato, y estabilidad en el período de post-irradiación. El rango en el que este dosímetro fue probado va desde 1 kGy a 150 kGy (0,10 a 15 Mrad), siendo el error del método no superior al 3%. La experiencia se realizó en la planta de irradiación semi-industrial del Centro Atómico Ezeiza y además en un irradiador tipo Gammacell-220.

1. INTRODUCCION

El estudio del dosímetro de nitrato/nitrito de potasio surgió en el año 1975 como una necesidad para controlar dosis del orden de los 10 kGy (1 Mrad) en los procesos de radioesterilización en la planta de irradiación semi-industrial del Centro Atómico de Ezeiza (CAE).

El hecho de que los cristales de NO_3^- sufren una reducción radiolítica es conocido desde 1963; Chen y Johnson [1] por un lado y Cunningham [2], por otro, fueron los que denunciaron una correlación entre dosis entregada al sólido y producción de NO_2^- según la siguiente reacción:

NO_3^- ——— $NO_2^- + O$

$NO_2^- + O$ ——— NO_3^-

$NO_3^- + O$ ——— $NO_2^- + O_2$

$O + O$ ——— O_2

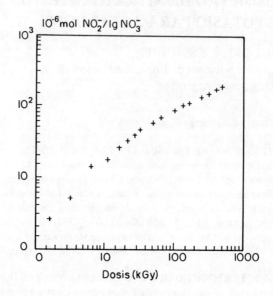

FIG.1. Correlación entre dosis entregada y producción de nitrito según los resultados obtenidos por Chen y Johnson (1963).

Chen y Johnson [1] llegaron a determinar que el nitrato de cesio era el más adecuado de los nitratos inorgánicos.

El ión nitrito fue analizado por el método de Shinn [3].

Sin embargo, analizando los resultados obtenidos por estos autores, se comprueba que la linealidad obtenida no fue buena (véase la Fig.1).

En nuestro trabajo se eligió el nitrato de potasio en estado sólido, y se estudió desde el punto de vista dosimétrico para cubrir rangos que van desde 1 kGy hasta 150 kGy (100 krad hasta 15 Mrad), mostrando muy buena linealidad.

Se determina el NO_2^- producido por valoración colorimétrica después de intervenir en una reacción de diazotación y posterior copulación en determinadas condiciones de estandarización.

El sistema $NO_3 K - NO_2 K$ ofrece muy buena estabilidad en el período de post-irradiación y su respuesta es independiente de la temperatura de irradiación.

2. FUENTE DE RADIACION (rayos gamma de ^{60}Co)

Se utilizó para las irradiaciones un Gammacell con fuente de ^{60}Co, con una actividad de $0,111 \times 10^{15}$ Bq (3000 Ci) y una velocidad de dosis de 1700 Gy/h.

Se trabajó además irradiando en la planta de irradiación semi-industrial con una actividad de 19,648 × 10^{15} Bq (532 000 Ci) de ^{60}Co y una velocidad de dosis de 2600 Gy/h a 1,50 m de la fuente.

3. MATERIALES

Se envasó el NO_3K seco (aproximadamente 1 g) en ampollas de vidrio selladas que fueron previamente lavadas y secadas.

3.1. Instrumento de medición

Espectrofotómetro Beckman de doble haz y cubetas de pyrex de 1 y 0,1 cm de paso óptico.

3.2. Reactivos

Se utilizaron los reactivos siguientes: nitrato de potasio Baker Analysed, nitrito de potasio Baker Analysed, ácido acético glacial Merck, sulfanilamida B & H, diclorhidrato de N-1 naftiletilendiamina Merck, ácido clorhídrico Merck, almidón Merck, ioduro de potasio Merck, MnO_4K 0,1 N tritisol Merck, $S_2O_3Na_2$ 0,1 N tritisol Merck, y ácido sulfúrico Merck.

3.3. Solución colorante

A) 2 g de sulfanilamida se disuelven en 1 litro de solución de ácido acético 30%.
B) 1 g de N-1 naftiletilendiamina se disuelven en 1 litro de ácido acético 30%.

En el momento de llevar a cabo la colorimetría se mezclan las soluciones A y B (5:1).

Las soluciones A y B se conservan cerradas en oscuridad a bajas temperaturas; en estas condiciones su duración es de 3 meses.

4. PROCEDIMIENTO

Para determinar NO_2^- en el NO_3K irradiado se procede a pesar 0,5 g de éste, se disuelve en 50 ml de agua contenida en un matraz de 100 ml y se agregan 20 ml de solución colorante. Transcurridos exactamente 10 min se enrasa y se lee la absorbancia de la solución a 546 nm (véase la Fig.2).

La coloración se basa en la formación de una sal de diazonio y posterior copulación con el clorhidrato de N-1-naftiletilendiamina en medio ácido.

La reacción se detalla a continuación.

5. CURVAS DE CALIBRACION

5.1. Curva de Lambert-Beer

Se utiliza una solución de NO_2K, aproximadamente 0,1 N, que se valora por la técnica de Klothoff, método Lunge.

De la solución anterior se prepara una serie de soluciones de concentración decreciente y se agrega a cada una 0,5 g de NO_3K sin irradiar (véase la Sección 5.1.2). Se debe cuidar de mantener el mismo volumen (50 ml) en todos los matraces al agregar la solución colorante. Esto se debe a varias razones, entre ellas la dependencia de la reacción con la acidez del medio (véase la Sección 6).

Se obtuvo un valor medio para el coeficiente de extinción molar de

$$40341 \ \frac{1}{mol \cdot cm}$$

Los datos se muestran en el Cuadro I y se representan en la Fig.3.

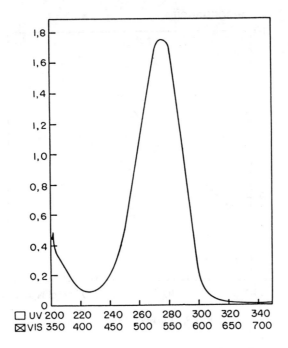

FIG.2. Curva de coeficiente de extinción molar.

CUADRO I. CURVA LAMBERT-BEER

Dispersión	Absorbancia	Conc. (M)
	0,000	0,000
0,003	0,064	$0,156 \times 10^{-5}$
0,006	0,157	$0,389 \times 10^{-5}$
0,010	0,315	$0,778 \times 10^{-5}$
0,026	0,628	$1,556 \times 10^{-5}$
0,030	0,788	$1,945 \times 10^{-5}$
0,066	1,569	$3,89 \times 10^{-5}$

$$\text{Pendiente } (0,0) = \epsilon \cdot b = 40341 \; \frac{1}{\text{mol} \cdot \text{cm}} \qquad r = 0,99999$$

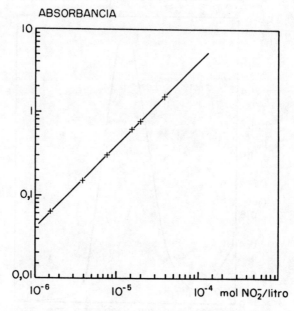

FIG.3. Curva de Lambert-Beer.

CUADRO II. INFLUENCIA DE LA PRESENCIA DE NO_3^- EN LA CURVA DE LAMBERT-BEER

Conc. (M)	Absorbancia	
	con 0,5 g NO_3^-	sin NO_3^-
$0,363 \times 10^{-5}$	0,139	0,130
$0,725 \times 10^{-5}$	0,282	0,264
$1,450 \times 10^{-5}$	0,557	0,527
$1,813 \times 10^{-5}$	0,702	0,657
$3,625 \times 10^{-5}$	1,404	1,317

5.1.1. *Influencia de la presencia de NO_3^- en la curva de Lambert-Beer*

Se comprobó que hay una variación en el coeficiente de extinción de un 6%, al trabajar con el agregado de 0,5 g de NO_3K.

En el Cuadro II se muestra una de las mediciones realizadas, donde se observa dicha influencia.

CUADRO III. RELACION ENTRE DOSIS ENTREGADA Y CONCENTRACION DE NO_2^- PRODUCIDO EN PRESENCIA DE 0,5 g DE NO_3K

Dosis (kGy)	Absorbancia	Dispersión	Moles NO_2^- prod./0,5 g NO_3^-
1,75	0,058	0,0021	$0,148 \times 10^{-6}$
3,50	0,115	0,0012	$0,294 \times 10^{-6}$
5,25	0,161	0,0031	$0,412 \times 10^{-6}$
7,00	0,208	0,0032	$0,532 \times 10^{-6}$
10,5	0,315	0,0104	$0,805 \times 10^{-6}$
21,0	0,609	0,0073	$1,556 \times 10^{-6}$
31,5	0,913	0,0247	$2,333 \times 10^{-6}$
50,8	1,435	0,0129	$3,667 \times 10^{-6}$
70,0	1,959	0,0274	$5,006 \times 10^{-6}$
83,0	2,292	0,0252	$5,858 \times 10^{-6}$
113,8	3,000	0,0600	$7,665 \times 10^{-6}$
148,2	3,968	0,0569	$9,800 \times 10^{-6}$

Pend. (0,0) = $0,6685 \times 10^{-7}$ $\dfrac{\text{mol } NO_2^- /0,5 \text{ g } NO_3^-}{\text{kGy}}$ r = 0,9992

5.2. Curva de dosis en relación con la concentración

Basándose en la curva de la Fig.3, se construyó otra de NO_2^- producido por irradiación de 0,5 g de NO_3^- a distintas dosis; los valores obtenidos se registran en el Cuadro III y se grafican en la Fig.4.

6. ESTUDIO DEL EFECTO DEL pH SOBRE LA REACCION DE COLORACION

La curva (véase la Fig.5) se construyó con una concentración de NO_2^- constante igual a $27,23 \times 10^{-5}$ M. De ella se deduce la importancia de trabajar con un pH entre 1 y 2,2 aproximadamente.

La acidez del medio estuvo dada por ácido clorhídrico o ácido acético en distintas concentraciones.

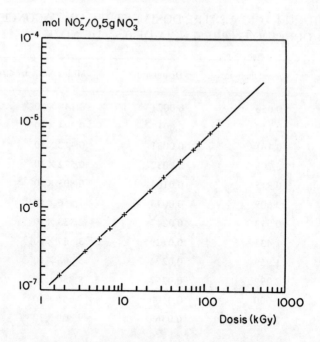

FIG.4. Curva de concentración en relación con dosis.

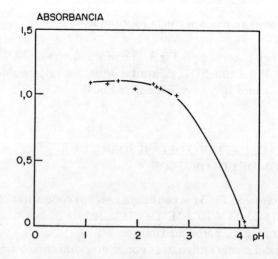

FIG.5. Curva del efecto del pH sobre la reacción de la coloración.

CUADRO IV. DETALLE DE LA MEDICION REALIZADA CON EL DOSIMETRO NO_3-NO_2K

Tiempo (min)	Absorbancia	Dosis (kGy)	$\dot{D}$ (Gy/min)
180	0,156	4,95	27,50
351	0,289	9,90	28,21
365	0,312	10,25	28,08
1017	0,795	27,2	26,75
1084	0,845	29,5	27,21
1449	1,125	39,5	27,26
1574	1,188	41,0	26,05
2340	1,796	64,0	27,35
4123	3,045	108,0	26,19

$$\dot{D} = 27,18 \text{ Gy/min} = 2,7\%$$

7. ESTUDIO DE LA RESPUESTA DEL DOSIMETRO A DISTINTAS TEMPERATURAS DE IRRADIACION

Se han hecho irradiaciones a temperaturas que van de 8°C a 100°C y no se han observado discrepancias que superan el error del método (véase la Sección 9).

Este hecho constituye una característica muy importante para la aceptación del dosímetro ya que en otros casos (p. ej. alanina) este parámetro es crítico.

8. ESTABLIDAD DE POST-IRRADIACION

No se observó decaimiento, habiéndose controlado la estabilidad post-irradiación del dosímetro hasta los tres meses.

9. PRECISION Y EXACTITUD

Se utilizó el dosímetro Fricke como patrón de calibración por extrapolación y simultáneamente se hicieron intercomparaciones con el dosímetro ferroso-cúprico en el rango de superposición de dosis (Cuadro IV).

Los valores de intercomparación de dosímetros fueron: Fricke, $\dot{D} = 27,93$ Gy/min; ferroso-cúprico, $\dot{D} = 27,16$ Gy/min; nitrato/nitrito de potasio, $\dot{D} = 27,18$ Gy/min.

En dicho cuadro se puede observar que el error del método es del 2,7%.

En los años 1977 y 1978 participamos en la encuesta de Intercomparación para Altas Dosis organizada por la OIEA. Se utilizó el dosímetro nitrato/nitrito de potasio, teniendo como dosímetro de referencia el método Fricke [4].

La discrepancia con el laboratorio GSF no superó el 2,5% para el año 1977 y el 3,3% para el año 1978.

Se inició en el año 1980, a través del CIEN, un acercamiento entre los países latinoamericanos, organizando una encuesta sobre intercomparación para altas dosis. Esto permitió tener conocimiento de las factibilidades con que contaban dichos países, y comprobar el comportamiento del dosímetro. Se realizaron entonces trabajos de intercomparación, calibración y orientación.

REFERENCIAS

[1] CHEN, Tungho, JOHNSON, E.R., Nucleonics (Abr.1963) 66.
[2] CUNNINGHAM, J., J. Phys. Chem. **65** (1961) 628.
[3] SHINN, M.B., Ind. Eng. Chem. Anal. Ed. **13** (1941) 33.
[4] ORGANISMO INTERNACIONAL DE ENERGIA ATOMICA, Manual of Food Irradiation Dosimetry, Colección de Informes Técnicos N° 178, OIEA, Viena (1977).

IAEA-SM-272/17

STANDARD MEASUREMENT OF PROCESSING LEVEL GAMMA RAY DOSE RATES WITH A PARALLEL-PLATE IONIZATION CHAMBER

R. TANAKA, H. KANEKO, N. TAMURA
Takasaki Radiation Chemistry Research Establishment,
Japan Atomic Energy Research Institute,
Takasaki, Gunma-ken

A. KATOH*, Y. MORIUCHI**
Electrotechnical Laboratory,
Sakura, Ibaraki-ken

Japan

Abstract

STANDARD MEASUREMENT OF PROCESSING LEVEL GAMMA RAY DOSE RATES WITH A PARALLEL-PLATE IONIZATION CHAMBER.

An accurate and practical method of measuring the processing level exposure rates of ^{60}Co gamma rays with an ionization chamber was studied to develop standardization of a dosimetry system with wide and flexible applicability for current gamma ray irradiation facilities. A parallel-plate, free-air cavity ionization chamber was designed with an aluminium wall and an electrode gap of 1 mm having some geometries for accurate measurement of high exposure rates. The active collection volume of the chamber was estimated within an accuracy of ±1%. As a result of experiments, the polarity effect on the saturation current was observed without beam collimation, and was found to have its origin in extraneous current generated in the signal cable, the chamber stem and the connector outside the ionization volume of the chamber. The net collection current can be evaluated by reversing polarity and taking the mean value of the measured currents. Other influences such as the saturation characteristics of the ionization chamber, gamma ray dose attenuation in the chamber wall, a scattered gamma ray and a temperature rise of the cavity gas during irradiation were also examined. For the standard ^{60}Co gamma ray radiation field of the order of 0.258 $C \cdot kg^{-1} \cdot h^{-1}$ (1 $kR \cdot h^{-1}$), the exposure rate values absolutely measured with the chamber agree with the standard values within an accuracy of ±1%; these agreements have also been maintained for as long as six years. In comparing high exposure rate measurements with different ionization chambers, good agreement (within ±2%) was obtained over the range from 5.16 $C \cdot kg^{-1} \cdot h^{-1}$ (20 $kR \cdot h^{-1}$) to 516 $C \cdot kg^{-1} \cdot h^{-1}$ (2 $MR \cdot h^{-1}$). It was proved experimentally that the parallel-plate cavity ionization chamber absolutely calibrated in the standardization field (even of the order of 0.258 $C \cdot kg^{-1} \cdot h^{-1}$ (1 kR/h)) can be applied with good reliability for processing level radiation fields up to 774 $C \cdot kg^{-1} \cdot h^{-1}$ (3 MR/h), even with broad beam geometry and an appreciable amount of scattered gamma rays.

---

* Present address: The Institute of Radiation Measurement, Naka-gun, Ibaraki-ken, Japan.
** Present address: Gifu College of Medical Technology, Seki, Gifu-ken, Japan.

1. INTRODUCTION

In the gamma ray irradiation facilities at the Takasaki Radiation Chemistry Research Establishment (TRCRE), Japan Atomic Energy Research Institute (JAERI), various types of radiation processes are frequently required. Therefore, wide ranges of dose and dose rate have to be covered. Two liquid chemical dosimeter systems (Fricke and cerium sulphate) are useful as reference (field standard) dosimeters, but unfortunately these systems cannot cover wide dose ranges and they require complicated handling. Although calorimetry is considered to be the most reliable calibration method, it is not practical in irradiation facilities, except in primary standard dosimetry laboratories.

The ionization chamber is basically the most simple, practical and reliable method used as reference dosimeters for wide dose and dose rate ranges of ^{60}Co gamma rays. However, ionization chambers have mostly been used in a range lower than about 2.58 $C \cdot kg^{-1} \cdot h^{-1}$ (10 $kR \cdot h^{-1}$).[1] Various thimble-type exposure rate meters are commercially available for therapy level exposure rates, but they are not clarified for processing level exposure rates higher than about 2.58 $C \cdot kg^{-1} \cdot h^{-1}$ (10 $kR \cdot h^{-1}$). Usually, considerable variation of exposure response is found among the exposure rate meters at processing level dose rates.

We tried to develop an accurate and practical reference dosimetry system using a cavity ionization chamber measuring a high exposure rate of up to a few megaröntgen per hour. For this purpose we adopted a parallel-plate ionization chamber with a thin air gap allowing good saturation characteristics at high dose rates. Through extensive investigation [1–4], we have examined various characteristics of the parallel-plate ionization chamber at the high dose rate region, and have ascertained its excellent performance as a reference dosimeter in ^{60}Co gamma ray irradiation facilities.

Some problems regarding the design of the parallel-plate ionization chamber are discussed and also the detailed experimental results that were obtained with the ionization chamber. Data include an intercomparison study against national exposure rate standard measurements.

2. METHOD OF MEASUREMENT

Figure 1 shows a cross-section of the basic structure of the parallel-plate, free-air cavity ionization chamber with a thin electrode gap. The thickness of the high potential electrode and the collecting electrode satisfies electronic equilibrium conditions. The high potential plate is supported by three insulating spacers to keep a uniform electrode gap and is adjusted by the thickness of the spacers.

---

[1] 1 R = 2.58 × 10^{-4} $C \cdot kg^{-1}$.

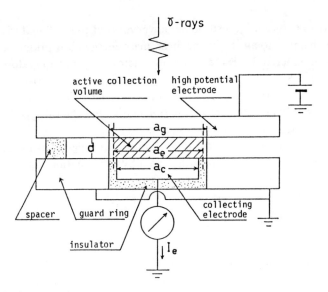

FIG.1. *Cross-section of the basic structure of the parallel-plate, free-air cavity ionization chamber with a thin electrode gap.*

If distortion of the electric field is negligible, the active collection volume is estimated to be $\pi(a_e/2)^2 d$, where a_e is the effective diameter of the active collection volume and d is the electrode gap. The effective diameter is given by $a_e = (a_c + a_g)/2$, where a_c is the diameter of the collecting electrode and a_g is the inner diameter of the guard rings. The electric field distortion in the parallel-plate ionization chamber is generally due to the edge effect of parallel-plate electrodes, the relative potential drop at the collector electrode caused by load resistance in the charge measurement system, and the electric charge of the gap insulator between the collector electrode and the guard rings. The edge effect is avoided by using a narrow electrode gap and a wide guard ring; the relative potential drop is negligible compared with the normal voltage (150 V) of the ionization chamber. The charging effect on the electric field distortion was minimized by narrowing the insulation gap to 0.3 mm.

With this method, the exposure rate, $\dot{X}$ (R/h), at the point of measurement in a uniform irradiation field is given by the formula

$$\dot{X} = 1.079 \times 10^{13} \frac{I_e}{v} \frac{T + 273.15}{273.15} \frac{760}{P} K_s K_\mu K_a K_r \tag{1}$$

where I_e is the collection current (A), v is the active collection volume (cm³), T is the atmospheric temperature (°C), P is the atmospheric pressure (torr), K_s is the mean electron mass collision stopping power ratio of the wall material of the ionization chamber to air, K_μ is the mean mass energy absorption coefficient ratio of air to wall material, K_a is the correction factor for gamma ray dose attenuation in the wall, and K_r is the correction factor for recombination loss of positive and negative ions that is the reciprocal of ion collection efficiency ($K_r \geq 1$).[2] The value of K_a should be determined experimentally, and the value of K_r equals unity if the recombination loss can be neglected.

When the recombination loss is small but not negligible in a parallel-plate ionization chamber, K_r is theoretically given by the equation [5]

$$K_r = 1 + \frac{m^2 d^4 q}{6 V^2} \qquad (2)$$

where m is a constant defined as the m-value, q is the ion density (esu cm^{-3}·s^{-1}) and V is the applied voltage (V). Since the second term of Eq.(2) is proportional to the fourth power of the electrode gap, to narrow the gap is the most effective way of reducing the recombination loss; it also results in high special resolution, especially in high dose rate dosimetry. We adopted 1 mm as the electrode gap so that the value of ($K_r - 1$) might be less than 10^{-3} at 1 MR/h for an applied voltage of 100 V in Eq.(2).

If the ionization chamber is calibrated in a standard ^{60}Co gamma ray radiation field, the exposure rate measured is expressed by the formula

$$\dot{X} = K_c \frac{T + 273.15}{273.15} \frac{760}{P} I_e \ (R/h) \qquad (3)$$

where K_c is the calibration factor for the measured current. It is assumed in Eq.(3) that the effect of the difference of photon energy spectrum from the standard field can be neglected.

Figure 2 shows (a) a cross-section and (b) a plan view of the parallel-plate free-air cavity ionization chamber (JTC-6) designed on the basis of the above considerations. We took 20 mm as the diameter of the collecting electrode, assuming that the lateral variation of exposure rate is of negligible order. We also prepared another chamber (JTC-2) of a similar type, which was calibrated in the standard ^{60}Co gamma ray field in the Electrotechnical Laboratory (ETL).

Although graphite is best suited for the wall material from the point of view of the equivalence to air and radiation resistance, it does not have good mechanical properties for precise cutting and also has a tendency to result in a

---

[2] 1 torr = 1.333 × 10^2 Pa.

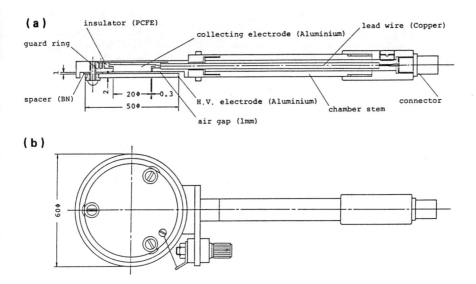

FIG.2. (a) Cross-section and (b) plan view of the parallel-plate, free-air cavity ionization chamber (JTC-6).

decrease of electric insulation of the narrow gap between the collecting electrode and the guard ring. For this reason we adopted aluminium as the wall material. The non-equivalence to air of aluminium may cause an error in exposure rate measurement in the presence of low energy photons, but the error is considered to be very small in usual ^{60}Co gamma ray radiation fields where primary gamma rays are predominant. The effect of a scattered gamma ray was examined by relative measurement with two ionization chambers with graphite and aluminium walls, as described in Section 5. The thickness of the high potential electrode is 2 mm. Polychlorofluoroethylene was used as the insulator molded into the 0.3 mm gap between the collecting electrode and the guard ring.

Boron nitride was used as the insulating material of the spacer because of radiation resistance and mechanical properties such as machining and hardness. The thickness of the spacers must be exactly the same as the electrode gap in order to evaluate accurately the small active volume. Therefore, surface finishing of the collector plate was carried out after molding the insulator to obtain complete flatness of the plate.

On the basis of the experimental study with JTC-6 and the theoretical examination, a trial standard ionization chamber with a 3 mm air gap and aluminium wall (PC-B) was designed and manufactured to establish a primary standard for processing level exposure rates [2, 3, 6]. For comparison with different types of ionization chambers, a cylindrical ionization chamber (CIC, graphite wall, active volume: 8.2 cm^3) was also prepared as a conventional chamber which is commonly used in lower dose rate regions [7].

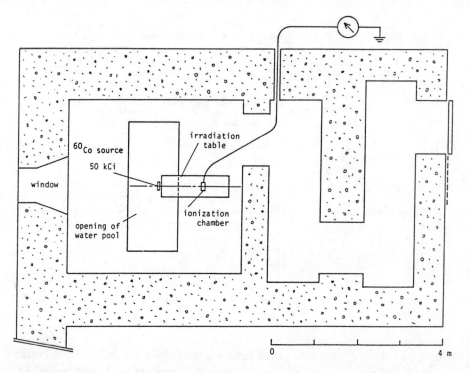

FIG.3. *Plan view of the ^{60}Co gamma ray irradiation room at JAERI.*

Figure 3 shows a plan view of the ^{60}Co gamma ray irradiation room at JAERI which was used for high exposure rate measurements. A 1.85 PBq (50 kCi) ^{60}Co plaque source (300 × 150 mm) was set at a fixed position for irradiation by lifting it from a storage position in a water pool.[3] The reproducibility of the source positioning is good enough to give a reproducible collection current even at the point of measurement near the source. The distance from source to ionization chamber (SID) was varied from 15 to 150 cm, corresponding to exposure rates from 5.16 $C \cdot kg^{-1} \cdot h^{-1}$ (20 $kR \cdot h^{-1}$) to 516 $C \cdot kg^{-1} \cdot h^{-1}$ (2 $MR \cdot h^{-1}$), with a reproducibility of positioning of the ionization chamber within ± 0.5 mm above the irradiation table. The lateral uniformity of exposure rate over the active collection volume of the ionization chamber is within ± 0.5%, even 15 cm from the source. The collection current was measured mainly with a vibrating reed electrometer (Takeda, TR-8411) with calibrated high load resistances and a digital pico-ammeter (Takeda, TR-8641D). As the signal cable, we used a 3C-2V low noise coaxial cable.

---

[3] 1 Ci = 3.70 × 10^{10} Bq.

Dose rate measurements were also carried out in a standard ^{60}Co gamma ray field in ETL with a well-collimated gamma ray beam [6]. The SID value was varied from 1 to 2 m, corresponding to exposure rates from about 0.5 to 2 kR/h, with good reproducibility of setting of the chambers (within ± 0.5 mm) and good lateral dose rate uniformity.

3. DETERMINATION OF THE VALUES OF CONSTANTS AND CORRECTION FACTORS FOR ABSOLUTE MEASUREMENT OF EXPOSURE RATES

3.1. Mean electron mass collision stopping power ratio and mass energy absorption coefficient ratio

The mean electron mass collision stopping power ratio, K_s, of aluminium to air is estimated to be 0.864 (Spencer-Attix theory) [8]. In this estimation, the mean path length of secondary electrons through the cavity is assumed to be twice the length of the electrode gap by using Birkhoff's calculation results on track-length distribution in an infinite slab [9].

The mean mass energy absorption coefficient ratio of air, K_μ, to aluminium was evaluated at 1.038 [10]. In this evaluation, the influence of scattered gamma rays was not taken into account, but it was experimentally checked, as shown in Section 5.

3.2. Active collection volume of the ionization chamber

The value of the air gap distance, d, was given by the mean value of the thickness of three boron nitride spacers which were measured within an accuracy of ± 0.005 mm. Variation in the measured thickness of the spacers was within ± 0.01 mm. The values of a_e and a_g were measured within an accuracy of ± 0.01 mm.

As a result of the measurements, the active volume of the ionization chamber was determined within an accuracy of ±1% on the assumption that the distortion of electric field in the ionized air space is negligible. The active volume was not influenced by mechanical pressure on the boron nitride spacers.

3.3. Correction factor for dose attenuation in the chamber wall

The correction factor for dose attenuation in the chamber wall is given by the equation

$$K_a = e^{\mu_d t} \approx 1 + \mu_d t \tag{4}$$

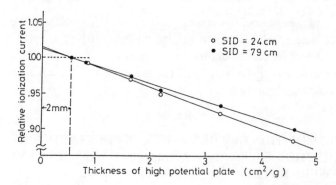

FIG.4. Decrease of the collection currents of the JTC chamber, with the thickness of the high potential aluminium plate relative to the collection current for a nominal thickness of 2 mm.

where μ_d is the dose attenuation coefficient in the high potential plate (cm²/g) and t is the wall thickness (cm). The value of μ_d was experimentally given as the decreasing rate of the collection current with the thickness of the high potential aluminium plate. Figure 4 shows the decrease in the collection currents of the JTC chamber, with the thickness of the high potential aluminium plate relative to the collection current for a nominal thickness of 2 mm. The values of μ_d were 0.025 and 0.028 cm²/g for SIDs of 79 and 24 cm, respectively. The dose attenuation coefficient is almost the same as the mass energy absorption coefficient for a large SID, but increases with decreasing SID for oblique incidence on the high potential plate.

The influence of oblique incidence on the dose attenuation coefficient can be calculated for simple irradiation geometries. The value of μ_d is 1% larger for a SID of 12 cm than for a very long SID. The parallel-plate cavity ionization chamber, however, cannot be applied for geometries with an extremely large oblique incidence because it is difficult to estimate accurately the dose attenuation in the chamber wall.

3.4. Saturation characteristics of collection current

Figure 5 shows typical saturation curves of the JTC chamber for positive and negative potentials at 18.1 C·kg⁻¹·h⁻¹ (70 kR/h). The collection current saturates at 10 V; at the highest exposure rate of 774 C·kg⁻¹·h⁻¹ (3 MR/h) it nearly saturates at 100 V. The most important result is that the negative saturation current is 3% higher than that of the positive one at 18.1 C·kg⁻¹·h⁻¹ (70 kR/h). The difference depends on the geometrical condition of measurement, but it is reproducible under the same geometrical condition. The ratio of the

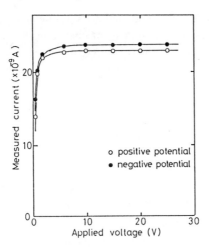

FIG.5. *Typical saturation curves of the JTC chamber for positive and negative potentials at 18.1 $C \cdot kg^{-1} \cdot h^{-1}$ (70 kR/h).*

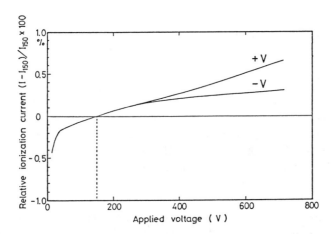

FIG.6. *Increase of the saturation curve of the JTC-6 chamber with an applied voltage of 51.6 $C \cdot kg^{-1} \cdot h^{-1}$ (200 kR/h) relative to the saturation current at 150 V. (+V and −V indicate positive and negative potentials, respectively. I_{150} and I indicate the ionization current at an applied voltage of 150 V and different value, respectively).*

negative to positive saturation current approaches unity with the increase of exposure rate. This polarity effect is discussed in detail in Section 4.

The saturation current still increases very slowly with the applied voltage. The increase of current with an applied voltage of 51.6 $C \cdot kg^{-1} \cdot h^{-1}$ (200 kR/h), relative to the saturation current at ±150 V, is shown in Fig.6. The positive saturation current is a little higher than that of the negative current in the voltage region higher than 300 V; this tendency does not depend on the exposure rate. The gradual increase may be due to a slight change in the active volume of the ionization chamber. However, the influence of the increase on determination of the exposure rate is considered to be negligible, since variation of the collection current within the range from 100 to 300 V, relative to that at 150 V, is smaller than ±0.2%.

4. STEM EFFECT

Regarding the polarity effect [5] on the saturation current, the influence of the difference of contact potential and space charge on the ion collection efficiency can be neglected. The polarity effect here is ascribed to two causes:

(1) Stem effect [11], i.e. radiation-induced current in the stem of the signal lead, which is generated outside the gas volume of the ionization chamber

(2) Polarity effect on the active volume due to space charge in the ionized gas and charging of the insulator.

To examine the contribution of the latter effect, the ionization chamber was sealed in a vacuum vessel, and positive and negative collection currents were measured under reduced air pressure. The change of the collection current with an applied voltage of 150 kR/h is shown in Fig.7. In all the saturation curves, there is a point symmetry about a common point $(0, -3.7 \times 10^{-11})$, with no dependence on air pressure. The negative current of 37 pA is ascribed to stem effect, but not to the ionization current from the gas volume. The change of the collection current with an applied voltage at 10^{-2} torr results from low energy electrons emitted from the chamber wall, and the sharp increase of the saturation current at low air pressures is presumably due to electron avalanche.

Figure 8 shows the changes in positive and negative saturation currents with air pressure in the ionization chamber. Both currents change linearly, with air pressure up to 760 torr, and both straight lines are symmetrical with respect to the axis of the current which is -3.7 pA. This shows that the real collection current of the ionization chamber can be evaluated by averaging the absolute values of both saturation currents; the stem current is obtained as the difference of both absolute values.

Most of the stem current is considered to originate from photo-Compton current, which is proportional to dose rate. In the ionization chamber system, however, the stem current is not proportional to the exposure rate at the point of measurement, since the total stem current is integrated along the stem which consists of the chamber stem and the signal cable; also, the integral dose rate over the total length of the stem usually increases sublinearly with the exposure rate at the point of measurement. The stem current basically depends on the geometrical arrangement of the source, the ionization chamber and the signal cable in the gamma ray radiation field.

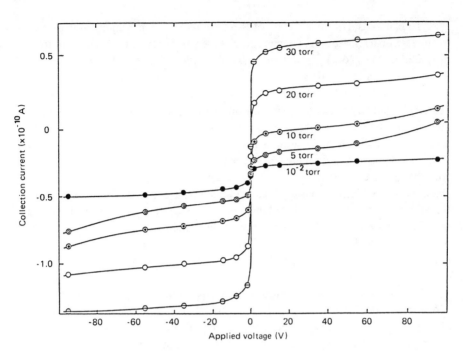

FIG.7. *Change of collection currents of the JTC chamber with an applied voltage of 150 kR/h under various reduced air pressures.*

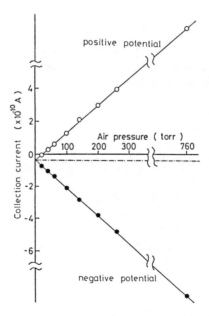

FIG.8. *Changes in the positive and negative saturation currents with air pressure in the JTC chamber.*

To clarify the dominant part of the dosimeter system which contributes to the stem current source, we measured the stem current under the condition of shielding a few local parts of the system with lead blocks. As a result of the measurement, a large part of the stem current was found to be radiation-induced current in the coaxial cable, but current generation from the chamber stem cannot be disregarded. The relative contribution of the current from the chamber stem increases with exposure rate at the point of measurement because the integral dose rate over the chamber stem nearly increases linearly with the exposure rate.

As origins of the stem current other than photo-Compton current, we also considered the current generation in the electric connector of the ionization chamber, probably due to a small additional air space around the joint, and the time-dependent radiation-induced current due to electric field that had formed in the insulator around the copper wire of the cable resulting from accumulation of excess charge during irradiation. The connector current was below 10^{-11} A, but it is not reproducible, depending on the irradiation condition and the ionization chamber.

We also studied in detail photo-Compton current from the coaxial cable and found that its origin can be divided into two contributions [12]; one is the displacement current due to electronic non-equilibrium in the insulator-conductor interface around the signal lead wire (interface current); the other is the displacement current due to the intensity gradient of gamma rays around the signal wire in an electronic equilibrium condition (bulk current). Both current generation rates are proportional to the dose rate integrated over the cable exposed to gamma rays. The interface current is inversely proportional to the thickness of the inner insulator of the coaxial cable in the range thicker than the mean projected range of Compton electrons, while the bulk current very nearly increases linearly with the thickness. Photo-Compton current also generally increases with the diameter of the core wire and is also influenced by the thickness of the outer insulator, depending on the contribution rate of contaminant electrons in the radiation field [13].

5. INFLUENCE OF A SCATTERED GAMMA RAY

Since aluminium is used as the cavity wall, the mean mass energy absorption coefficient ratio, K_μ, and the mean mass collision stopping power ratio, K_s, in Eq.(1) are considered to be influenced by scattered gamma rays from the wall of the irradiation room, the irradiation table, the source, etc. In most ^{60}Co gamma ray fields in which primary gamma rays are predominant, however, the effect of a scattered gamma ray is very small and it is not easy to determine the effect quantitatively.

We examined the overall effects of scattered gamma rays on the collection current of aluminium and graphite wall ionization chambers of the same type

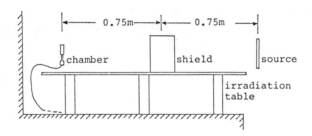

FIG.9. Irradiation geometry to examine the effects of a scattered gamma ray on the collection current of the aluminium wall ionization chamber (JTC).

under the irradiation conditions shown in Fig.9. The incidence of a primary gamma ray on the cavity of the ionization chamber was shielded by thick lead blocks. The ratios of the ionization current in the presence of the shield to that in the absence of it were 0.087 and 0.082 for aluminium and graphite wall, respectively. Since the value of K_μ and K_s for the graphite wall are almost unity, even for a low energy scattered gamma ray, the 0.5% difference represents the difference of the values of K_s and K_μ between a primary gamma ray and all the incident gamma rays.

In the geometry of measurement shown in Fig.9, gamma rays scattered at a small angle from the source and the irradiation table are prevented from impinging on the ionization chamber by the lead blocks, but this influence on energy dependence is considered to be negligible partly because a gamma ray scattered at a small angle has relatively high energy. The ratio of a scattered to a primary gamma ray increases with relative SID to the source to wall distance. Therefore, the contribution of a scattered gamma ray at the point of measurement shown in Fig.9 is relatively large in the irradiation room. Under similar geometries of measurements, it can be assumed that the systematic error due to a scattered gamma ray does not exceed about 0.5%.

6. EFFECT OF TEMPERATURE RISE DURING EXPOSURE

While measuring the high exposure rate with an ionization chamber, energy absorption of gamma ray and heat radiation from the source cause a temperature rise in the cavity gas and the aluminium walls, which results in a decrease in the collection current. Assuming an adiabatic condition, the increasing rate of temperature in a small target (where the exposure rate is $\dot{X}$) is given by the relation

$$\Delta T = 3.47 \times 10^{-8} \frac{\dot{X}}{K_s C} \quad (^\circ C/min) \tag{5}$$

where C is the specific heat (cal·°C^{-1}·g^{-1}) of the target.[4] Assuming that the temperature of the cavity gas is the same as that of the chamber wall, the calculated value of ΔT for the aluminium wall is about 0.5°C/min at 774 C·kg^{-1}·h^{-1}(3 MR/h). This value agrees with the initial increasing rate of the temperature observed with a thermocouple, although after some time the temperature reaches an equilibrium value at which the heat loss from the chamber is equal to the energy input. The value of the initial increasing rate results in a decreasing rate of 0.2%/min of the collection current.

7. RESULTS OF ABSOLUTE AND RELATIVE MEASUREMENTS OF EXPOSURE RATE

On the basis of examination of the various factors described above, we tried to carry out comparison studies of high exposure rate measurements in different ^{60}Co gamma ray radiation fields as follows:

(1) Measurements in the standard radiation field of 0.5 to 2 kR/h in the ETL
(2) Measurements in the processing level radiation field above 5.16 C·kg^{-1}·h^{-1} (20 kR/h) at JAERI with different ionization chambers.

Exposure rate measurements were carried out in two ways: absolute measurement with the JTC-6 chamber and measurements with two calibrated ionization chambers (JTC-2 and JTC-6). For absolute measurement, the value of the constants in Eq.(1) was as follows: $v = 0.325$ cm^3, $K_s = 0.864$, $K_\mu = 1.038$ and $K_r = 1$.

Table I shows the results of the exposure rate measurements in the standard radiation field which were made four times over six years. The ratios of the exposure rate values obtained with the calibrated ionization chamber JTC-2 ($\dot{X}_{JTC2}$) and those absolutely determined with JTC-6 ($\dot{X}_{JTC6}$) to the standard values ($\dot{X}_{STD}$) are shown for different SIDs. The JTC-2 chamber was calibrated at 10 kR/h at JAERI with the standard cylindrical ionization chamber of ETL (active volume: 62.5 cm^3). Each value was obtained from the average of three or four repeated measurements. The values obtained by absolute measurement agree very well with the standard values (within about ± 1%), with good reproducibility. The measured values with JTC-2 are 0 to 2% higher than the standard values, but the difference is within the experimental error and the ratio has good reproducibility (within ±1%). These agreements have been maintained for as long as six years. A small irregular variation between the ratios is chiefly due to current noise for a very low collection current of the order of 10 pA.

Figure 10 shows a comparison of the measured values between the JTC-6 and PC-B chambers in the high exposure rate region in the irradiation room at

---

[4] 1 cal = 4.184 × 10^0 J.

TABLE I. COMPARISON OF EXPOSURE RATE VALUES ($\dot{X}_{JTC2}$ AND $\dot{X}_{JTC6}$) MEASURED WITH THE CALIBRATED JTC-2 AND ABSOLUTELY DETERMINED WITH JTC-6, RESPECTIVELY, AGAINST THE PRIMARY STANDARD VALUES $\dot{X}_{STD}$ IN THE ELECTROTECHNICAL LABORATORY

	SID	1.0 m (2376 R/h)	1.5 m (1054 R/h)	2.0 m (582 R/h)
$\dfrac{\dot{X}_{JTC2}}{\dot{X}_{STD}}$	(1)	1.013	1.002	1.016
	(2)	1.010	1.006	1.009
	(3)	1.016	1.009	1.023
	(4)	1.021	1.008	1.013
$\dfrac{\dot{X}_{JTC6}}{\dot{X}_{STD}}$	(1)	0.997	0.987	0.996
	(2)	1.005	--	1.004
	(3)	1.007	0.991	0.998
	(4)	0.994	--	--

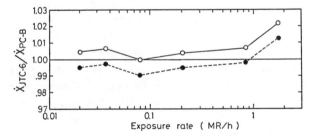

FIG.10. Comparison of the measured values between the JTC-6 and PC-B chambers in the high exposure rate region in the irradiation room at JAERI (see Fig.3). ((○) = the ratio of exposure rate values calibrated with JTC-6 to those calibrated with PC-B; (●) the ratio of exposure rate values absolutely determined with JTC-6 to those calibrated with PC-B.)

JAERI (see Fig.3). The lines show the ratios of exposure rate values calibrated with JTC-6 to those calibrated with PC-B and the ratios of exposure rate values absolutely determined with JTC-6 to those calibrated with PC-B. The ratios were corrected for the difference of sensitivity between the charge measurement systems. In both cases, i.e. calibration in the standard radiation field and absolute measurement with JTC-6, the values measured with JTC-6 and PC-B agree within a variation of ±1%, except for the highest exposure rate. The relatively large difference (about 516 C·kg⁻¹·h⁻¹ (2 MR/h)) was considered to have resulted from the setting error of the ionization chamber at the short distance from the source.

In the case of CIC, the correction factor for recombination loss, K_s, was empirically given on the basis of Eq.(2), since the ion collection efficiency is lower than unity at high exposure rates. The measured values with CIC and JTC-6 agree within ±2% when both ionization chambers were calibrated in the standard radiation field.

We obtained satisfactory results from comparisons of exposure rate measurements in the standard radiation field and the radiation field of high exposure rate. The parallel-plate free-air cavity ionization chamber was found to be reliable as a reference dosimeter for high exposure rate ranges in the ^{60}Co gamma ray irradiation facilities; a primary standard for processing dose rates higher than 2.58 C·kg⁻¹·h⁻¹ (10 kR/h) can also be established with the parallel-plate ionization chamber for gamma ray fields where primary photons from a ^{60}Co source are predominant. The uncertainty in determination of the exposure rate measured with PC-B was estimated to be within ±1.7% in quadratic addition of each uncertainty.

8. DISCUSSION AND SUMMARY

The JTC chamber has not yet been applied for measurement of an exposure rate higher than 774 C·kg⁻¹·h⁻¹ (3 MR/h). It does not seem to be difficult to measure 10 MR/h or more with JTC considering that the value of $(K_s - 1)$ is inversely proportional to V^2, but it requires proper correction for a rapid temperature rise in the cavity gas. It is expected that the graphite wall chamber will be applied as the reference dosimeter for high intensity and high energy X-rays with a continuous energy spectrum and unsteady dose rate depending on beam scanning.

The good agreement between JTC and CIC in the measured values in the high exposure rate region shows that cylindrical-type ionization chambers can be used for measurement of high exposure rates of the order of 0.258 C·kg⁻¹·h⁻¹ (1 MR/h) on the basis of accurate estimation of the correction factor for ion recombination. A practical and accurate method for measuring high exposure rates under poor saturation conditions has already been suggested by Moriuchi et al. [14] and Moriuchi [15]. Small cylindrical ionization chambers will be useful as reference dosimeters in the geometries of measurements with a very wide distribution of the gamma ray incidence angle for which the parallel-plate ionization chambers cannot be applied.

Conversion from exposure rate to absorbed dose rate in the medium, which is finally needed in reference dosimetry, is obtained from the following relation, assuming that electronic equilibrium exists

$$\dot{D}_m = 0.869 \frac{(\mu_{en}/\rho)_m}{(\mu_{en}/\rho)_{air}} \dot{X} f_a \tag{6}$$

where D_m is the absorbed dose rate in the medium, $(\mu_{en}/\rho)_m$ and $(\mu_{en}/\rho)_{air}$ are the mass energy absorption coefficients of the medium and air, respectively, which are given for a primary ^{60}Co gamma ray, and f_a is the correction factor for gamma ray dose attenuation from the incidence surface of the medium to the point of interest in the medium ($f_a < 1$). The effect of a scattered gamma ray on the ratio of mass energy absorption coefficients is considered to be almost negligible from the experimental results with different chamber walls. In this case, the size of the medium is small compared with the mean free path of the primary gamma ray and the depth at the point of interest in the medium is not so great as to change the gamma ray spectrum; the atomic number of the medium is also smaller than that of aluminium.

In the radiation field where a primary ^{60}Co gamma ray is not predominant, graphite ionization chambers are recommended. However, it is difficult to determine accurately the small active volume of the chamber, and, unlike the aluminium wall chamber, much more careful maintenance and frequent calibrations seem to be necessary.

The JTC is the most convenient chamber for calibration of the exposure rate in processing level ^{60}Co gamma ray fields and routine dosimeters. At the local point where the JTC cannot be set, accurate measurement of the dose rate can be carried out with precise routine dosimeters such as the alanine dosimeter or the cobalt-glass dosimeter, instead of Fricke and cerium sulphate dosimeters.

As a result of the above studies and comparisons, it can be concluded that the accuracy of exposure rate determinations up to a few megaröntgen per hour can be estimated to be about ± 1–2% with parallel-plate air-cavity ionization chambers for the radiation field with broad beam geometry and with an appreciable amount of scattered gamma rays. Uncertainty in absorbed dose determinations can be estimated to be within about a few per cent. The reliability of reference dosimetry in the TRCRE at JAERI is maintained by periodic checks of traceability against national exposure rate standard measurements in ETL.

ACKNOWLEDGEMENTS

The authors wish to thank S. Tajima and M. Kawai of the Japan Atomic Energy Research Institute for their important experimental work. They also wish to thank I. Yamachi and J. Naoi of the Electrotechnical Laboratory for their calibration work.

REFERENCES

[1] MORIUCHI, Y., et al., Accurate Methods for Measuring High Dose Rates Radiations, Research Committee on High-level Radiation Dosimetry, Irradiation Development Association (1978) 11 (in Japanese).

[2] MORIUCHI, Y., "Current work on dosimetry standards in Japan", National and International Standardization of Radiation Dosimetry (Proc. Symp. Atlanta, 1977), Vol.1, IAEA, Vienna (1977) 139–158.

[3] MORIUCHI, Y., KATOH, A., YAMACHI, I., NAOI, J., TAMURA, N., TANAKA, R., Proc. Annual Meeting on Radioisot. Phys. Sci. Ind., Japan Radioisotope Association, Tokyo (1979) 54 (in Japanese).

[4] TANAKA, R., KANEKO, H., TAMURA, N., KATOH, A., YAMACHI, I., NAOI, J., MORIUCHI, Y., ibid. 55 (in Japanese).

[5] BOAG, J.W., Radiation Dosimetry (ATTIX, F.H., ROESCH, W.C., Eds), Vol.2, Academic Press, New York (1966) 1.

[6] TAMURA, N., Radiat. Phys. Chem. **18** (1981) 281.

[7] KATOH, A., YAMACHI, I., Bull. Electrotech. Lab. **47** 9&10 (1983) 52 (in Japanese).

[8] BURLIN, T.E., Radiation Dosimetry (ATTIX, F.H., ROESCH, W.C., Eds), Vol.1, Academic Press, New York (1968) 331.

[9] BIRKHOFF, R.D., et al., Health Phys. **19** (1970) 1.

[10] HUBBELL, J.H., Int. J. Appl. Radiat. Isot. **33** (1982) 1269.

[11] MORIUCHI, Y., Oyo Butsuri **34** (1965) 435 (in Japanese).

[12] TANAKA, R., TAMURA, N., IEEE Trans. Japan 100-A (1980) 141 (in Japanese).

[13] BURLIN, T.E., CHAN, F.K., Int. J. Appl. Radiat. Isot. **22** (1971) 73.

[14] MORIUCHI, Y., KATOH, A., TAKATA, N., TANAKA, R., TAMURA, N., "Practical, accurate methods for measuring the m-values of ionizing gases", National and International Standardization of Radiation Dosimetry (Proc. Symp. Atlanta, 1977), Vol.2, IAEA, Vienna (1979) 45–63.

[15] MORIUCHI, Y., Estimation of General Ion Recombination Loss in High Intensity Radiation Dosimetry, Research of the Electrotechnical Laboratory, Report No.739 (1973).

IAEA-SM-272/39

Invited Paper

PROGRESS IN ALANINE/ESR TRANSFER DOSIMETRY*

D.F. REGULLA, U. DEFFNER
Gesellschaft für Strahlen- und
  Umweltforschung mbH München,
Neuherberg, Federal Republic of Germany

Abstract

PROGRESS IN ALANINE/ESR TRANSFER DOSIMETRY.

    High accuracy of dose measurement requires a reliable dosimetry method and established traceability of measurement to national standards by means of reference dosimetry besides routine dosimetry; however, at the moment there is no calibration service available for high doses at primary standard dosimetry laboratories. Considering in particular international trade with processed food, the different doses required and the variety of dosimetry systems applied without primary standards there is an urgent need for international standardization in high-level dosimetry. In contrast to traditional dosimetry methods that are based on radiation effects in inorganic materials, e.g. ionization in air (ion chamber dosimetry), temperature rise in graphite or metal (calorimetry), coloration of chemical solutions (Fricke dosimetry) and induction of luminescence capability (thermoluminescence, photoluminescence), the dosimetry method described here uses solid organic compounds, in particular crystalline alanine CH_3-$CHNH_2$-COOH. For measurement, the effect of radiation-induced free radicals is used whose concentration is determined. Alanine is an organic compound which is closely tissue-equivalent with respect to atomic composition and specific gravity. Additionally, the generation of free radicals by radiation, quantitatively accessible by ESR spectroscopy, is part of the chain of cellular damage in biological materials. These properties together with high precision in the determination of spin concentration make the alanine/ESR dosimetry system useful for the envisaged task. After 8 years of experience gained under various climatic and transport conditions the new dosimetry method may be considered an off-line tissue-equivalent chamber dosimetry. A report is given of the most recently evaluated physical and dosimetrical properties of the Gesellschaft für Strahlen- und Umweltforschung alanine/ESR dosimetry system, characterized by a wide dose range from 0.1 Gy to 0.5 MGy and negligible energy dependence of response. The evaluation system and procedure are described including the computer-assisted dose assessment. Figures on precision and inter-specimen scattering are given together with data on influence parameters. Overall uncertainty is estimated and compared with recent data from intercomparisons with primary and secondary standard dosimetry laboratories.

---

    * Research carried out with the support of the IAEA under Research Contract No. 2819/RB.

1. INTRODUCTION

The Gesellschaft für Strahlen- und Umweltforschung (GSF) research activities in the field of high-level photon dosimetry are traceable back to the years 1973 and 1974 when they were dealing with the development of both solid-state effects and appropriate calibration procedures [1]. A method using solid organic compounds revealed most promising results; this work referred to the earlier efforts of Bradshaw et al. in 1952 [2] and Bermann et al. in 1971 [3]. It was in 1976 that the first rather accurate results in the dose range 100 Gy to 100 kGy were reported by the GSF [4].

In 1977, the GSF was invited by the IAEA to join an international group working on the programme High Dose Standardization and Intercomparison for Industrial Radiation Processing with the ultimate intention of providing an international dose assurance service. For this purpose, it was proposed that dosimeters for high-level dosimetry should be tested in different organizations with the aim of finding the most appropriate system for transfer dosimetry.

In parallel, the IAEA established a Co-ordinated Research Programme with the activities concentrated, among others, on the study of influence quantities affecting reliability and accuracy, in which the GSF also participated. The common IAEA and GSF activities were bound into a Research Agreement in 1978 and 1979; from 1980 onwards co-operation has been based on Research Contracts.

The joint IAEA/GSF research project aimed at developing and settling precision and reliability in alanine/ESR metrology for the purpose of transfer dosimetry has now been finalized. The objectives successfully achieved are as follows:

(1) Analysis of the ESR equipment to reduce or eliminate sources of uncertainties
(2) Development of software to evaluate ESR signals in terms of dose quantities, its implementation and performance tests
(3) Assurance of quality in alanine sample preparation and production standardization
(4) Calibration of the alanine/ESR system with respect to dose evaluation
(5) Performance tests with the alanine/ESR system for practical transfer dosimetry
(6) Investigation of influence quantities during sample preparation, exposure, transport and evaluation.

The state of dosimetry of the alanine/ESR method is compiled in Table I. A more detailed description of the method is given in Ref.[5].

TABLE I. SELECTED CHEMICAL-PHYSICAL AND DOSIMETRIC PROPERTIES OF THE GSF ALANINE/ESR TRANSFER DOSIMETRY SYSTEM

Composition	L-α-alanine CH_3-$CHNH_2$-COOH (90%), paraffin (10%)
Effective atomic number	$Z_{eff} \cong 7.2$
Effective specific gravity	1.22 g/cm^3
Dimensions	4.9 mm dia. × 10 mm length (photons, neutrons)
	4.9 mm dia. × 1 mm length (electrons)
Dose range	0.1 Gy–0.5 MGy
Time behaviour of fading	$(1 - S/So) \leqslant 0.01$ for 3 years, at 22°C and 40–60% relative humidity
Inter-specimen scattering	$s \leqslant \pm 1.5\%$
Precision	$\leqslant 0.5\%$ (2s)
Overall uncertainty	$\leqslant 3\%$ (2s)

2. SYSTEM COMPONENTS

The alanine/ESR transfer dosimetry system is based on the following components: alanine samples, an ESR spectrometer, computer/software, registration and calibration facilities.

2.1. Hard and software

The ESR spectrometer for alanine sample evaluation is of the Varian E9 type. Software is derived from a data acquisition system. Data and results are displayed on graphic and alphanumeric screens; hard copies are made on an XY recorder and a matrix printer (Fig.1). The whole system is operated under air-conditioning.

2.2. Dosimeters

The samples are made of 90% L-alanine and 10% paraffin. The latter acts as a bonding agent and does not contribute zero reading or radiation-induced signals. For reasons of better handling, the samples are solid and, because of the dimensions of the measuring cell, are of a cylindrical shape (Fig.2).

Since response, zero reading and reproducibility of the ESR signal differ with the preparation procedure of the samples, standardization and quality control are mandatory. The GSF has set up laboratory equipment for preparation of the alanine samples.

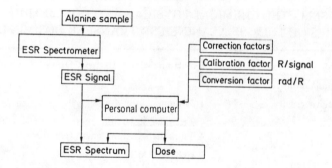

FIG.1. Block diagram of alanine/ESR dose evaluation (1 R = 2.58 × 10^{-4} C/kg; 1 rad = 1.00 × 10^{-2} Gy).

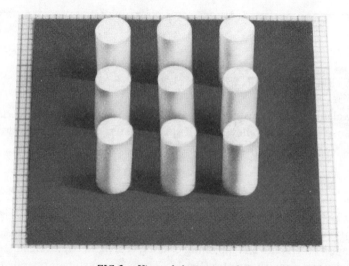

FIG.2. View of alanine samples.

2.3. Calibration

Calibration of the alanine/ESR system is based on secondary standard ionization chamber dosimetry, using a ^{60}Co teletherapy source (Atomic Energy of Canada Limited (AECL), Eldorado), in the range 0.1 Gy to 1 kGy. Traceability to the primary dosimetry standards of the National Physical Laboratory (NPL), Teddington, and the Physikalisch-Technische Bundesanstalt (PTB), Braunschweig, is provided (Fig.3). In the range 0.1 kGy to 0.1 MGy, a ^{60}Co

IAEA-SM-272/39   225

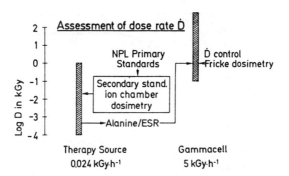

FIG.3. *Scheme of high-dose ^{60}Co calibration.*

AECL Gammacell unit serves for calibration with a dose rate determined by alanine/ESR dosimetry and cross-checked by Fricke dosimetry.

3. SCIENTIFIC PROGRESS

Progress is reported in terms of hard and software, samples, calibration and influence quantities.

3.1. Spectrometer specifications

Quantitative ESR measurements are affected by a number of uncertainties resulting mainly from spectrometer stability, spectrometer settings and sample positioning.

It was shown that the ESR spectrometer needs a warm-up time of approximately 2 hours for precision measurements. For evaluating the warm-up behaviour, an alanine sample was measured 200 times successively, the experiment starting immediately after switching on the spectrometer (Fig.4). The measurement group numbers as indicated in the figure represent the mean value of 10 successive measurements. Within the first 100 measurements, which take about 2 hours, the spectrometer sensitivity increases by 1.5%. Then it reaches a plateau with a slope of $(\Delta S/S)/\Delta t = 0.001$ hours.

The gain steps marked at the spectrometer console are not exactly linear, due to the tolerances of the resistors in the voltage divider chain. Pre-factor and decade factor settings of gain deviate from linearity by approximately ±5%. Respective correction factors for the gain settings have been evaluated (Fig.5); the precision of corrections is ±0.25%.

Reproducibility of the ESR spectrometer has been tested with a sample irradiated to 1 kGy and evaluated successively 20 times. In the first series of

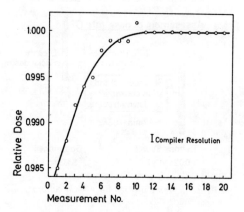

FIG.4. *Relative sensitivity of the alanine/ESR system after starting the spectrometer. Absorbed dose 8.5 kGy. Each measurement group represents the mean value of 10 readings.*

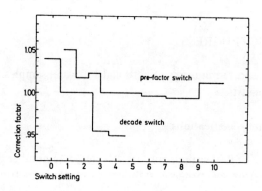

FIG.5. *Correction factors for the gain pre-factor and decade factor settings of ESR spectrometer.*

tests, the spectrometer settings were not changed and the sample was not taken out of the cavity. A standard deviation of ±0.13% was found for the readings.

In a second series of reproducibility tests, the sample was taken out of the cavity after each measurement and inserted again upside down; furthermore, the settings of the spectrometer were re-adjusted before each measurement. The standard deviation of the readings was ±0.23% (Fig.6).

3.2. Software for ESR spectrum analysis

According to software developed at the GSF (Fig.7), the dose evaluation of the ESR signal can be performed either by double integration of the ESR

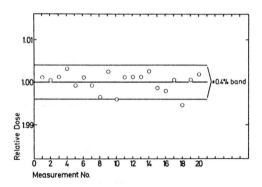

FIG.6. Reproducibility of an individual alanine sample removed from the sample cell after each readout procedure and inserted again upside down.

spectrum or by measuring the maximum peak-to-peak amplitude. Double integration is preferred for several reasons, but needs comparatively long measuring times.

Integration is performed either over the whole scan range (200 gauss), or a partial range of the spectrum only (approximately 130 gauss). The latter procedure leads to improved precision and is time saving.

The main peak-to-peak amplitude can either be assessed as part of the integration procedure, or by execution of a quick full-range scan during which the spectral position of the central peak is localized and marked on a proper time-scale.

Apart from these procedures, the main peak-to-peak amplitude can be localized by the G-value of alanine. This technique is advantageous, since there is no need to evaluate or scan the whole ESR spectrum, only the central peak.

3.3. Multiple ESR spectrum scan

For both dose evaluation methods, i.e. double integration or assessment of the main peak-to-peak amplitude, the software provides the possibility of noise reduction by a multiple-scan procedure, including calculation of the arithmetic mean of the ESR amplitude values.

Performance tests revealed the multiple-scan method to be principally applicable, particularly for assessment of 0.1 Gy doses, but extended measuring time due to repeated reading cycles is necessary.

3.4. System calibration

The shape of the dose response curve was found to be independent of the batch of alanine samples. As a consequence, the experimentally determined

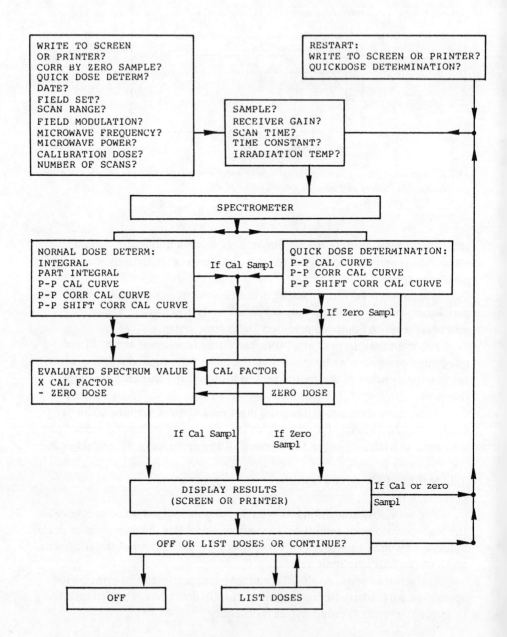

FIG.7. Flow scheme for computer evaluation of ESR signals (peak-to-peak or integral).

TABLE II. REPRODUCIBILITY OF ALANINE/ESR MEASUREMENTS AT THE GSF, SHOWN AT 12 CONSECUTIVE READINGS OF AN INDIVIDUAL SAMPLE

Number of measurement	Reading value
16	858.310
17	858.310
18	858.310
19	858.310
20	858.970
21	858.310
22	858.970
23	858.310
24	858.310
25	858.310
26	858.970
27	858.310

relationship between dose and response is replaced by a mathematically described function. This was achieved by a mathematical fit with exponential spline functions up to the saturation level. The temperature dependence of the curve is also fitted by mathematical functions. Furthermore, amplification settings of the spectrometer are automatically considered. The calibration curve can be adjusted to eliminate minor changes in spectrometer sensitivity and gain or for alanine samples of different size and alanine content.

The programme provides the possibility of considering the zero reading of an unirradiated sample which is not radiation induced and cannot be annealed. Prereading is regularly considered for dose computation, i.e. for doses <0.1 kGy.

A check of the dose response curve is performed by a single calibration alanine sample, usually at a level of 1 kGy, at the beginning of each measurement series. Based on the evaluated dose of this sample the dose response curve is shifted and adjusted for the minor changes of spectrometer sensitivity. These day-to-day shifts of the system's sensitivity are in the order of ≤1%.

The corrections applied to the ESR signal for gain setting, sensitivity shift of the ESR spectrometer, zero reading of an unirradiated sample, irradiation temperature, saturation of the dose effect curve at higher doses and base line drift are automatically taken into account for dose evaluation.

Since the ESR spectrum is compiled digitally there is a small quantization error of 0.05% if the main amplitude of the spectrum is spread over the whole ordinate (Table II).

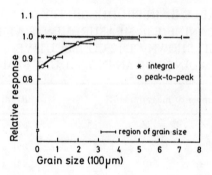

FIG.8. *Relative response of alanine samples 4.9 mm dia. × 10 mm, as a function of grain size.*

3.5. Quality assurance of the alanine sample production

Preparation of the samples is done in three steps: mechanical processing of the crystalline alanine, blending of the two components alanine and paraffin, and pressing of the pellets. Crystalline L-alanine from different manufacturers (Fluka, Merck, Serva) revealed no significant difference in the dosimeter properties.

From the results of previous experiments it was known that batch uniformity depends on the quality of production, e.g. sample mass and other parameters. Recent findings showed that different sample responses may result from varying grain sizes of the basic alanine material. The present experiments include five different grain sizes between around 50 and 700 μm; separation of grain sizes was achieved by applying sieves of different mesh numbers. Then each powder sample was mixed with paraffin. As a result, response increases with grain size (Fig.8). This holds only for the peak-to-peak method, while no effect of grain size on sensitivity has been found when using the integration method. The reason is that the shape of the ESR spectrum varies slightly with grain size.

Some thousands of dosimeters have been prepared. The alanine mass of 250 mg used has been reproduced within 1% on an almost 3s basis (Fig.9).

Batch uniformity of the individual responses was found in a band of ±1.5% when normalized to the mean of the relative readings (using 32 alanine samples with a grain size of around 100 μm); the relative standard deviation is 0.7% (Fig.10). Compared with earlier results the batch uniformity has thus significantly been improved.

Interbadge scattering has been found to be within ±1%.

3.6. Detection limits

Alanine/paraffin samples show a zero reading corresponding to about 0.6 Gy (Fig.11). The lower limit of detection is determined by the reproducibility of the

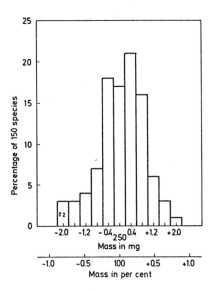

FIG.9. Mass distribution of 150 alanine samples.

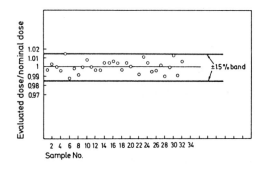

FIG.10. Batch uniformity of response; 32 alanine samples; grain size around 100 μm.

zero reading value and by the noise of the spectrometer. The signal-to-noise ratio for the zero reading is of the order of 10. These two limiting factors result in a scattering of the zero reading of about ±0.1 Gy which, for the moment, represents the lower limit of detection.

To complete the dose response curve for high doses, the alanine samples were irradiated with electrons (2.5 MeV) up to doses of 5 MGy. At a dose level of 0.5 MGy the ESR signal shows saturation (Fig.12). For doses above 0.5 MGy the ESR signal decreases again, which holds for both reading methods, i.e.

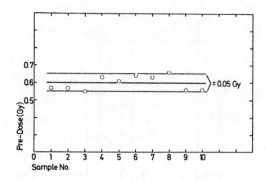

FIG.11. Zero reading of 10 different alanine samples.

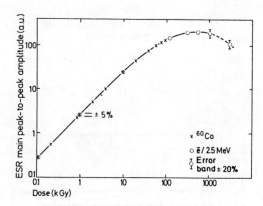

FIG.12. Dose effect curve of alanine/ESR samples in the range 0.1 kGy to 5 MGy.

peak-to-peak amplitude assessment and integration. The alanine/paraffin samples change structure, soften and deform above about 2 MGy thus loosing their capability for precise dosimetry. Above 0.3 MGy, however, precision is limited anyway, since response saturates in this range.

3.7. Study of effects from influence quantities

Previous findings showed an increase in the yield of free radicals with irradiation temperature. This increase is linear up to 80°C and can be characterized by a slope of $(\Delta S/S)/\Delta T = 0.0018°C$. To explore whether this phenomenon depends on the irradiation atmosphere, the experiments were repeated with the difference that the dosimeter samples were exposed in pure oxygen or pure nitrogen atmosphere.

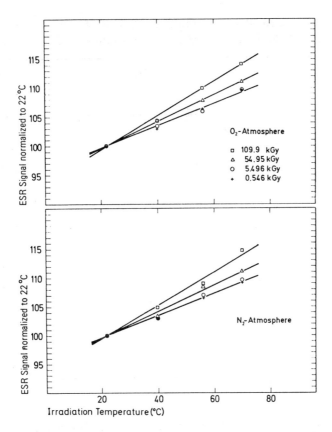

FIG.13. Dependence of ESR signal on irradiation temperature in oxygen and nitrogen atmosphere. Parameter absorbed dose.

The results were identical with those from previous measurements. Moreover, the dose effect curves obtained for oxygen, nitrogen and air showed no noticeable difference (Fig.13). It becomes evident that the partial pressure of oxygen and nitrogen during irradiation has practically no influence on the yield of free radicals. Therefore, paramagnetic oxygen molecules from the ambient gas cannot explain the increased rate of spin concentration.

4. INTERCOMPARISON WITH PRIMARY STANDARD DOSIMETRY LABORATORIES

Early in 1984 a set of transfer dosimeters was exposed at the NPL (for reference calibration), with the doses unknown to the GSF. The evaluated results are given in Table III. For the ten doses applied between 1 kGy and roughly

TABLE III. RESULTS OF 1983 NPL/GSF CALIBRATION DOSE INTER-COMPARISON BY THE IAEA (irradiation temperature 30°C)

Dosimeter Serial No.	NPL nominal dose (kGy)	GSF evaluated dose (kGy)	NPL/GSF ratio nominal/ evaluated dose
651	40.8	40.7	1.002
652	48.9	49.1	0.996
653	1.03	1.03	1.000
654	40.8	40.8	1.000
655	48.9	48.7	1.004
656	1.03	1.03	1.000
657	9.6	9.6	1.000
658	9.6	9.5	1.011
659	18.7	18.8	0.995
660	18.7	18.8	0.995
$\bar{x}$			1.0003
S_{abs}			0.0048
S_{rel} %			0.48

50 kGy, the mean of the NPL/GSF dose ratios is 1.0003 with a relative standard deviation of 0.48%. The figures stem from transfer dosimeters which have been mailed several times and evaluated 3 months after having left the GSF. The results represent a further improvement compared with earlier intercomparisons between GSF and NPL, where there was a maximum spread of ±2.5% [5].

5. CONCLUSIONS

The radiation-induced free radical generation in alanine, evaluated quantitatively by ESR spectroscopy, has proved applicable for remarkably accurate dosimetry, e.g. precision ≤0.5%, overall uncertainty ≤5% (see Table III).

Hence, the alanine/ESR technique approaches to a certain extent the figures of uncertainty and reliability known from ion chamber dosimetry. Based on good dosimetric properties and the low effects of influence quantities, which in addition can widely be corrected for, the method developed by the GSF lends itself to the envisaged task of transfer dosimetry. In the dose range 1 Gy to 0.5 MGy, the alanine/ESR transfer dosimetry may serve in industrial radiation processing but also for medical therapy. The new technique seems capable of complementing or even replacing Fricke dosimetry in transfer dosimetry. An extension of the lower detection limit is probable.

The alanine/ESR dosimetry may also be useful for transfer or reference dosimetry in neutron beam therapy, and incidental radiation protection based on an almost rad-equivalent response to photons and high-energy betas, electrons and neutrons.

We are pleased that quite a number of research institutions, among them primary and secondary standard dosimetry laboratories, have started research and application of alanine/ESR dosimetry, stimulated by the GSF and IAEA activities and results.

Apart from transfer dosimetry, there are colleagues of many scientific and dosimetric centres world-wide who have expressed interest and offered co-operation in an international dose library system based on alanine and ESR, a project which has tentatively been proposed by the authors.

REFERENCES

[1] REGULLA, D.F., SCHURMANN, G., SUESS, A., "Dosimetry at a high-activity ^{60}Co waste treatment facility", Radiation for a Clean Environment (Proc. Symp. Munich, 1975), IAEA, Vienna (1975) 465–476.

[2] BRADSHAW, W.W., CADENA, D.G., CRAWFORD, E.W., SPETZLER, H.A., Radiat. Res. **17** (1952) 11.

[3] BERMANN, F., De CHOUDENS, H., DESCOURS, S., "Application à la dosimétrie de la mesure par résonance paramagnétique électronique des radicaux libres créés dans les acides aminés", Advances in Physical and Biological Radiation Detectors (Proc. Symp. Vienna, 1970), IAEA, Vienna (1971) 311–325.

[4] REGULLA, D.F., DEFFNER, U., "Radiation for pollution abatement", (Proc. 1st Int. Conf. ESNA Working Group on Waste Irradiation), European Society for Nuclear Methods in Agriculture (GRONEMAN, A.F., Ed.), Wageningen (1976) 21.

[5] REGULLA, D.F., DEFFNER, U., Int. J. Appl. Radiat. Isot. **33** (1982) 1101.

IAEA-SM-272/12

DOSIMETRY OF ELECTRON AND GAMMA RADIATION WITH ALANINE/ESR SPECTROSCOPY*

M.K.H. SCHNEIDER, M. KRYSTEK,
C.C.J. SCHNEIDER
Physikalisch-Technische Bundesanstalt,
Braunschweig,
Federal Republic of Germany

Abstract

DOSIMETRY OF ELECTRON AND GAMMA RADIATION WITH ALANINE/ESR SPECTROSCOPY.
    A new method for the preparation of alanine dosimeters was investigated. The absorbed dose response of these dosimeters was demonstrated for 10 MeV electron and ^{60}Co gamma radiation in the range from 20 Gy to 1.1 kGy. Concentration of the irradiation-induced free radicals in the alanine was determined by ESR spectroscopy. In addition to measurements at ambient temperature, the alanine dosimeters were also subjected to thermal treatment during irradiation (up to about 50°C) in order to assess their performance characteristics under extreme conditions which might arise in future technical applications. The results show that under normal conditions the alanine calibration curves are linear, whereas at higher temperatures the dosimeters require a correction of 0.3%/K for absorbed doses above 200 Gy.

1. INTRODUCTION

Free radicals produced in alanine by irradiation and measured using ESR seem to be a very promising dosimeter system to cover the range from the therapy level (~ 40 Gy), where the Fricke (ferrous sulphate) dosimeter is a reference dosimeter, to high doses in radiation processing, 1 MGy or more [1, 2]. The main advantages are small size, low fading, small energy dependence of response with respect to the quantity water-absorbed dose produced by gamma and electron radiation, and easy readout with good repeatability. However, preparation of the dosimeter samples as described in Ref.[1] is not easy.

    Alanine (CH_3-CH(NH_2)-COOH) is a simple amino acid. On irradiation at room temperature predominantly free paramagnetic radicals of the type CH_3-CH-COOH are produced which can be measured quantitatively using ESR. To avoid the orientation problems of single crystals in the magnetic field alanine powder is used, yielding an ESR spectrum averaged over all directions due to the

---

* Research carried out with the support of the IAEA under Research Agreement No. 3815/CF.

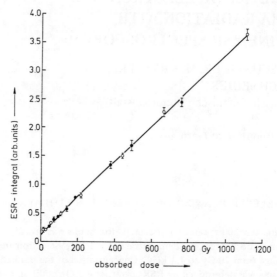

FIG.1. ESR calibration of alanine dosimeters. Double-integrated ESR signal in arbitrary units versus absorbed dose for ^{60}Co gamma (⦶) and 10 MeV electron (⦶) irradiation.

randomly oriented crystallites. Since determination of absolute spin concentration is very difficult, a relative method was applied. Representative samples of a batch of pellets were irradiated with photons or electrons, giving different absorbed doses under calibration conditions. These pellets are kept sealed in polyethylene foil at 6°C, a temperature at which almost no fading occurs [1]. They are evaluated by ESR analysis to give a calibration curve of radical concentration in arbitrary units versus absorbed dose.

The subject of this investigation is to obtain calibration curves for ^{60}Co gamma and 10 MeV electron radiation as well as to study the behaviour of alanine irradiated by electrons under higher temperature conditions.

2. SAMPLE PREPARATION

The dosimeter material was pure polycrystalline alanine powder (Merck No. 963/1046) which may be used in special encapsulation during irradiation. The powder was dried in vacuum at approximately 40°C for 7 hours until no change of mass could be detected in a sample. Contrary to Ref.[2] solid pellets are preferred for practical reasons. In the Physikalisch-Technische Bundesanstalt (PTB) they are produced by compressing 200 mg of the pure powder without admixture of any binding substance to form a cylinder of 5 mm diameter and 8 mm length under a pressure of about 12 000 N/cm². For encapsulation these

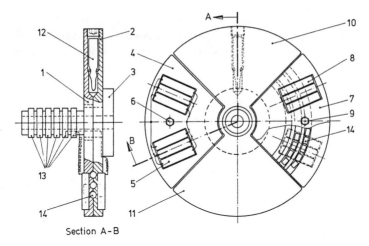

FIG.2. *Irradiation apparatus for alanine pellets subjected to different temperatures. 1+2: front side and rear side sectored aluminium disc; 3: flange on the motor axis; 4: resistor-heated sector; 5: heating resistor; 6+9: NTC resistors for temperature measurement; 7: non-heated sector with resistors 8 to balance electron scattering in both complementary sectors; 10+11: sectors bearing Fricke ampoules 12; 13: ring contact for heating and temperature measurement with NTC resistors 6 and 9; 14: alanine pellets.*

pellets were then covered either with wax or paint, both of which yield no significant ESR signal after irradiation. The final sample mass was 200 ± 3 mg plus 20 ± 10 mg wax or paint; the mass density was 1.20 g/cm^3. The sample dispersion obtained from the individual results for a batch of pellets irradiated under the same conditions is expressed by the bars (see Fig.1). At doses of 640 Gy the relative standard deviation of four samples is 3.7% and increases to about 12% at 20 Gy due to the deteriorating signal-to-noise ratio.

3. IRRADIATION CONDITIONS

Three different types of irradiation set-up were used:

(1) Cobalt-60 gamma radiation was used as a reference source, with a dose rate of about 0.6 Gy/min at a distance of 35.2 cm from the source. The dosimeters were positioned at a depth of 2.5 cm in a Perspex phantom. The axes of the pellets were perpendicular to the beam. Ampoules with Fricke solution were irradiated at the same depth in Perspex and used as a reference for electron beam calibration.

(2) Ten MeV electrons emitted from an accelerator for medical purposes (Philips SL 75-20) lead in the treatment position to a dose rate of approximately 2 Gy/min. The pellets were irradiated in sets of four inserted in a Perspex tube (1.5 mm wall thickness) 2.0 cm deep in a water phantom perpendicular to the horizontal beam. The dimensions of the water phantom are 23 × 30 × 22 cm; it is placed with its front surface 1 m from the source. The field size was 10 × 10 cm. At the same depth 2 cm away from the alanine samples on each side an ampoule with Fricke solution was installed for calibrating the dosimeters with respect to the reference cobalt beam.

(3) Electron beam dose rates of about 600 Gy/min were produced by the same accelerator using the experimental beam outlet, where the electrons are not deflected by magnets, and without any treatment tubes. As the beam shape and homogeneity were not well defined, the irradiations for intercomparisons of heated and ambient alanine samples were performed alternately in fast sequence. The phantom comprised a circular aluminium disc rotating at a frequency of about 400/min around an axis in the beam direction (see Fig.2). The pellets were placed horizontally in two complementary 90° sectors at a depth of 9 mm. One of the sectors was thermally insulated and electrically heated by resistors. The temperature of the sector holding the alanine samples was measured during rotation using calibrated NTC resistors.[1] Heating by electric current took place until the desired temperature was reached, then the power was reduced to keep this temperature constant for about 1 hour to warm up the pellets uniformly. It was found that there was still some thermal conduction from the hot sector to the cooler region. This was taken into account by measuring the temperature of the cooler alanine sector as well. In each of the other sectors between the sectors containing the alanine there were up to three ampoules with Fricke solution for calibration.

4. READOUT PROCEDURE

The ESR measurements were performed in the PTB using an AEG XT20 spectrometer operated in the X-band range (9.1 GHz). The magnetic field setting was at 0.3240 T with a scan of 20 mT. The RF field modulation amplitude, amplification and microwave power had to be adjusted for each sample individually according to the absorbed dose of the sample. For the measured dose range only one integration time setting was necessary, except for the lowest absorbed doses where the integration time was increased tenfold, while the magnetic field sweep was reduced by the same factor. The detector bias current,

---

[1] NTC resistors = resistors with a negative temperature coefficient of resistivity versus temperature.

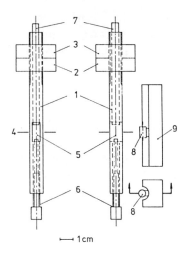

FIG.3. *Sample holder for alanine pellets in the ESR microwave cavity. 1: Teflon rod 10 mm dia. with thread M10 × 1 on top; 2 + 3: Teflon nuts for adjustment of the height of the holder in the microwave (MW) field; 4: centre line for adjustment of the centre of pellet 8; 5: rectangular aperture for the MW flux in the cylindrical cavity for the pellets; 6: lower rod with thread M5 to move the pellet into the centre; 7: upper rod to hold the pellet by pressure; 8: alanine pellet; 9: insertion aid for pellets.*

which influences the signal intensity, was kept constant at 140 µA for all measurements. The samples were inserted into the microwave resonator in a Teflon holder with a cylindrical cavity open to both sides (see Fig.3) such that the sample was perpendicular to the magnetic field. The alanine pellets were taken from their storage temperature (6°C) and allowed to warm up for several hours to be measured at room temperature. Owing to the technique of the ESR apparatus, the first derivative of the alanine paramagnetic absorption spectrum as a function of the magnetic field sweep is the output signal. It was recorded on a chart strip recorder and simultaneously digitized at 600 points by a digital voltmeter. The first and last 50 points of the spectrum were used to correct the base line.

5. EVALUATION OF THE ESR SIGNALS

(1) Direct method

Assuming that the shape of the recorded curve is independent of the spin concentration (this is satisfied for high absorbed doses > 1 kGy), the difference between the maximum and minimum amplitudes of the ESR spectrum on the chart strip can be taken as a direct measure of the spin concentration. This method was applied to all samples irradiated with doses higher than 80 Gy; for smaller doses the signal-to-noise ratio became so poor that determination of the peak amplitudes was unsatisfactory. It was found that the standard deviation of a batch of irradiated samples was 1.2 times (1 kGy) to 3 times (80 Gy) higher than that obtained using the following method.

(2) Double integration method

The data digitized during the magnetic field sweep are assimilated by the computer and stored on a floppy disc for subsequent evaluation. Since the data acquisition occurred in equidistant field intervals, in the on-line evaluation all the ESR amplitudes measured can be simply summed twice to give the integral of the absorption. Small displacements of the base line of the ESR spectrum can strongly affect the second integral and so the first summation was used for correction. The value of the first summation was divided by 600 (number of data points) and the result subtracted from each input data value. After this co-ordinate transformation, the second summation divided by the amplification factor yielded the free radical concentration for each irradiated pellet in arbitrary units.

In addition to this simple method of integration a more complex one was applied to a representative sample of results using cubic spline interpolation and subsequent double integration. The results agreed to within 0.5% and hence the straightforward summation was preferred to save computer time.

6. RESULTS

Determination of unknown absorbed doses using the ESR method requires calibration of the pellets and readout devices. For this purpose pellets having ESR absorption of the same order of magnitude are used. In this way global calibration of the apparatus is achieved and the calibration curve in Fig. 1 can be used for interpolation. This calibration curve is not corrected for zero dose reading. The error bars referring to 4 to 9 pellets indicate a statistical interspecimen dispersion of about 3% at 600 Gy to 12% at 20 Gy, the latter due to deterioration in the signal-to-noise ratio. This is in good agreement with Ref.[1], where a maximum deviation of 4% for a dose of 4500 Gy is reported. Error may be caused during preparation of the samples as well as by the positioning in the microwave cavity. The relative change in response as a function of increased temperature during electron irradiation is shown in Fig.4. At absorbed doses of 200 and 420 Gy, produced by 10 MeV electrons, an increase of the temperature from 20 to 47°C increases the ESR signal by about 8.7%, which corresponds to a temperature dependence of 0.3%/K. The uncertainty of these measurements is the same as obtained for the ambient and calibration samples. Nevertheless, despite the relatively large value of this uncertainty (anticipated for low doses),

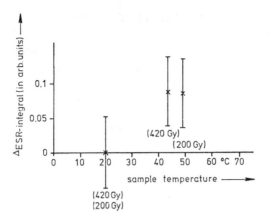

FIG.4. *Increase of alanine double-integrated signal versus temperature during irradiation. The absorbed doses were in the range from 200 to 420 Gy.*

this effect is still significant, in contradiction to Ref.[1] where for absorbed doses smaller than 10^4 Gy a value of 0.18%/K is reported, while for doses above 10^5 Gy the value of 0.31%/K is given; no uncertainty statement was made.

7. CONCLUSIONS

Radiation-induced free radical production in alanine with ESR evaluation has been found to be a very practicable system to bridge the gap between therapy level dosimetry and technical dosimetry. While the high-dose measurements are described in Ref.[1], we have investigated the behaviour of the alanine/ESR system at low doses for ^{60}Co gamma and electron irradiation. Since the calibration results for these radiations are equivalent, they are combined in one calibration curve. Currently, the overall uncertainty of measurement is predominantly due to dispersion of the ESR results as dose assessment is done by comparison with calibrated samples, the uncertainty of which is small compared with the standard deviation of the ESR dose readings. Dispersion of the results is largely caused by two effects: the positioning of the pellets in the sample holder in the resonance cavity introduces the smaller error component, whereas pellet preparation leaves much improvement to be desired for future work since it is here that the predominant error contribution arises.

Concerning the temperature during irradiation by electrons, a coefficient of 0.3%/K for low doses was found. This is about two times higher than that in Ref.[1] for low doses, but it is equal to the value in Ref.[1] for high doses, these doses being generated by ^{60}Co gamma irradiation.

ACKNOWLEDGEMENTS

The authors are deeply indebted to L. Bliek for the use of his laboratory and for his support, and to W. Knittel for selfless and unfailing assistance throughout the ESR measurements.

REFERENCES

[1] REGULLA, D.F., DEFFNER, U., Dosimetry by ESR spectroscopy of alanine, Int. J. Appl. Radiat. Isot. **33** (1982) 1101–1114.
[2] BRADSHAW, W.W., CADENA, D.G., CRAWFORD, G.W., SPETZLER, H.A.W., The use of alanine as a solid dosemeter, Radiat. Res. **17** (1962) 11–21.

IAEA-SM-272/41

APPLICATIONS OF ALANINE-BASED DOSIMETRY

A. BARTOLOTTA, B. CACCIA, P.L. INDOVINA,
S. ONORI, A. ROSATI
Laboratorio di Fisica,
Istituto Superiore di Sanità,
Rome, Italy

Abstract

APPLICATIONS OF ALANINE-BASED DOSIMETRY.
    Alanine-based radiation dosimetry and related dosimeters developed at the Istituto Superiore di Sanità, Rome, Italy, and capable of providing high accuracy absorbed dose determination by ESR are presented. Overall uncertainty is shown to be ± 3.9% in the 10 Gy to 3 kGy range. Possible applications to radiotherapy and industry are discussed. Percentage depth dose values and dose profiles measured with alanine dosimeters in phantom are presented.

1. INTRODUCTION

    The accuracy with which a radiation quantity such as absorbed dose can be measured depends on the existence of a proper primary standard to which the measuring instrument used must be traceable by means of the so-called 'metrological chain'. It is well known that no such primary standard is yet available for high-dose dosimetry. Hence, field dosimeters cannot always be calibrated in the dose range for which they will be used, i.e. in industrial plants for the sterilization of medical supplies or the preservation of food.
    Regulla and Deffner [1] discuss the proposal made by the IAEA in 1977 that an international comparison of various dosimetric systems used in different plants seemed to be the right approach to check the reliability of the dosimetric systems themselves. One such system is the alanine-based dosimetry developed at the Gesellschaft für Strahlen- und Umweltforschung, Neuherberg, and already used for the international high-dose photon dosimetry intercomparison organized by the IAEA between 1978 and 1981 [2].
    Because of the substantial increase in the industrial use of ionizing radiation in Italy, the Italian health authorities have become very interested in this field. It is, in fact, well known that even though ionizing radiation is effective in achieving sterilization and long lasting food preservation, it could give rise to such modifications in food as lowering its nutritional capacity and triggering the formation of substances detrimental to health; hence, the importance of checking that definite lower and higher dose limits be complied with during product

irradiation [3]. The Istituto Superiore di Sanità (ISS), Rome, which is the technical and scientific organ of the Italian Health Service (SSN), is by statute responsible for assessing "the validity and conformity of the dosimetric methods to be used in irradiation plants, and for supervising compliance with individual treatment exposures, i.e. that doses are kept to within statutory limits".

At this point, it may be worth remembering that absorbed dose can only be evaluated indirectly because a suitable tool for its determination in the product itself is still lacking.

Another open problem in high-dose dosimetry is dose assessment in patients undergoing radiotherapy, the success of which depends on strict compliance with optimal dosage (to within ± 5%). The major sources of error in the evaluation of absorbed dose in the target volume depend on quality of irradiation, reproducibility and accuracy of measuring each single dose. Owing to the consistency of these errors, it is therefore desirable to assess the dose actually absorbed by the patient throughout treatment. This would enable total dose in the target volume to be evaluated at any stage of treatment.

On the basis of work in free-radical dosimetry by Regulla and Deffner [1, 2], the ISS is now conducting a programme aimed at developing alanine-based dosimetry for use in both industry and radiotherapy.

This paper reports experimental data on the behaviour of the alanine dosimeters made at the ISS, as well as results on percentage depth dose values and dose profiles measured in phantom irradiated with ^{60}Co γ-rays.

2. MATERIALS AND METHODS

A semi-automatic process for making solid-state alanine dosimeters was developed. Currently, the process consists of obtaining solid cylinders (4.8 mm in diameter, various lengths) by pressing blended L-alanine (80%) and paraffin (20%). Figure 1 shows some such dosimeters.

For calibration purposes, dosimeters were irradiated with a ^{60}Co source at a controlled temperature of 20°C in the Laboratorio di Metrologia delle Radiazioni Ionizzanti, ENEA, Casaccia. To achieve electronic equilibrium conditions, dosimeters were placed in a proper Perspex phantom. Exposure was measured at dosimeter sites inside the phantom by means of a secondary standard (National Physical Laboratory, Teddington, 'Therapy Level' ionization chamber).

A 'Mix-D' tissue-equivalent phantom was used to check alanine dosimeter accuracy in determining tissue dose distributions for ^{60}Co radiotherapy. Irradiation was performed at the Istituto per la Cura e la Prevenzione dei Tumori 'Regina Elena', Rome.

Alanine ESR spectra were recorded with an X-band ESR Varian E112 spectrometer. Suitably sized quartz holders were used to situate dosimeters

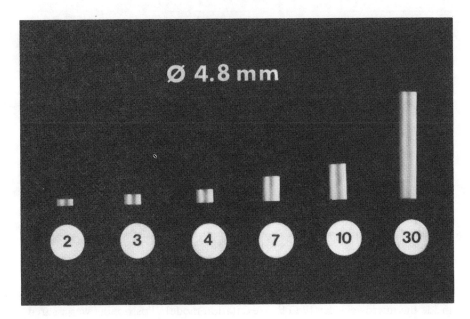

FIG.1. Alanine dosimeters of various lengths (2, 3, 4, 7, 10 and 30 mm) made at the ISS.

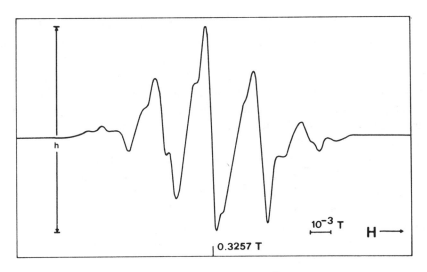

FIG.2. ESR spectrum of alanine dosimeter irradiated with ^{60}Co γ-rays (D_w = 97 Gy); recording conditions are to be found in text (H = magnetic field strength).

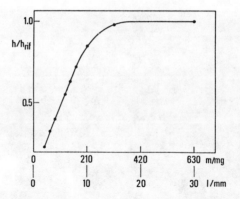

FIG.3. ESR peak-to-peak signal height, h, versus mass (or length, l) of dosimeters. Data normalized to h_{rif}, obtained with a 30-mm long dosimeter. For all dosimeters absorbed dose $D_W = 97$ Gy.

inside the microwave cavity (TE_{011} excitation mode). Positioning was assured using Perspex spacers of various thicknesses.

The signal: noise ratio of low-dose irradiated alanine dosimeters (absorbed dose in water, $D_W < 10$ Gy) was improved with a Varian data acquisition system linked to a P 9835B computer, all of which enabled ESR spectra to be accumulated.

3. MAJOR FEATURES OF ISS ALANINE DOSIMETERS

Figure 2 shows the ESR spectrum of an irradiated alanine dosimeter ($D_W = 97$ Gy). The signal is due to the microwave energy which has been absorbed by the long living free radicals that are produced in alanine when it is struck by ionizing radiation. Since the line shape of the alanine ESR spectrum does not depend on the concentration of free radicals over a wide range of absorbed doses, the absolute number, N, of free radicals is proportional to peak-to-peak ESR signal height, h. Once calibration has been completed, the absorbed dose in alanine can then be evaluated by measuring h.

Hence, the first step was to analyse the accuracy with which the ISS facility could record ESR spectra. If and how much each instrumental parameter contributed to the overall uncertainty in measuring h has now been experimentally studied. The following set-up enables h to be determined to within an uncertainty of 0.8% at the 95% confidence level: microwave power, 1 mW; modulation amplitude, 2.5×10^{-4} T; time constant, 0.25 s; scan time, 240 s; scan range, 2×10^{-2} T. This 0.8% uncertainty value is due to short- and long-term instrumental fluctuations as well as to minor errors in positioning the dosimeter inside the microwave cavity. There are, however, also other sources of uncertainty with the

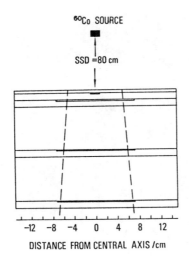

FIG.4. *Configuration of the ^{60}Co source and phantom during irradiation. Dosimeter housings are indicated by the double line.*

ESR recording parameter set-up mentioned. These uncertainties, estimated at the 95% confidence level, in the evaluation of absorbed dose with 30 mm long dosimeters in the 10 Gy to 3 kGy range are:

(1) *Fading*. Dosimeters did not exhibit fading after 12 months of post-irradiation storage in darkness at 15 to 25°C and 40 to 60% relative humidity.

(2) *Batch homogeneity (differences among dosimeter masses)*. Twelve dosimeters, randomly chosen from a batch of 100, were given the same dose and then measured; the interspecimen scattering uncertainty was ±1.1%.

(3) *Recording temperature*. This must be controlled because the ESR peak-to-peak signal height was seen to be temperature dependent (0.23%·°C^{-1} in the 15 to 25°C range). Since all spectra were recorded at 18°C, uncertainty due to recording temperature was ±0.1%. Similar figures would be obtained if spectra were recorded at different temperatures and normalized to 18°C.

(4) *h/dose conversion*. Straight line equation ($\chi^2 = 15.5$, $\nu = 15$) obtained in the 0.5 Gy to 3 kGy range enabled h/dose conversion to be obtained with an uncertainty of ±1.1%.

(5) *Zero dose*. Dosimeters showed no signal for zero dose. However, the lowest absorbed dose detectable was about 0.1 Gy.

(6) *Energy dependence*. No energy dependence was observed for X-ray energies higher than 120 keV. Doses in the lower energy range are underestimated.

(7) *Calibration dose*. As reported in Section 2, exposure in the phantom was measured with a secondary standard. D_W calculated by means of a conversion factor of 37.6 Gy/(C·kg^{-1}) was seen to have an overall uncertainty of ±3.5%.

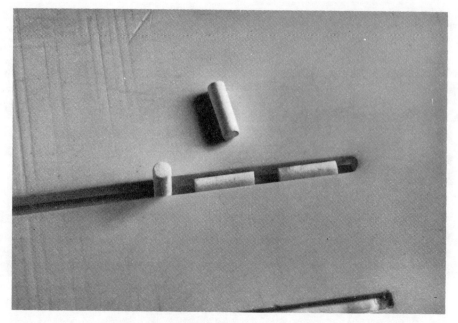

FIG.5. *Dosimeter housing with samples in phantom.*

A quadratic combination of all uncertainties for 30-mm long dosimeters gives an overall uncertainty of ±3.9% at the 95% confidence level. For shorter dosimeters overall uncertainty in absorbed dose could quite possibly be different mainly because high batch homogeneity is much more difficult to achieve than for 30-mm long dosimeters. As a result (see Fig.3), at any given dose a minor error Δm in the proper preselected mass, m, of the dosimeter leads to a major variation Δh in h. This does not appear to hold true for dosimeters longer than 15 mm.

4. APPLICATIONS

The ISS intends to use its ESR dosimetric facility for: (1) assessment of both absorbed dose in products irradiated in industrial plants in Italy and absorbed dose homogeneity ratio, and (2) monitoring radiation in vivo during radiotherapy. A programme for point (1) is about to start. A set of dosimeters will be sent to each plant for irradiation. Nominal doses will be compared with those measured with alanine dosimeters. This will also enable comparison of the various dosimetric systems for standardization purposes. Regarding point (2), percentage depth doses in phantom were measured.

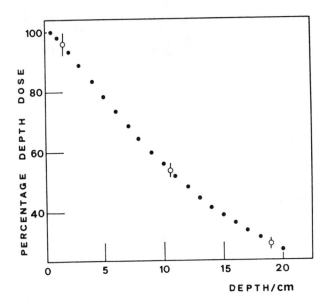

FIG.6. *Percentage depth doses in phantom along central axis experimentally measured with alanine dosimeters (○), together with calculated values (●) [4]. Data expressed as percentage of build-up dose.*

Figure 4 shows the geometrical arrangement of the ^{60}Co source and phantom during irradiation. A rectangular field (10×10 cm^2) and a build-up dose of 50 Gy were used. The phantom was made in such a way as to enable dosimeters to be located at various depths; 16-mm long dosimeters were chosen for this purpose. Figure 5 is a detailed view of a dosimeter housing with samples inside the phantom. Figure 6 plots both central axis percentage depth doses measured with alanine dosimeters and calculated values taken from the literature [4]. Differences between calculated and experimental values were less than ±1%.

To verify the accuracy of phantom doses measured with alanine dosimeters, a dose profile study was conducted. Figure 7 shows the results of this study. Three different depths were chosen (1.44, 10.44 and 19.44 cm). Alanine-measured dose values were compared with calculated values [4] and with those obtained in a similar study conducted with TLDs [5]. On the basis of these results, it was concluded that alanine dosimeters seem to be a suitable tool for radiotherapy. What is more, in spite of their length (16 mm), they nonetheless enable the penumbra region to be drawn accurately. As for relative values, h/dose conversion uncertainty need not be taken into account in the evaluation of overall uncertainty, which thereby decreases to only ±2%.

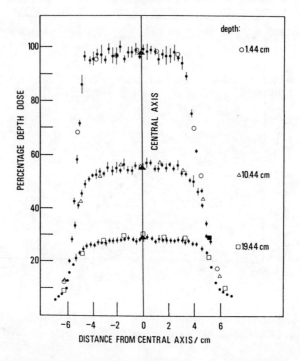

FIG.7. *Dose profiles measured with alanine dosimeters at three different depths. The calculated values (▲) [4] and TLD measured values (●) are reported for comparison. All data are expressed as percentage of build-up dose.*

Hence, we feel that alanine-based dosimetry can be proposed for use in monitoring cumulative radiotherapy doses by measuring a patient's skin dose at each and every stage of treatment. This particular use is possible because alanine dosimeters accumulate absorbed dose, do not fade and their reading is non-destructive.

In conclusion, the data reported in this paper confirm that the alanine dosimeters made at the ISS could be used either with, or as a substitution for, other conventional techniques in high-dose dosimetry.

REFERENCES

[1] REGULLA, D.F., DEFFNER, U., "Standardization in high-level photon dosimetry based on ESR transfer metrology", Biomedical Dosimetry: Physical Aspects, Instrumentation, Calibration (Proc. Symp. Paris, 1980), IAEA, Vienna (1981) 391–404.

[2] REGULLA, D.F., DEFFNER, U., Dosimetry by ESR spectroscopy of alanine, Int. J. Appl. Radiat. Isot. **33** (1982) 1101.
[3] WORLD HEALTH ORGANIZATION, Wholesomeness of Irradiated Food (Rep. Joint FAO/IAEA/WHO Expert Committee), Technical Report Series 659, WHO, Geneva (1981).
[4] JOHNS, H.E., CUNNINGHAM, J.R., The Physics of Radiology, Charles C. Thomas Publications, Washington, DC (1977).
[5] D'ANGELO, L., Verifica dei piani di trattamento radioterapeutici con dosimetria a termoluminescenza, Thesis, Rome University, 1984.

IAEA-SM-272/30

APPLICATION OF COMMERCIAL SILICON DIODES FOR DOSE RATE MEASUREMENTS

M.S.I. RAGEH, A.Z. EL-BEHAY
National Centre for Radiation Research
and Technology,
Cairo

F.A.S. SOLIMAN
Nuclear Materials Co-operation,
Cairo

Egypt

Abstract

APPLICATION OF COMMERCIAL SILICON DIODES FOR DOSE RATE MEASUREMENTS.
  Dose rate measurements using PN junction diodes have been carried out in two types of radiation facilities: a Gammacell-220 and an industrial irradiator J-6500. Measurements inside the Gammacell gave isodose curves similar to those reported by the manufacturer. The measured values of dose rates at different distances from the source of the industrial irradiator, using the silicon diode type BY100, were compared with the values measured by radiochromic dye films. Good agreement between these values was obtained. The experimentally measured values were compared with theoretical calculations. The results obtained recommend use of silicon diodes as solid state dosimeters.

1. INTRODUCTION

Interest has recently been raised in studying the effects of ionizing radiation on the physical properties of semiconductors and insulating materials. Within this frame, use of diodes as radiation monitors was considered.

It is known [1, 2] that the value of the current flow in a PN junction depends on the mode of biasing. When the junction is operating under reverse biasing conditions, the current is due to minority charge carriers only and a much lower conductivity occurs. Investigations have shown that gamma irradiation of diodes increases the reverse current significantly, while the effect on the forward current is negligible [3, 4]. Detailed studies have also shown that similar diodes do not yield similar currents under the same reverse bias conditions [3]. This behaviour was also observed during irradiation. It should be noted that before using a diode as a radiation monitor its photocurrent-dose rate dependence must first be obtained.

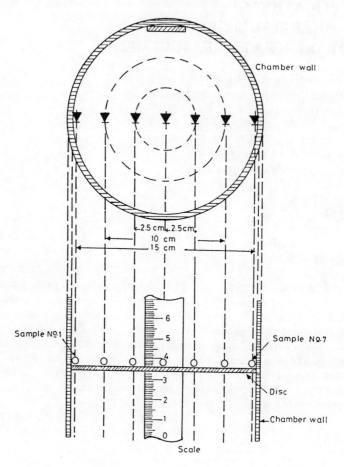

FIG.1. General arrangement of diodes inside the irradiation chamber of Gammacell-220.

One of the important properties of the silicon diodes used in this work is the linear dependence of its photocurrent on the delivered dose rate [3]. Such a property, together with some other advantages, such as high sensitivity, small size and ease of collection of these diodes, recommend their application for dose rate measurements.

2. EXPERIMENTAL PROCEDURES

2.1. Measurements inside the Gammacell-220 chamber

Figure 1 shows the general arrangement used to map the dose rate values in the sample chamber (diameter, 15.5 cm; length, 20 cm) of the Gammacell. The

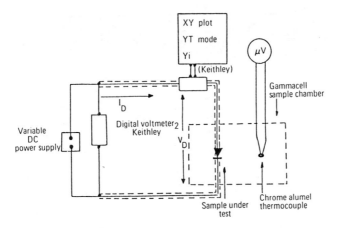

FIG.2. *Circuit used in recording the effect of gamma radiation on PN junction diodes.*

source elements are in a fixed position cylindrically surrounding the sample chamber. The chamber automatically moves up and out of the source area at the end of exposure time. The circuit used in the given experiment is shown in Fig.2, where the diode used was a commercial type BY100. A reverse bias voltage of 125 volts was applied to the diode according to the recommendations of Ref. [3].

Although fabrication of semiconductor devices has progressed significantly over the last twenty years, it is still difficult to obtain two similar devices of the same type. This fact necessitates normalization procedures if more than one diode is used in the same experiment. In the given case seven diodes of the above-mentioned type were fixed to a cardboard disc (Fig.1) and were used in mapping the chamber volume in order to study the distribution of dose rate values inside it.

Since the value of the photocurrent of each diode corresponds to the dose rate value at its position, vertical variation of the cardboard disc provides information on the variation of dose rates along the length of the chamber. Dose rate values along the radius of the chamber were obtained by means of the seven diodes located along the radius. The correlated radial change of dose rate values was measured by moving the central diode (diode No. 4) along the diameter of the chamber at the mid-height position of the cardboard disc. The values obtained at certain points were used for normalization of the results of the other diodes located at these points. It is normal practice to relate the nominal dose rate values at various points to that of the geometrical centre of the irradiation chamber. If I_p represents the net photocurrent, it can be expressed as

$$I_p = I_t - I_i \tag{1}$$

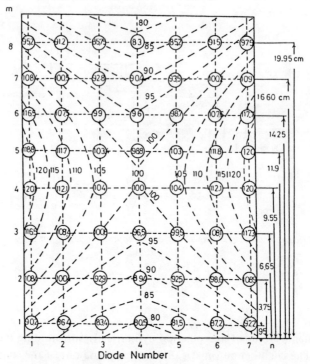

FIG. 3. Isodose curves inside the irradiation chamber of Gammacell-220 measured by silicon diodes.

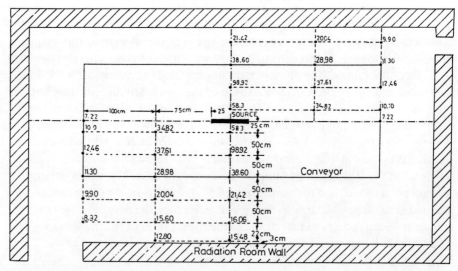

FIG. 4. Dose rate distribution inside the radiation room of the industrial facility (values in rad/s).

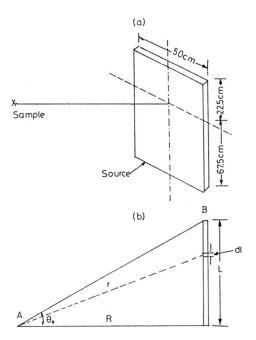

FIG.5. Geometry of source used in the radiation room: (a) total source position; (b) geometry used in theoretical calculations.

where I_t represents the total current measured at position (n, m) during exposure to radiation and I_i represents the initial current measured far from the radiation field; n represents the diode position along the chamber diameter (it can have values between 1 and 7) and m represents the vertical position (it can have values from 1 to 8).

The value of relative dose rate per cent, $D_r\%$ (relative to that of the geometrical centre), point (4,4), is given by the expression

$$D_r\% = D_{n,m}(N) \times 100/D_{4,4}(N) = XY \times 100 \qquad (2)$$

The parameters X and Y are used to correlate the measured photocurrent values of different diodes at any position (n, m) to that of the geometrical centre, point (4,4) (see Fig.3). They are given by the following expression

$$Y = I_{p_{n,m}}(N)/I_{p_{n,4}}(N) \qquad (3)$$

$$X = I_{p_{n,4}}(4)/I_{p_{4,4}}(4) \qquad (4)$$

where N represents the diode number.

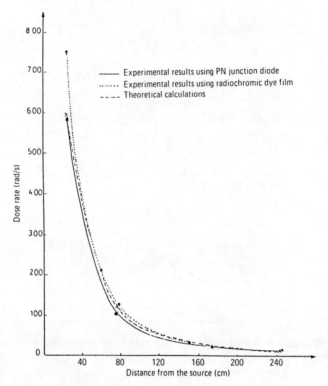

FIG.6. Dependence of dose rate on the perpendicular distance from the source.

The results obtained by the above method are shown in Fig.3. The isodose curves constructed using the measured values are in good agreement with those reported by the manufacturer.

Absolute values of dose rates can be obtained if the photocurrent dose rate dependence of the central diode is already known. Dependence may be obtained by calibration in a known radiation field.

2.2. Measurements inside the irradiator radiation room

The dose rates of the industrial ^{60}Co gamma irradiator were measured at different points inside the irradiation room using commercial diodes of the same type as above.

To convert the obtained values of photocurrent at any position to dose rate values, the sensitivity of the used diode was measured in the Gammacell-220. The initial diode reverse saturation current, I_i, and its total reverse current, I_t, during irradiation at a known dose rate were measured and used to calculate the sensitivity, S, according to the following relation

$$S = I_p/D \qquad A/(rad \cdot s^{-1}) \qquad (5)$$

where D is the dose rate and I_p is defined by Eq. (1).[1]

The results obtained from scanning of the irradiation room are given in Fig.4, where the source is represented by a dark line in the centre of the irradiation room.

To check the results obtained, another well-established dosimetry system, i.e. radiochromic dye film, was used. This system is known to be very stable and has an accuracy which is better than 3% [5, 6]. For comparison, the dose rate values were measured by diodes and films at several positions.

The dose rate at some points could also be calculated theoretically in the first approximation according to the following simple formula

$$P_A = K(C/L)(1/R)\arctan(L/R) \qquad (6)$$

where P_A is the exposure dose rate at a given point, A, K is the gamma constant of the ^{60}Co, C is the source activity in curies, L is the length of the rod source, R is the minimum distance between point A and the rod source (Fig.5).[2]

Figure 6 illustrates the values of dose rates along a perpendicular distance AB obtained from theoretical calculations, experimental results using the PN diode type BY100 and results obtained from radiochromic dye film dosimeters. It is seen that good agreement between the results is obtained at distances up to 40 cm from the source. Errors in measurements at distances less than 40 cm are due to the high dose rate, large scattering effects and approximations in the formula.

3. CONCLUSIONS

Diode type BY100 has been successfully employed for measurement of dose distribution inside two different radiation facilities. Measured values were compared with measurements using radiochromic dye films and also with calculated values. Full agreement was obtained between experimental and theoretical results.

---

[1] 1 rad = 1.00×10^{-2} Gy.
[2] 1 Ci = 3.70×10^{10} Bq.

The results obtained in this work recommend the application of silicon diodes for dose rate measurements. It should be noted that only special diode types can be employed in such measurements. These diodes must have high sensitivity to radiation, high breakdown voltage and low leakage currents.

REFERENCES

[1] SZE, S.M., Physics of Semiconductor Devices, John Wiley, New York (1969).
[2] CHAFFIN, R.J., Microwave Semiconductor Devices: Fundamentals and Radiation Effects, John Wiley, New York (1973).
[3] RAGEH, M.S.I., EL-BEHAY, A.Z., SOLIMAN, F.A.S., Effect of radiation on PN junction diodes (in preparation).
[4] SCHARF, K., Exposure rate measurements of X- and gamma rays with silicon radiation detectors, Health Phys. **13** (1967) 575–586.
[5] McLAUGHLIN, W.L., et al., Radiochromic plastic films for accurate measurements of radiation absorbed dose and dose distribution, Radiat. Phys. Chem. **10** (1977) 119–127.
[6] MILLER, A., Radiochromic Dye Film Dosimetry, Risø National Laboratory, Roskilde, Rep. IFUNAM 79-24 (1979).

IAEA-SM-272/2

RAPID DETERMINATION OF OPTIMAL IRRADIATION CONDITIONS ON LARGE RADIONUCLIDE SOURCES

M. PEŠEK
Institute for Research, Production and
　Application of Radioisotopes,
Prague, Czechoslovakia

Abstract

RAPID DETERMINATION OF OPTIMAL IRRADIATION CONDITIONS ON LARGE RADIONUCLIDE SOURCES.
　A description is given of the determination of optimal conditions for the irradiation of decorative hollow glass facade elements for a new extension to the National Theatre in Prague (called the 'New Scene') which was carried out on the irradiation sources PERUN I and RADEGAST I at the Institute for Research, Production and Application of Radioisotopes. The exposure rates were determined with Czechoslovak semiconductor PN elements. With these semiconductor gauges it is possible to measure exposure rates in large radionuclide sources; moreover, it is possible to use these PN elements for detection of the main direction of incident ionizing radiation. The current generated in PN elements is not greatly dependent on system temperatures in the range from 20 to $50°C$. For dosimetric purposes it is possible to use measurements of current and generated voltage. For measuring exposure rates or doses of radiation it is also possible to use some polymer films, e.g. cellulose triacetate or other dosimetric systems. Various positions of the hollow glass facade elements in the irradiation chambers and various cobalt source arrangements were calculated on the computer PDP 11/23. The mathematical procedure is based on the summing up of various exposure rates at individual points, taking into account the radiation absorption in shielding capsules and irradiated material. Various linear and spatial arrangements of sets of points were incorporated into a computational program. Results have shown that this combined procedure enables rapid determination of the optimal irradiation conditions, even in flat construction parts such as facade elements. Moreover, it has been found that by using PERUN I and RADEGAST I irradiation can successfully be carried out in the required time and with the required quality.

1. INTRODUCTION

Most irradiation work performed with large radiation sources and especially with experimental radiation sources requires frequent determination of the optimal irradiation conditions. For example, it is necessary to find the most advantageous irradiation time under irradiation conditions that are as uniform as possible. In addition, some large radiation sources require that an optimal arrangement of radionuclide sources be found. Simple computational procedures make the solution of these problems relatively easy.

Fast determination of exposure rates can be done, for example, by means of semiconductor PN elements [1]. For our measurements large semiconductor elements manufactured in Czechoslovakia were used. For exposure or dose rate measurements other dosimetric systems can also be used, e.g. foils of cellulose triacetate [2], radiochromic dosimetric films [3], etc. However, semiconductor PN elements provide immediate data on the exposure rate at a measured point and evaluation can be made by means of simple amperometric measurements. Evaluation of exposure rates can also be performed on the basis of generated voltage.

For exposure rate evaluations, a series of computational procedures were worked out which enabled more or less accurate evaluation of the exposure rates in radionuclide sources. Many problems require practical, fast and relatively simple computational procedures. On the basis of such computations it is possible to determine rapidly the optimal conditions for various irradiation requirements.

Two radiation sources at the Institute for Research, Production and Application of Radioisotopes, PERUN I and RADEGAST I, were used for our irradiation work and optimization studies. To ensure acoustic insulation, and also from the aesthetic point of view, architects proposed covering the outer walls of a new extension to the National Theatre in Prague (called the 'New Scene') with hollow glass facade elements (60 × 80 × 20 cm) made from the Czechoslovak glass Simax. The original unsuitable green tint of Simax glass was changed to a smoky brown colour by ^{60}Co gamma irradiation using the Institute's ÚVVVR radiation sources. The absorbed dose that ensured the required colour was found to be 750 Gy, which has no effect on the technical and mechanical properties of the glass elements. A study of the temperature stability of these colour centres revealed that at 25°C they disappear. These colour centres have a half-life of approximately 18 to 20 years [4], which appears to be sufficient considering the lifetime of the facade; however, UV radiation decreases the colour intensity by another 10% on the side exposed to the sun.

As nearly 5000 of these glass elements had to be irradiated, it was necessary to find the optimal irradiation conditions so that irradiation could be performed in the required time and with the required quality; the possibility of irradiation by both radiation sources with sufficient uniformity also had to be taken into account.

2. EXPERIMENTAL PROCEDURES

2.1. Radiation sources

Two large radiation sources at the Institute were used for irradiating the glass facade elements: the experimental radionuclide source PERUN I containing 1570 TBq of ^{60}Co, and the more simple radiation source RADEGAST I containing

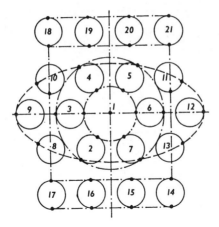

FIG.1. Arrangement of rods with cobalt sources for PERUN I – ground plan. The source contains 21 rods into which the cobalt sources can be placed. Rods 2 to 21 are placed eccentrically in turnable cylinders so that they can be arranged in various configurations.

411 TBq of ^{60}Co. PERUN I enables ^{60}Co sources to be placed in 21 vertically movable rods; their position regarding the irradiation chamber centre is changeable within certain limits (Fig.1). In RADEGAST I, ^{60}Co sources are fixed in one rod only, which is placed in the middle of the irradiation chamber.

The floor area of the irradiation chambers is the same for both irradiation sources (3 × 3 m). It was assumed that most of the irradiation work would be done in PERUN I and that RADEGAST I would serve mainly as a reserve source. It was therefore necessary to find irradiation conditions under which the glass elements could be irradiated in both sources at the same (possible) exposure rate distributions.

2.2. Measurement of exposure rates by means of semiconductor PN elements

Semiconductor PN elements with a diameter of 20 mm were used for determining the exposure rates on the front surface of the glass facade elements. The exposure rate was determined on the basis of measurement of the generated current. The current generated by these semiconductor elements depends on the direction of the incident ionizing radiation. This direction dependence may be strengthened by additional toroidal shielding around the semiconductor PN element. Therefore, these simple, flat semiconductor elements cannot be used in certain cases, e.g. between several sets of sources, because the results would be erroneous. The direction dependence of semiconductor PN elements can be limited by combining several systems into properly arranged sets, e.g. if two or more semiconductor PN elements are arranged into a 'cross set' (at an angle of 90°), direction

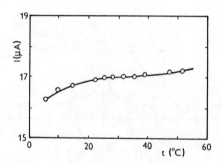

FIG.2. *Temperature dependence of the current generated in the PN element; measured at an exposure rate of 25.2 mA/kg.*

dependence is significantly suppressed [5]. By turning such a set of semiconductor PN elements in the direction of the ionizing radiation, the differences of generated current were found to be $< \pm 2\%$. Dependence of the generated current on the exposure rate is linear from 0 to 40 mA/kg, independent of the direction. Calibration of semiconductor PN elements and PN systems was performed by means of ionization chambers and an integrating ratemeter, model 555 RADOCON II Victoreen; some measurements were also performed using a ferrous sulphate dosimeter.

Current efficiency measurements with PN elements of different areas (with different diameters) revealed that there are no substantial differences between the generated current values if these are taken per unit effective area. The average current efficiency for PN elements with a diameter of 20 mm was found to be 0.352 $\mu A/cm^2$ and for PN elements with a diameter of 40 mm, 0.35 $\mu A/cm^2$; in both cases the exposure rate was 1 mA/kg. Decrease of the current efficiency of the semiconductor PN elements during long-term irradiation with high energy radiation reveals a certain disadvantage. After a total exposure of 10 kC/kg at an exposure rate of 24.4 mA/kg the generated current decreases by 30.5%. A slight increase in generated current (voltage) for a short time after the beginning of irradiation is another unfavourable phenomenon which slightly complicates measurements. Although this increase totals only 1.5% of the measured values for irradiation at an exposure rate of 23.2 mA/kg, it nevertheless decreases the accuracy of measurement.

If, however, the generated current or voltage is measured immediately after placing the PN element into the radiation field, reproducible values can be obtained. This slight increase in generated current or voltage is probably connected with the temperature dependence of the PN element response (Fig.2).

For generated current or voltage measurement simple measuring instruments such as a milliammeter or millivoltmeter with high input resistance can be

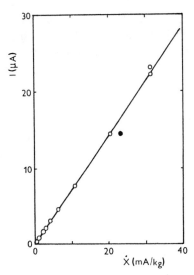

FIG.3. *Dependence of generated current on exposure rate ($\dot{X}$) for semiconductor PN elements with a diameter of 20 mm ((●) obtained in irradiation chamber of Gammacell-220).*

used. In our measurements the universal measuring instrument UNI-10 (VEB Messtechnik Mellenbach, German Democratic Republic and the picoammeter BM 545 (Tesla Brno, Czechoslovakia) were used.

For dosimetric purposes it is possible to use the results obtained from generated current measurements and generated voltage. Generated current is linearly dependent on an exposure rate up to 40 mA/kg, as can be seen in Fig.3. On the other hand, generated voltage is not directly proportional to the exposure rate in the given range. These results were obtained using a millivoltmeter with high inner resistance. However, even these measurements can be used, e.g. at lower exposure rates.

For exposure rate measurements at the front surface of the glass facade elements the simple semiconductor PN element was used, i.e. the same instrument as that used for measuring the generated current; on the basis of the readings data on exposure rates were obtained.

These semiconductor elements and systems, together with a simple evaluating device, can be used to measure the radiation fields of large radiation gamma sources. It is advantageous to combine the measurements with simple computational procedures. Different optimization studies can also be treated in this way.

2.3. Computations of radiation field distributions in radiation sources

For exposure rate computations, only simple expressions were used which could easily be processed by means of simple computers.

In the first step, the computational program was written for a linear set of sources, the radiation of which is absorbed by the casing material and, if required, by the irradiated material itself. The results are obtained both in digital and graphic form, the latter being especially important in order to obtain an immediate, very rapid concept of the exposure rate values. The dependence of exposure rates on the height at the chosen point of the irradiation chamber (assuming that the cobalt sources are arranged in a linear set located vertically in the irradiation chamber) can be calculated as the sum of exposure rate contributions of individual cobalt sources given by the following expression

$$\dot{X} = K_F K_1 \frac{A_i}{a_i^2 + b_i^2} \exp{-\frac{\sqrt{a_i^2 + b_i^2}}{a_i}} [\mu_2 a_2 + \mu_3 (a_i - a_1 - a_2)]$$

The total exposure rate at the calculated point is

$$\dot{X}_{total} = \sum_i \dot{X}$$

where A_i is the activity of the i-th cobalt source
 K_F is the correction factor
 K_1 is the gamma constant of the isotope (radionuclide) used
 μ_2 is the absorption coefficient of the shielding material of cobalt sources (shielding tubes)
 μ_3 is the absorption coefficient of the irradiated material.

The meaning of other symbols is evident from Fig.4. In these computations backscattering from the walls or other parts of the radiation chamber is not considered. The contributions of these scattered photons may be included in the computations [6]. However, it is difficult to provide a true picture of the actual conditions in the majority of cases. It is always necessary to introduce certain simplifications, because the irradiation chamber also contains, along with concrete walls, other different metal parts, often rather bulky, e.g. metal bushings, various control mechanisms, etc., so that even by means of complicated mathematical expressions it is impossible to describe all the effects and factors contributing to the total exposure rate.

For these reasons we used simplified procedures to facilitate and accelerate the computations. The calculated exposure rates are in good agreement with the measured values, with the correction factor value, K_F, approaching unity. However, absorption in a larger amount of irradiated material was not taken into consideration.

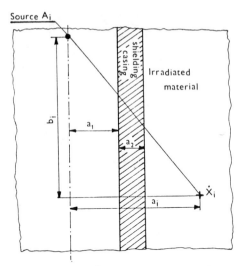

FIG.4. *Computation of exposure rates ($\dot{X}$) from individual cobalt sources with activity (A_i) at different points of the irradiated material, considering radiation absorption in the casing and the irradiated material.*

The computational procedures were modified for larger rod sets of cobalt sources located in parallel and vertical positions at different points of the irradiation chamber so that this computational method could be used for different radiation source constructions. It is evident that computations can be made for an arbitrarily distributed set of cobalt sources in the irradiating chambers. The exposure rates are then calculated and graphically registered at individual points on the straight lines parallel to the individual linear cobalt source sets or at individual positions in the irradiation chamber. At the beginning, computations were done with the computer HP (Mod. 9101A), later with the computer VARIAN 620/L, and at present with the computer PDP 11/23.

2.4. Determination of optimal irradiation conditions for the glass facade elements

Most of the irradiation assignments on large radiation sources and especially on experimental radiation sources require determination of the optimal irradiation conditions. Such a procedure was used to determine the optimal irradiation conditions for irradiation of the glass facade elements used for the 'New Scene' of the National Theatre in Prague.

First, the exposure rates on the front surfaces of the glass facade elements were determined by means of large semiconductor PN elements for a chosen distance in PERUN I. The exposure rates at the same points were calculated using the computational procedure described in subsection 2.3. In this way the

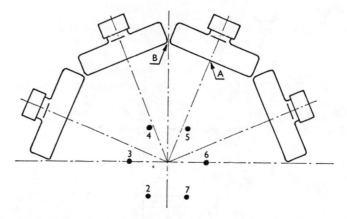

FIG.5. Illustration of the glass facade elements located in one half of the PERUN I irradiation chamber around rods 2 to 7 with cobalt sources. The distance from the central points of the glass elements to the centre of the irradiation chamber is 74 cm.

correction factor values were specified for subsequent computation. The exposure rate values on the surface of the glass facade elements were 5 to 10% higher than those calculated without facade elements. This was to be expected considering backscattering from the relatively massive glass.

Other exposure rate values at different arrangements of the glass facade elements and cobalt sources were calculated by the more precise computational procedure. Computations for the PERUN I source were performed for rods 2 to 7 arranged either in a small ring (Fig.5) or in a partially extended circle, and for three cobalt sources placed in the rods at heights of 35, 55 and 75 cm above the floor. Another variant was also considered, i.e. with the cobalt sources in the rods placed at heights of 25, 55 and 85 cm above the floor. Examples of exposure rate computations for PERUN I with the glass facade elements placed at distances of 52.5, 74 and 93 cm from the centre of the irradiation chamber are illustrated in Figs 6 to 8, respectively. The cobalt sources for these variants were placed at heights of 35, 55 and 75 cm. The results obtained for cobalt sources placed at heights of 25, 55 and 85 cm and glass facade elements placed at 52.5, 74 and 93 cm are given in Figs 9 to 11, respectively. Examples of computations for RADEGAST I with the glass facade elements placed at a distance of 74 cm from the central cobalt source are given in Fig.12.

At a distance of 52.5 cm six glass facade elements (placed vertically, i.e. on the shorter side) can be placed around the 'small' ring of rods in PERUN I. At a distance of 74 cm eight glass facade elements (Fig.5) can be irradiated; at 93 cm this number increases to ten. The horizontal position of the glass facade elements (on the longer side) was also considered; variants when the ring of rods 2 to 7

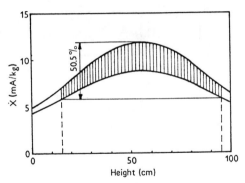

FIG.6. Exposure rate ($\dot{X}$) distribution on the surface of irradiated glass elements placed 52.5 cm from the centre of the irradiation chamber in PERUN I. Cobalt sources were placed in the rods at heights of 35, 55 and 75 cm.

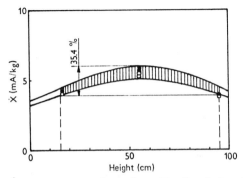

FIG.7. Exposure rate ($\dot{X}$) distribution on the surface of irradiated glass elements placed 74 cm from the centre of the irradiation chamber of PERUN I. Cobalt sources were placed in the rods at heights of 35, 55 and 75 cm. (●) = exposure rate values measured by a semiconductor dosimeter (PN element) at the central points of the facade elements (perpendicular to point A in Fig.5); (○) = exposure rate values measured by a semiconductor dosimeter (PN element) at the marginal points of the facade elements (perpendicular to point B in Fig.5).

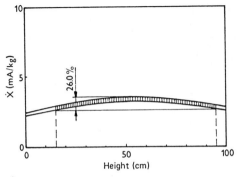

FIG.8. Exposure rate ($\dot{X}$) distribution on the surface of irradiated glass elements placed 93 cm from the centre of the irradiation chamber of PERUN I. Cobalt sources were placed in the rods at heights of 35, 55 and 75 cm.

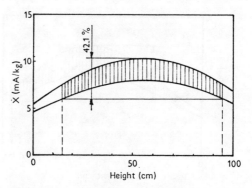

FIG.9. Exposure rate ($\dot{X}$) distribution on the surface of irradiated glass elements placed 52.5 cm from the centre of the irradiation chamber of PERUN I. Cobalt sources were placed in the rods at heights of 25, 55 and 85 cm.

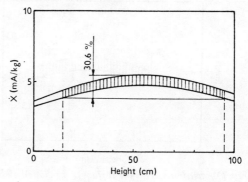

FIG.10. Exposure rate ($\dot{X}$) distribution on the surface of irradiated glass elements placed 74 cm from the centre of the irradiation chamber of PERUN I. Cobalt sources were placed in the rods at heights of 25, 55 and 85 cm.

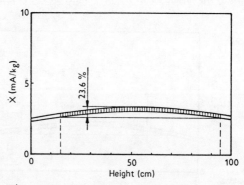

FIG.11. Exposure rate ($\dot{X}$) distribution on the surface of irradiated glass elements placed 93 cm from the centre of the irradiation chamber of PERUN I. Cobalt sources were placed in the rods at heights of 25, 55 and 85 cm.

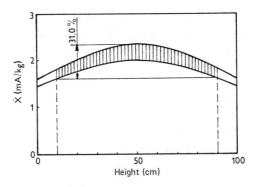

FIG.12. *Exposure rate ($\dot{X}$) distribution on the surface of irradiated glass elements placed 74 cm from the centre of the irradiation chamber of RADEGAST I.*

was extended, or variants with a different distribution of cobalt sources in rods, and also other variants were considered. The computational procedures that were used enabled a very fast and relatively precise conception on exposure rate distributions to be achieved for various arrangements of cobalt sources and glass facade elements. Figures 6 to 12 present the calculated exposure rates on the front surface of the glass facade elements and their vertical distribution at different points of the facade elements. The height of the glass facade elements is 80 cm and their most advantageous position in PERUN I is at heights of 15 to 95 cm above the steel plate floor of the source (with the centre at a height of 55 cm). For RADEGAST I (Fig.12), the most advantageous position is at heights of 10 to 90 cm above the steel plate floor of the source (with the centre at a height of 50 cm).

In Figs 6 to 12 the upper curves limiting the hatched areas give exposure rates at the glass facade element points nearest to the centre of the chamber (point A in Fig.5), and the lower curves limiting the hatched areas give the lowest exposure rates at the most distant points of the glass facade elements (point B in Fig.5). In the area limited by upper and lower curves (the hatched area) there is the region of exposure rates that are used for the irradiation of individual parts of the glass facade elements. Maximal exposure rates are in the central points of the glass elements nearer to the cobalt sources; minimal exposure rates are in the marginal points.

If the cobalt sources were placed at heights of 25, 55 and 85 cm, irradiation would be more uniform; however, the irradiation time would increase and more extensive modifications of the source would be necessary. At the same time this modification would influence unfavourably other irradiation work performed in the source. The arrangement with the cobalt source placed at heights of 35, 55 and 75 cm was therefore used for irradiation of the facade elements. Moreover, the exposure rate distribution on the surface of the elements in this arrangement, i.e. irradiation at a distance of 74 cm in PERUN I, is very similar to that of RADEGAST I (see Figs 7 and 12), which was considered a reserve source. Measurements, computation and visual colour assessment of the irradiated facade elements have shown that irradiation homogeneity was sufficient with cobalt sources placed at heights of 35, 55 and 75 cm, i.e. when the glass facade element centres were placed at a distance of 74 cm from the centre of the ring of cobalt sources, the difference between exposure rates in the centre of irradiated elements and on their edge is 35.4%. If the elements were irradiated at a distance of 52.5 cm, the exposure rate on the edge of the elements would be 50.5% lower; if elements were irradiated at a distance of 93 cm from the centre of the ring of rods, the exposure rate at the edge would be only 26% lower than the exposure rates in the centres of individual elements. On the basis of these computations it is then possible to determine the productivity in individual radiation sources and to choose the most suitable variant by means of which it is possible to ensure the required quality of work in the required time and to choose the variant providing sufficiently uniform irradiation.

REFERENCES

[1] OSVAY, M., STENGER, V., FÖLDIÁK, G., "Silicon detectors for measurement of high exposure rate gamma rays", Biomedical Dosimetry (Proc. Symp. Vienna, 1975), IAEA, Vienna (1975) 623-632.
[2] PUIG, J.R., LAIZIER, J., SUNDARI, F., "Le film «TAC», dosimètre plastique pour la mesure pratique des doses d'irradiation reçues en stérilisation", Radiosterilization of Medical Products 1974 (Proc. Symp. Bombay, 1974), IAEA, Vienna (1975) 113–134.
[3] BUENFIL-BURGOS, A.E., URIBE, R.M., de la PIEDAD, A., McLAUGHLIN, W.L., MILLER, A., Radiat. Phys. Chem. 22 3–5 (1983) 325.
[4] PRÁŠIL, Z., ŘEŘICHOVÁ, M., Sklář Keramik 34 1 (1984) 15.
[5] PEŠEK, M., JANŮ, M., Radioisotopy 25 1 (1984) 149.
[6] PEŠEK, M., HOŠPES, M., ČECHÁK, T., KLUSOŇ, J., Radioisotopy 23 4 (1982) 509.

IAEA-SM-272/11

INVESTIGATIONS OF THE USE OF LiF CRYSTALS FOR ROUTINE HIGH-LEVEL DOSIMETRY AT CERN

B. BAEYENS, F. CONINCKX,
P. MAIER, H. SCHÖNBACHER
European Organization for Nuclear Research
 (CERN),
Geneva

Abstract

INVESTIGATIONS OF THE USE OF LiF CRYSTALS FOR ROUTINE HIGH-LEVEL DOSIMETRY AT CERN.
  The possible use of optically clear LiF crystals for the routine European Organization for Nuclear Research (CERN) high-level dosimetry programme was investigated. Two types of commercially available crystals, isotopically enriched ^{7}LiF and natural abundance LiF, were irradiated in a ^{60}Co gamma source and in a nuclear reactor up to 10^8 Gy. By absorption measurements at wavelengths from 200 to 900 nm using a high-resolution photospectrometer a dose range of 10^3 to 10^8 Gy can be covered. For the upper part of this range (from 5×10^6 to 10^8 Gy) the colour centres below 700 nm are saturated and mainly a peak appearing at 780 nm is used. It was shown that an important increase in absorbance of this peak could be obtained by heat treatment of the crystals for 1 hour at 180°C. Comparisons were made of LiF crystals with silver-doped radiophotoluminescent (RPL) glass and hydrogen pressure dosimeters. Irradiations were carried out around high-energy proton and electron accelerators at CERN and the Deutsches Elektronen-Synchroton (DESY). The results have shown agreement within a factor of 2. LiF crystals did not show any energy dependence in the low photon energy radiation field at DESY. Follow-up of photospectrometer readings over two years did not give evidence of any fading. It can be concluded that the simplicity of readout, the small size of the dosimeter, the possibility of re-use and the applicability in a dose range of over five decades are very desirable characteristics for the use of LiF crystals as a long-term integrating high-dose dosimeter for both existing and future CERN installations.

1. INTRODUCTION

   At the European Organization for Nuclear Research (CERN), radiation dose measurements are carried out around the high-energy particle accelerators using radiophotoluminescent (RPL) glass detectors and detectors based on optical absorption in special phosphate glass [1, 2]. These high-level dosimetry (HLD) data have given reliable information on radiation ageing of accelerator components and have provided an indication of their projected lifetime [3, 4].
   The main radiation spectrum at present encountered at CERN is a mixed radiation field within the accelerator tunnels caused by losses of protons from

the accelerator beam at energies up to 450 GeV. A large electron positron (LEP) collider is at present under construction. The circulating electron beams in this machine will produce synchrotron radiation with a mean energy of about 1 MeV [5] when the beam is curved by a magnetic field.

The aim of this paper is to investigate the possibility of using commercially available, optically clear LiF crystals for HDL in existing and future CERN accelerators, using photospectrometric measurements of radiation-induced colour centres. The dosimeters are compared with RPL and hydrogen pressure dosimeters by exposing them in different radiation fields in a dose range of 10^3 to 10^8 Gy. The feasibility of LiF as a potential routine dosimeter for long-term high-dose integration with respect to fading and re-use is discussed.

2. DOSIMETER TYPES AND DOSIMETRY METHODS

Two different forms of optically clear LiF crystal dosimeters are used: (1) natural abundance LiF (Monokristalle-Kristalloptik, Korth, Kiel, Federal Republic of Germany), and (2) isotopically enriched ^{7}LiF (The Harshaw Chemical Company, Cleveland, Ohio, United States of America). The LiF dosimeters are either cleaved crystals or cleaved polished crystal chips.

Colour centres that are formed in LiF subjected to ionizing radiation are similar to those observed in the more frequently studied alkali halides. In the literature, the absorption bands produced by these colour centres have been labelled F, F′, M, R_1, R_2, N_1 and N_2 [6–8]. The wavelengths corresponding to the maximum absorption of the different radiation-induced colour centres for LiF, already used for dosimetry [9–12], are F-centre (λ_{max} = 247 nm), M-centre (λ_{max} = 443 nm), R_1-centre (λ_{max} = 315 nm), R_2-centre (λ_{max} = 374 nm), N_1-centre (λ_{max} = 517 nm) and N_2-centre (λ_{max} = 547 nm).

The model centres in LiF are described by trapped electrons at the negative ion vacancies in various combinations in the crystal lattice. Because of these centres, electronic transitions may occur if LiF is exposed to light in the UV and visible region of the electromagnetic spectrum. The absorbance A = log Io/I at selected wavelengths is a function of absorbed dose.

The basis of the RPL dosimeters, used for comparison with the LiF, is the formation of stable luminescent centres by electrons liberated in silver-doped phosphate glass during irradiation and trapped at vacancies or impurities [13]. Post-irradiation excitation by means of UV light releases a measurable quantity of fluorescent light which is a function of the absorbed radiation dose.

For hydrogen pressure dosimeters the principle of measurement is the formation of hydrogen gas (H_2) in polyethylene under irradiation. The dosimeter, utilizing hydrogen pressure build-up in a small, sealed glass capsule, has been developed at the Rutherford and Appleton Laboratories [14].

Table I summarizes the characteristics of the different dosimeter types, together with measurement techniques.

TABLE I. HLD RADIATION DETECTORS

Type	Dimensions	Useful range (Gy)	Measurement technique
LiF crystals		$10^3 - 10^8$	Absorption at selected wavelength.
Harshaw ^{7}LiF (USA)	6 × 6 × 2 mm^3		Reading instrument: PYE UNICAM PU 8800 UV/VIS spectrophotometer PHILIPS
Korth LiF (FRG)	6 × 6 × 1.5 mm^3		
RPL glass (Toshiba or Schott)	1 mm ϕ × 6 mm	$10^{-1} - 10^6$	Radiophotoluminescence stimulated by UV. Reading instrument: Toshiba FGD-6
Hydrogen pressure	0.25 g polyethylene in 5.8 cm^3 glass tube	$10^4 - 10^7$	Hydrogen pressure measurement

3. DOSIMETER CALIBRATION AND IRRADIATION

All dosimeter types are calibrated by means of a ^{60}Co gamma source at doses of up to 10^6 Gy. For higher doses (5×10^6 to 1×10^8 Gy) LiF crystals were irradiated in the 7 MW ASTRA research reactor (Austria) [15]. In the selected irradiation position in the reactor pool more than 95% of the dose contribution is due to gamma rays.

Figure 1 shows a semi-log plot of response at selected wavelengths of optical grade ^{7}LiF. It is seen that a dose range of 10^3 to 10^8 Gy is covered using the absorbance at different wavelengths corresponding to the colour centres given in Section 2. The visual colour changes of LiF crystals are also indicated in this figure. This feature of LiF makes it possible to estimate the order of magnitude of absorbed dose without any measurement using sophisticated apparatus.

Furthermore, the three dosimeter types were irradiated in the stray radiation field of the existing CERN proton accelerators and synchrotron radiation of the electron positron storage ring PETRA was carried out at the Deutsches Elektronen-Synchrotron (DESY), Hamburg. The first test was for comparison with our currently used high-dose dosimeters and the second was a representative test of the suitability of LiF crystals as a dosimeter for LEP. This is of particular interest as it is known that the currently used RPL dosimeters considerably overestimate at the low photon energy range of 0.01 to 0.3 MeV [13, 16].

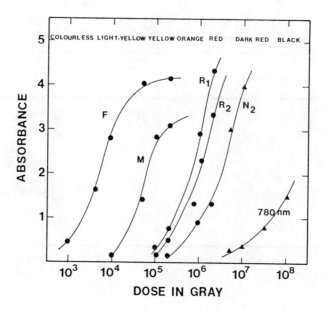

FIG.1. Cobalt source (●) and reactor irradiation (▲) response curves for the different colour-centre band absorption maxima of ^{7}LiF crystals having dimensions of $6 \times 6 \times 2$ mm^3 (absorbance $A = \log Io/I$).

4. RESULTS

4.1. UV and visible absorption spectra measurements in LiF crystals

Figure 2 shows the absorption spectrum, measured between 200 and 900 nm, of the different radiation-induced colour centres in LiF at an absorbed dose of 1 MGy obtained by irradiation at a ^{60}Co gamma source. The absorbance is plotted as a function of wavelength (λ). The full line represents the spectrum of a ^{7}LiF crystal that was stored at room temperature after irradiation. The F and M peaks are saturated at 1 MGy; R_1 and R_2 peaks are well developed. The N_1 peak at 517 nm is not resolved, while the N_2 peak appears as a shoulder.

In analogy with the theory of thermoluminescence electrons can be liberated from their traps by heating [7, 13]. This may be the explanation of the effect shown in Fig.2 where, after heating the same sample for 1 hour at 180°C (dotted line), the reorientation of electrons leads to a slight decrease of absorbance for the R_1 and R_2 peaks and a much better resolved N_2 peak.

At absorbed doses above 1 MGy a characteristic increase of absorbance is obtained at a wavelength of 780 nm. Figure 3 shows the spectrophotometric response, measured between 700 and 900 nm, for LiF irradiated in the reactor at a dose range varying between 5×10^6 and 10^8 Gy. For this dose range the

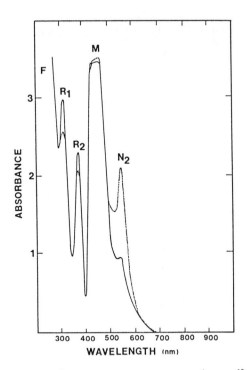

FIG.2. *Absorption spectrum of ^7LiF after exposure to 1×10^6 Gy in ^{60}Co gamma radiation. The full line shows the spectrum at room temperature; the dotted line shows the spectrum after thermal treatment of 1 hour at 180°C.*

band spectra at lower wavelengths (<700 nm) are saturated and are not suitable for dose evaluation. The effect of heating, again 1 hour at 180°C, on the absorption, measured between 700 and 900 nm, is also given in Fig.3 (broken line). A rather important increase in absorbance is caused by the heat treatment. This may be due to an analogous effect of reorientation of electrons, as described above. A hypsochromic shift (to shorter wavelength λ_{max} = 770 nm) is observed in all cases. The advantage of this heat effect for HLD is more accurate dose evaluation in the megagray range using the absorbance at 770 nm.

4.2. Intercomparison of LiF crystals with RPL and hydrogen pressure dosimeters

Intercomparisons of LiF single crystals with radiophotoluminescent (RPL) and hydrogen pressure dosimeters are carried out after irradiation around high-energy proton and electron accelerators at CERN and DESY, respectively. The absorbed doses are evaluated using ^{60}Co gamma and reactor calibrations (Section 3).

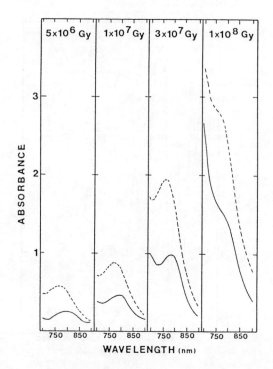

FIG.3. Effect of a short-term post-irradiation heat treatment on ^7LiF irradiated at the reactor. Full line: before thermal treatment; broken line: 24 hours after thermal treatment (1 hour at 180°C).

4.2.1. High-energy proton accelerator

Irradiations at the CERN 450 GeV super proton synchrotron (SPS) accelerator are carried out at different positions, the dose rates covering a range from 0.006 to 2.4 Gy·s^{-1}. Figure 4 shows the comparison of the different HLD detectors in squared symbols (CERN irradiations) by plotting the dose indicated by the routine RPL dosimeter in the abscissa and the dose indicated by LiF (filled symbols) and the hydrogen pressure dosimeter (open symbols) in the ordinate. The experimental data in the dose range 2×10^4 to 1×10^7 Gy show for all three types of dosimeter an agreement within a factor of 2 or better.

4.2.2. Synchrotron radiation

The aim of irradiations at PETRA (DESY) is to test the performance of LiF crystals for dosimetry of synchrotron radiation that occurs in the curved sections of electron accelerators where the beam particles undergo transverse deflection.

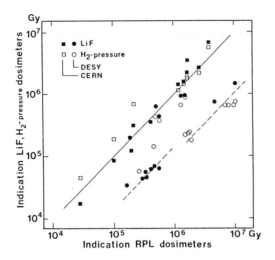

FIG.4. *Comparison of LiF and H_2-pressure dosimeter readings with the routine RPL dosimeter results (for irradiation at CERN and at PETRA/DESY).*

This is a representative test for LEP since the photon energy spectrum is very similar at PETRA, namely from tens of keV up to a few MeV [5, 16].

The intercomparison results of LiF, RPL and H_2-pressure dosimeters irradiated at PETRA are shown in round symbols (Fig.4). For a large number of LiF and hydrogen pressure dosimeters the dose is 7 to 10 times lower than that for RPL dosimeters. This is explained by the known overestimate of RPL at low energies that is strongly dependent on the silver content of the RPL glass. The maximum photon energy sensitivity is between 40 and 50 keV, while the minimum is between 0.3 and 2 MeV [13]. The few cases of good agreement indicate a higher energy spectrum of the PETRA synchrotron radiation caused by a recent increase of the electron beam energy or by low energy cut-off by shielding or machine components. This experience clearly shows the risk of dose overestimation by RPL dosimeters in a synchrotron radiation field. This is not the case for LiF or hydrogen pressure dosimeters, which appear to be energy independent.

5. FADING

The degree of stability of LiF over a period of about 2 years was determined by the follow-up of the photospectrometer readings. The samples were stored in darkness at room temperature. It was observed that for both natural LiF and enriched ^{7}LiF crystals irradiated with ^{60}Co gamma radiation in a dose range of 10^3 to 10^6 Gy a very good correspondence exists between the spectra measured a few weeks after irradiation and after two years. Moreover, at 1 MGy the colour-centre band at 547 nm is somewhat more pronounced in the aged sample. The storage of irradiated LiF crystals at room temperature over long periods of time leads to a reorientation of electrons in the crystal lattice. This kinetic process is strongly accelerated when the crystal is thermally treated after irradiation (see subsection 4.1).

6. ANNEALING – RE-USE

Annealing of LiF dosimeters was tested for a series of samples irradiated with ^{60}Co gamma rays and in the reactor. Thermal treatment at 550°C (±10°C) for 1 hour resulted in a spectrometric response of the annealed crystals that was comparable with the background absorption of non-irradiated samples in all cases. Irradiations of new and regenerated samples in the reactor in a dose range of 5×10^6 to 3×10^7 Gy showed nearly identical absorption spectra.

7. CONCLUSIONS

Solid-state dosimetry based on radiation-induced colour centres in pure crystalline LiF appears suitable for radiation environments where high-level doses (up to more than 10 MGy) are expected. The results obtained from ^{60}Co gamma and reactor calibrations showed the potential of LiF to measure absorbed doses in a dose range covering five decades. A thermal treatment for 1 hour at 180°C results in more accurate calibration curves, especially in the very high dose region (1 to 100 MGy), where a characteristic increase in absorbance at 770 nm was observed.

A comparison of the LiF with other high-level dose dosimeter systems such as RPL and H_2-pressure irradiated in the CERN high-energy accelerator tunnels shows good agreement within a factor of 2 when related to ^{60}Co gamma and reactor calibrations. On the other hand, irradiations of RPL dosimeters and LiF crystals to a representative low photon energy radiation field at DESY showed a 7 to 10 times too high dose reading for RPL but no energy dependence for LiF crystals. This makes LiF crystals also attractive as HLD for LEP.

The follow-up of the photospectrometer readings over 2 years did not show any fading, which shows that LiF is suitable for long-term dose integration.

The simplicity of the non-destructive photometric readout, the small size of the dosimeter and the possibility of re-use are very desirable characteristics of LiF. The possible use of this dosimeter in areas where dose levels to be measured extend over many orders of magnitude are interesting for both existing and future CERN installations as well as in other fields of application. The distinct colour changes of the crystals allow a visual estimate of the order of magnitude of the dose received, without having to carry out measurements.

REFERENCES

[1] IZYCKA, A., SCHÖNBACHER, H., "High-level dosimetry results for radiation damage studies at high-energy accelerators", Reactor Dosimetry (Proc. 3rd ASTM-EURATOM Symp. Ispra, 1979), Rep. EURA 6812, Ispra (1980) 316.

[2] CONINCKX, F., et al., High-Level Dosimetry Results for the CERN High-Energy Accelerators, European Organization for Nuclear Research, Rep. CERN HS-RP/060 (1981).

[3] SCHÖNBACHER, H., Material Irradiation Tests at CERN, European Organization for Nuclear Research, Rep. CERN TIS-RP/115/CF (1983).

[4] MAIER, P., STOLARZ, A., Long-Term Radiation Effects on Commercial Cable-Insulating Materials Irradiated at CERN, European Organization for Nuclear Research, Rep. CERN 83-08 (1983).

[5] FASSO, A., et al., Radiation Problems in the Design of the Large Electron Positron Collider (LEP), European Organization for Nuclear Research, Rep. CERN 84-02 (1984).

[6] SEITZ, F., Color centers in alkali halide crystals, I and II, Rev. Mod. Phys. **18** 3 (1946) 384; **26** 1 (1954) 7.

[7] SCHULMAN, J.H., COMPTON, W.P., Color Centers in Solids, Pergamon Press, Oxford (1963).

[8] KAUFMAN, J.V.R., CLARK, C.P., Identification of color centers in lithium fluoride, J. Chem. Phys. **38** 6 (1963) 1388.

[9] VAUGHAN, W.J., MILLER, L.O., Dosimetry using optical density changes in LiF, Health Phys. **18** (1970) 578.

[10] McLAUGHLIN, W.L., et al., Electron and gamma-ray dosimetry using radiation-induced color centers in LiF, Radiation Processing (Trans. 2nd Int. Meeting Miami, 1978) (SILVERMAN, J., Ed.), Radiat. Phys. Chem. **14** (1979) 467.

[11] McLAUGHLIN, W.L., et al., Radiation-induced color centers in LiF for dosimetry at high absorbed dose rates, Nucl. Instrum. Methods **175** (1980) 17.

[12] McLAUGHLIN, W.L., URIBE, R.M., MILLER, A., Megagray dosimetry for monitoring of very large radiation doses, Radiation Processing (Trans. 4th Int. Meeting Dubrovnik, 1982) (MARKOVIĆ, V.M., Ed.), Radiat. Phys. Chem. **22** (1983) 333.

[13] BECKER, K., Solid State Dosimetry, CRC Press, Cleveland (1973).

[14] MORRIS, A., et al., A remote reading integrating radiation dosimeter for the range $10^4 - 10^9$ rad, Rutherford High Energy Laboratory, Harwell, Rep. RHEL (1966) 132.

[15] SCHÖNBACHER, H., VAN DE VOORDE, M., BURTSCHER, A., CASTA, J., Study on radiation damage to high-energy accelerator components by irradiation in a nuclear reactor, Kerntechnik **17** 6 (1975) 268.

[16] DINTER, H., TESCH, K., Some measurements of absorbed dose due to synchrotron radiation in the PETRA tunnel, Deutsches Elektronen-Synchrotron, Rep. DESY D3/34 (1981).

IAEA-SM-272/32

USE OF 'MEMORY EFFECT' OF Al_2O_3 TL DETECTORS IN HIGH EXPOSURE DOSIMETRY

M. OSVAY, F. GOLDER
Institute of Isotopes,
Hungarian Academy of Sciences,
Budapest, Hungary

Abstract

USE OF 'MEMORY EFFECT' OF Al_2O_3 TL DETECTORS IN HIGH EXPOSURE DOSIMETRY.
  The phototransferred thermoluminescence (PTTL) characteristics of Al_2O_3 TL dosimeters were examined to investigate the possibility of re-estimation and to extend their measuring range. For 366 nm UV light, it was found possible to extend the measuring range by two orders of magnitude over the saturation value observed by usual TL readout techniques. No fading of the PTTL peak was observed during one month of keeping the dosimeters covered with dark polyethylene at ambient temperature.

1. INTRODUCTION

Easy handling, small size, cheapness, mechanical resistance, etc. have led to a great increase in the application of TL detectors over the last 20 years, but one of the disadvantages is that the signal is destroyed in the process of annealing. This means that after irradiation the detector can be evaluated only once; in some cases the information might be lost by error and it is not possible to repeat the evaluation.

Phototransferred thermoluminescence (PTTL) enables dose re-estimation utilizing the 'memory effect' in TL materials.

The phenomenon of PTTL is due to the retention of TL information — on exposure to UV light — from TL materials previously exposed to ionizing radiation and partially annealed. High temperature peaks, which are not thermally annealed during normal readout, enable re-estimation of the dose.

Application of PTTL for purposes of dosimetry only started in the last ten years. Mason and McKinlay et al. examined the possible use of PTTL for the re-estimation of LiF TL detectors [1, 2]. Among others, Caldas and Mayhugh [3] used this new technique for $CaSO_4$:Dy to extend its measuring range over the saturation value. Most of the common TL materials exhibit PTTL properties; Al_2O_3 is one of the most promising because the peaks on the glow curve are far from one another. The phototransfer characteristics of Al_2O_3 have been investigated in order to establish whether it can be applied in UV dosimetry [4].

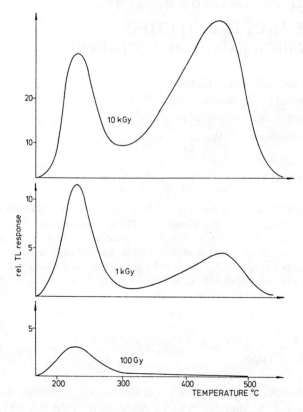

FIG.1. *TL glow curves of Al_2O_3:Mg,Y irradiated by a ^{60}Co gamma radiation source.*

An account is given of our PTTL investigations carried out on Al_2O_3 with the aim of utilizing the information that remained in the material after annealing for re-estimation of the dosimeters.

2. EXPERIMENTAL

The Al_2O_3 dosimeters were made of high purity powder by adding 0.1% MgO and 0.1% Y_2O_3 activators [5]. The diameter of the prepared ceramic disc was 6 mm and its thickness 1 mm. In Fig. 1 the glow curves of TL detectors irradiated by different doses in the 10^2 to 10^4 Gy dose range can be seen. The detectors were annealed by a Harshaw 2000 AB TL reader. The measurements were carried out at a heating rate of 10°C/s and the maximum temperature of 320°C was reached in each readout cycle. The UV source was provided by Swiss Camag-type low pressure mercury lamps with filters to give 254 and 366 nm monochromatic light. UV exposure was determined by actinometry using a ferricoxalate system. The exposure

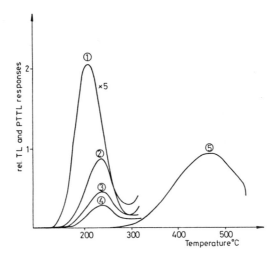

FIG.2. TL and PTTL responses of Al_2O_3 after 10^2 Gy exposure.

rate for 254 nm was 2×10^{15} photon·cm^{-2}·s^{-1}, whereas for 366 nm it was 4.8×10^{15} photon·cm^{-2}·s^{-1}. Irradiation before photostimulation was carried out by a ^{60}Co source. The absorbed doses were checked by chlorobenzene dosimeters [6].

3. RESULTS AND DISCUSSION

The gamma-radiation-exposed Al_2O_3 exhibits two TL glow peaks: one at 250°C, the other at 470°C. The latter can be observed only above 100 Gy, but with increasing gamma exposure its growth is quicker than that of the 250°C peak (see Fig.1).

The UV radiation-induced glow curve also shows a peak at 250°C; no higher peaks were observed because of the thermal background of the TL reader. The TL response of the dosimeters after 1 min illumination at 254 nm UV light (without previous gamma exposure) corresponds to 1 Gy gamma exposure. The UV exposure time of 1 min was found to be the optimum for re-estimation.

Figure 2 shows the TL and PTTL responses of the Al_2O_3 detector after 10^2 Gy exposure. After readout of the dosimetry peak (see curve 1 in Fig.2) the detectors were exposed to UV light. Curves 2, 3, and 4 in Fig.2 represent the phototransfer responses which were read out up to 300°C after successive UV irradiations. The detector can be re-evaluated several times, although as a result of successive UV irradiations the amount of information obtained decreases. Curve 5 in Fig.2 shows the residual TL response of the 470°C peak. To investigate the PTTL after very high gamma doses, six Al_2O_3 detectors were irradiated in the 100 to 10^5 Gy

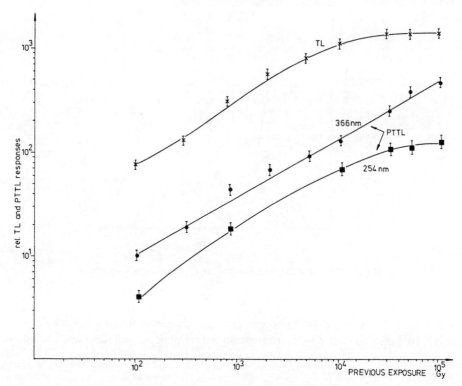

FIG.3. *TL and PTTL responses as a function of previous gamma doses.*

dose range. After each irradiation the detectors were read out up to 300°C. The readout cycle was repeated to check whether the traps corresponding to the peak at 250°C had been completely emptied. The detectors were then illuminated by 254 and 366 nm UV light separately and the PTTL responses were read out up to 300°C.

In Fig.3 the mean values of the TL and PTTL responses can be seen for 254 and 366 nm as a function of the previous gamma doses. Before every irradiation the detectors were heated to 600°C for 2 hours. In Fig.3, curve 'TL' represents the TL response of the Al_2O_3 obtained by the usual TL technique. This curve shows saturation at 5×10^3 Gy doses. We obtained similar saturation after UV exposures with a 254 nm wavelength, whereas with a wavelength of 366 nm no saturation was observed up to 10^5 Gy.

4. SUMMARY

The work described here indicates that the PTTL technique may be utilized for dose re-estimation of Al_2O_3 TL detectors; the threshold of re-estimation was found to be 1 Gy due to the UV sensitivity of the dosimeters. Phototransfer with 366 nm UV exposures not only enables re-estimation but also increases the measuring range by nearly 2 orders of magnitude.

REFERENCES

[1] MASON, E.W., Phys. Med. Biol. **15** (1970) 79.
[2] McKINLAY, A.F., BARTLETT, D.T., SMITH, P.A., Nucl. Instrum. Methods **175** (1980) 57.
[3] CALDAS, L.V.E., MAYHUGH, M.R., Health Phys. **31** (1976) 451.
[4] MEHTA, S.K., SENGUPTA, S., Phys. Med. Biol. **23** (1978) 471.
[5] OSVAY, M., BIRO, T., Nucl. Instrum. Methods **175** (1980) 60.
[6] KOVÁCS, A., STENGER, V., FÖLDIÁK, G., in Proc. 7th Int. Cong. Radiation Research, Vol. E, Nijhoff, Amsterdam (1983) 2.

DOSE STANDARDIZATION AND CALIBRATION, PHYSICAL ASPECTS

(Session 3)

Chairmen

A. MILLER
Denmark

A. KOVÁCS
Hungary

SOME PARAMETERS AFFECTING THE RADIATION RESPONSE AND POST-IRRADIATION STABILITY OF RED 4034 PERSPEX DOSIMETERS

B. WHITTAKER, M.F. WATTS,
S. MELLOR*, M. HENEGHAN*
UKAEA Atomic Energy Research Establishment,
Harwell, Didcot, Oxfordshire,
United Kingdom

Abstract

SOME PARAMETERS AFFECTING THE RADIATION RESPONSE AND POST-IRRADIATION STABILITY OF RED 4034 PERSPEX DOSIMETERS.
  Red 4034 Perspex in the form of 30 × 11 mm dosimeters is a radiation-sensitive dyed acrylic material used in gamma radiation processing dosimetry. Of the factors which could influence the radiation response characteristics of the material, absorbed water has been shown to be of primary importance. To optimize stability of response, water concentration requires careful adjustment. To maintain the water concentration the dosimeters are sealed in individual sachets before use. Other factors which could have a minor influence on the use of red 4034 dosimeters in industrial plant conditions, such as dose rates and temperature during irradiation, are discussed.

1. INTRODUCTION

All gamma radiation dosimetry techniques used for practical measurements in the 1 to 50 kGy radiation-processing dose range depend on the measurement of some chemical or physical quantity which relates indirectly to absorbed dose in water-equivalent material. For an ideal technique the quantity measured would depend on absorbed dose only and no other factor, an ideal which is difficult to achieve in practice but can be considered to be one of the ultimate goals of all development work on new radiation-sensitive systems for dosimetry. Practical methods in current usage depend for their continuing success on the following:

(i) Correct initial choice of a system in which the influence of factors other than dose is small.

---

* Visiting students from the University of Salford, Salford, Greater Manchester, United Kingdom.

(ii) Continuing research to define and quantify these factors with the aims of defining limitations of the system in practical applications and enabling small corrections to be made if necessary.

(iii) Regular intercomparison work involving national standards laboratories, giving traceabililty of the system to national standards of absorbed dose[1].

Red 4034 Perspex (formerly called red 400 Perspex) is a radiation-sensitive dyed acrylic material first applied to radiation-processing dosimetry in the U.K. prototype cobalt-60 irradiation plant (PIP) in 1960[2]. PIP and its back-up systems, including dosimetry, were developed by the Harwell organization to promote radiation-processing as a new industry[3]. The industry is now well established and expanding. This has created a growing requirement for red 4034 dosimeters, and Harwell continues to operate a service dedicated to their manufacture, calibration and supply.

On irradiation to doses above 1 kGy red 4034 Perspex progressively darkens owing to the formation of an optical absorption band peaking at 615 nanometers (nm) and extending from the dye-absorption edge (occurring at 600 nm) to beyond 700 nm. At wavelengths in the region of 630 to 650 nm post-irradiation changes in absorbance in properly prepared material are very small. In use, 640 nm is commonly used as the readout wavelength[4].

Red 4034 is supplied in the form of 30 x 11 x 3 mm rectangles pre-packed in individual sealed sachets. The purposes of these sachets are firstly to prevent scratching or other surface damage which could affect the accuracy of spectrophotometric readout and secondly to help maintain an optimized level of water-concentration gained in the dosimeters during production. In use, the dosimeters are irradiated in their sealed sachets and only removed from the sachets immediately prior to readout.

Each irradiated red 4034 dosimeter is measured on a good quality spectrophotometer having a bandwidth of 2 nm or less at the selected readout wavelength (nominally 640 nm), preferably against air as a reference. The practice of reading radiation-induced absorbance directly by measuring against an unirradiated control[4] is being phased out because different laboratories could use slightly different controls. Air-reference readout has become the agreed practice in Harwell/National Physical Laboratory (NPL) inter-comparison work[5]. Specific absorbance (640 nm), i.e. absorbance per unit thickness, is the dose-dependent quantity,

obtained by dividing absorbance by dosimeter thickness. Part
of the quality control used in the manufacture of the dosimeters is aimed at achieving a specific absorbance readout
self-consistency at 25 kGy dose for a given batch of dosimeters of better than ± 2%. For dose readout a calibration
curve or tables relating specific absorbance to cobalt-60
gamma dose is used. Specific absorbance/dose data are
supplied by Harwell but the users are encouraged to establish
their own curves by reading sets of dosimeters irradiated to
accurately known doses as this removes systematic errors due
to spectrophotometry and other causes. The Harwell data are
based on irradiations in a standardized 8 kilocurie cobalt-60
facility specially designed for this purpose and on cross-
calibrations in the 10 to 30 kGy range*. Calibration-curves
supplied cover the useful dose range of the system, namely
5 kGy to 50 kGy [1].

2. FACTORS AFFECTING RESPONSE AND STABILITY

Red 4034 (400) Perspex was initially chosen as a result
of an in-depth study of radiation-sensitive materials
available in the late 1950s because of its relative insensitivity to factors other than dose[2]. Most of the factors
influencing dose readout were identified in the early work
and have since been studied in some detail. Also,
manufacturing technique and quality control have been
improved with the aim of producing a highly consistent
material. At the present time the main factors known to have
an influence on radiation-response and pre- and post-
irradiation stability are as follows.

2.1 Batch Number

The sheet material from which dosimeters are made is
produced on a batch basis. The dye-formulation is accurately
reproducible, therefore general characteristics such as
absorption spectra and sensitivity to water and air are
reproduced from batch to batch. Sensitivity to radiation
however depends on dye concentrations and these for technical
reasons are less easily reproduced [6]. However, dye
concentrations and the general composition of the material
have been more carefully controlled in recent years and this
has resulted in a much more consistent product [1]. To
obtain accurate dose readout each batch is calibrated
separately.

---

*1 Ci = 3.70 x 10^{10} Bq.

2.2 Water Concentration

Polymethyl methacrylate, marketed in the U.K. under the trade name of Perspex, is slightly hygroscopic and can in fact absorb up to 2% by weight (2% w/w) of water if exposed to humid air. Absorbed water concentration was identified as a parameter which could influence spectrophotometric response, and hence readout dose in the early work on red Perspex. No attempt was made to quantify these changes in terms of water concentration, but a conditioning process was used in practice which produced dosimeters with a water content of about 1% w/w with good readout reproducibiliy and insensitivity to external atmospheric conditions, provided that they were sealed in individual sachets [2]. A major problem in the early work was that water concentrations in Perspex could not be accurately measured. An infrared analysis technique was used but the accuracy depended on standards with known concentrations which could not be readily made. Following an increasing usage of red 4034 dosimeters in the industry I.G. Jones at Harwell investigated water analysis techniques and presented a viable method depending on weight changes in 1977[7]. Olejnik affirmed the importance of water concentration in a 1979 paper in which it was shown that in the case of three early batches of red 4034 dosimeters differences between local and manufacturers response curves could be linked to this factor[8]. Further studies were reported by Barrett *et al.* in 1981 concluding that the water concentration achieved in manufacture was non-ideal and that a higher value (greater than 0.4% w/w) would be preferable[9].

Recent work at Harwell has been directed to the examination of water concentration as a parameter in a series of experiments. These were aimed at accurately defining the relationship between specific absorbance (640 nm) and water concentration, for water concentrations from 0.1 to 1.4% w/w and fixed dose levels from 15 to 35 kGy in steps of 5 kGy. The relationship between post-irradiation changes (640 nm) and water concentration was also studied over 15 day intervals. Because of the large number of spectrophotometric and water concentration measurements needed to accurately define these relationships, only one batch of red 4034 dosimeters, namely batch W, has been exhaustively studied.

Irrespective of dose, specific absorbance fell rapidly with increasing water concentration in the 0.1 to 0.4% w/w range. From 0.4 to 0.6% water concentration, specific absorbances fell marginally, by approximately 2%. At all five doses studied a virtually water concentration

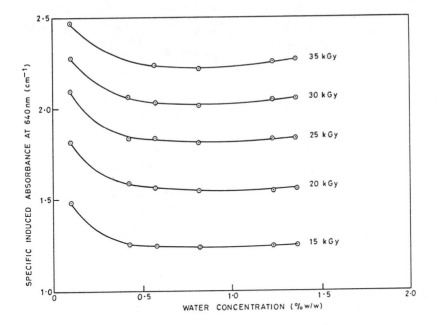

FIG.1. Response versus water concentration (red 4034 W).

independent region was observed extending from 0.6 to 1.0% w/w water. Above 1.0% w/w small increases in specific absorbances were seen (Fig. 1). Post-irradiation changes were at a minimum, within ± 2% in 15 days, for water concentrations in the range of 0.7 to 1.1% w/w. The 25 kGy post-irradiation/water concentration characteristics are shown in Fig. 2. Similar characteristics were found at the other doses studied.

The absorption of water by Perspex is reversible, i.e. exposed red 4034 dosimeters absorb or desorb water depending on atmospheric conditions of temperature and humidity. In temperate climates (for example the U.K.) serious dehydration of these dosimeters (to below 0.4% w/w) is improbable but clearly reliable international application must depend on some form of containment for the optimized water concentration imposed during production if the calibration curves supplied by Harwell are to be used. This containment is provided by sealed sachets which effectively form a part of the dosimetry system. To achieve good readout accuracy it is important that the dosimeters be kept in these sealed packs both before and during irradiation.

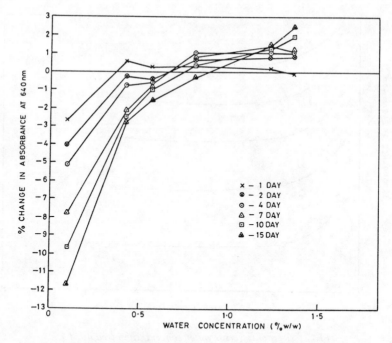

FIG.2. *Fading characteristics (red 4034 W); dose 25 kGy.*

2.3 Dose Rate

Early investigations with a pulsed beam of 4 MeV electrons showed no significant dose rate effect up to 10^5 Gy.s^{-1} [4]. Low dose rates were not investigated because it was thought that, based on radiochemical theory, significant effects would only appear at high rates, due to radical recombination reactions.

Chu and Antoniades reported an apparent low dose rate effect in 1975 [10]. In the 1983 laboratory intercomparison studies involving Harwell and NPL a small consistent disparity in data, about 2%, was attributed to a dose rate effect[1]. Recently dose rate effects have been examined with batch W dosimeters at Harwell over a range 0.14 to 3.25 Gy.s^{-1} with the aim of properly quantifying the response/dose rate relationship.

Because the effect to be examined was very small, of the same order as possible post-irradiation changes, the experiment was carefully designed to isolate dose rate as a parameter. Post-irradiation changes being wavelength-dependent the philosophy used was that if the effect could be duplicated at several

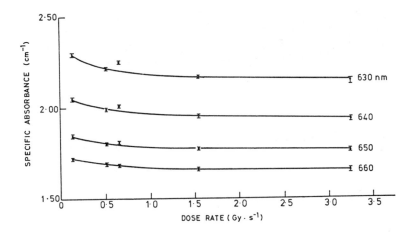

FIG.3. *Specific absorbance versus dose rate for readings taken close to end of irradiation (error bars represent one standard error).*

different readout wavelengths then it could be assigned to a true dose rate effect. As an extension of this approach the irradiated dosimeters were measured at intervals up to 20-25 days after irradiation.

The dose rates available were 3.252 (5.3.84), 1.547 (6.3.84), 0.662 (14.3.84), 0.525 (19.3.84) and 0.142 (6.3.84) $Gy.s^{-1}$, dates of measurement by means of dichromate dosimeters supplied by NPL being given in brackets. The dose rates were cross-checked by local Fricke dosimetry at Harwell with agreement within ± 1%. In the red 4034 dosimeter irradiations these rates were first corrected for the decay of ^{60}Co ($T_{\frac{1}{2}}$ 5.27 years). Sixteen red 4034 dosimeters were irradiated to 25 kGy at each dose rate. These were measured at 630, 640, 650 and 660 nm as soon as possible after irradiation (within 15 hours) and reading was repeated at intervals to define the post-irradiation effects. For readings taken shortly after irradiation a general trend was observed at four wavelengths from 630 to 660 nm, response increasing marginally with decreasing dose rate (Fig. 3). The 0.662 $Gy.s^{-1}$ data points were however anomalous for 630-650 nm readout. This anomaly was attributed to differential post-irradiation effects and the probability that the processes causing these effects occur during as well as after irradiation. Accordingly all the dosimeters were re-read approximately 48 hours after the middle of the respective irradiation periods. The resulting curves showed a wavelength-independent relationship between specific absorbance and dose rate with no anomalous points (Fig. 4). Based

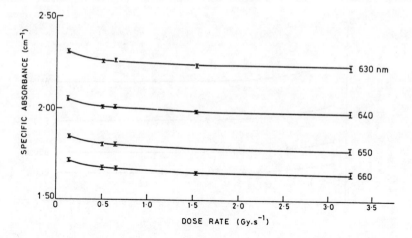

FIG.4. Red 4034 W specific absorbance versus dose rate for readings taken between 37 to 46 hours after midpoint of irradiation time.

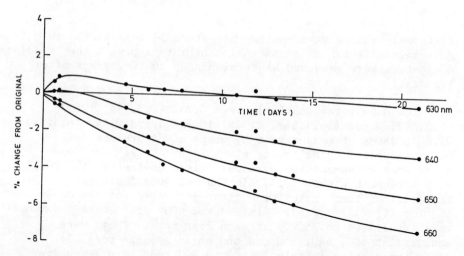

FIG.5. Red 4034 W post-irradiation changes versus wavelength; dose rate 0.142 $Gy \cdot s^{-1}$.

on these curves, no significant rate effect was observed between 1.55 and 3.25 $Gy.s^{-1}$.

The 20-25 day post-irradiation studies unexpectedly showed a regular trend indicating that the readout wavelength for minimum post-irradiation changes was also slightly dose-rate-dependant. At the highest rate used 640 nm readout was found to be acceptable, specific absorbance changes falling within ± 2% during 25 days, whereas at the lowest rate 630 nm readout was a better choice, with ± 1% in 21 days (Figs. 5-9).

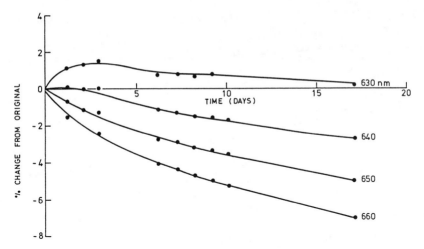

FIG.6. *Red 4034 W post-irradiation changes versus wavelength; dose rate 0.525 Gy·s^{-1}.*

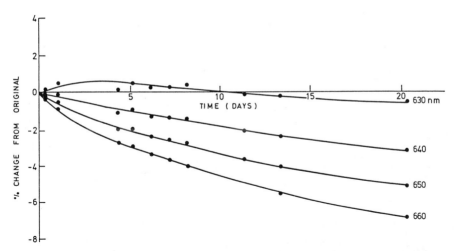

FIG.7. *Red 4034 W post-irradiation changes versus wavelength; dose rate 0.662 Gy·s^{-1}.*

2.4 Temperature during Irradiation

In the original work the response of red Perspex was found to be independent of irradiation temperature from 0 to 30°C [2]. Miller *et al.* and Miller and McLaughlin have reported temperature studies on Red 4034 dosimeters over a range of -80 to +80°C finding little dependence (less than 2% change in response) from -20 to +40°C but significant increase in sensitivity below -30 and above +50°C [11,12].

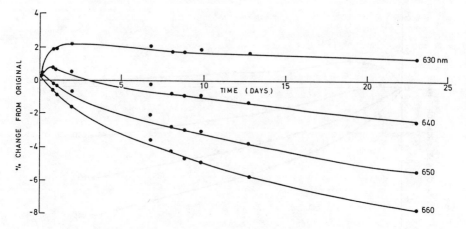

FIG.8. Red 4034 W post-irradiation changes versus wavelength; dose rate 1.55 Gy·s^{-1}.

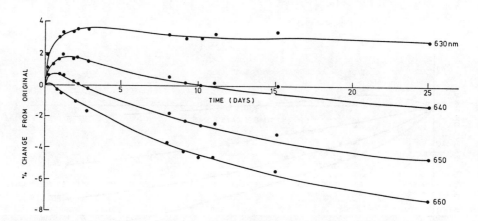

FIG.9. Red 4034 W post-irradiation changes versus wavelength; dose rate 3.25 Gy·s^{-1}.

Using a different batch of dosimeters (H) but a narrower temperature range Barrett et al. found only 1% increase in sensitivity from +25 to +43°C but 4% increase at +50°C[13].

2.5 Elapsed Time before Readout

At the time of presenting this paper the dose rate used at Harwell for calibrating Perspex dosimeters is 1.40 Gy.s^{-1}. Irradiations are carried out using an automatic timing system set for five accurately known dose levels in the 5 to 50 kGy range[1]. The irradiations are carried out overnight and the dosimeters read the following morning, i.e. within 20 hours from the end of the low dose irradiations. This exercise is carried out twice for each calibration of a batch of dosimeters to provide a 10 point response curve, each point representing the mean specific absorbance (640 nm) of four dosimeters.

In the practical application of these dosimeters the best dosimetric accuracy is to be expected for readings made shortly after irradiation, i.e. simulating to some extent what is done in the calibration. This is not always possible in practice, particularly in the case of exchange and intercomparison of dosimeters irradiated at different establishments. This raises the question of possible dose readout errors as a function of time lapse before readout.

The time-dependence of specific absorbance is non-linear and depends on water concentration, dose, dose rate and readout wavelength (Figs. 2, 5-9). Post-irradiation storage temperature is also known to be important[13] but for practical purposes normal temperature storage (15-25°C) as used in the present studies need only be considered. In view of the evident complexity of post-irradiation processes it is unlikely that fading corrections are of much practical value and it is perhaps better to define the maximum limits of dosimetric inaccuracy expected within a given elapsed time. Readout within two or three days of irradiation is not necessarily preferable to readout within a longer interval, say 10 days, as shown by the 640 nm curve in Fig. 9. The present results (for batch W) indicate that at 640 nm changes in absorbance fell within $\pm$ 2% in 10 days irrespective of dose rate (0.14 to 3.25 $Gy.s^{-1}$) and dose (15 to 35 kGy).

3. IRRADIATION PLANT DOSIMETRY

Miller *et al.* pointed out that the irradiation conditions used in practice can be quite different from those used in calibration and that this could be a source of dosimetric inaccuracy[11]. In most irradiation plants the dosimeters are exposed to changing dose rates and temperatures, whereas in calibration steady conditions are normally used. However, variable dose rate conditions have been used for calibration in a standardized irradiation facility at Risø[11,12].

The present work suggests a slight enhancement of response at the start of plant irradiation cycles when the dose rates are low. In most plants a large part of the integral dose is delivered at medium to high dose rates so the overall effect of the dose rate dependence could be very small. Miller et al. using the Risø facility found no significant change in the red 4034 specific absorbance dose curve when variable dose rates were used[11].

Information on the temperatures encountered in radiation processing is sparse; however, a temperature rise from 18 to 43°C in the centre of a product box has been reported as typical[12]. Within this range red 4034 dosimeters have, in fact, been shown to be less temperature sensitive than some commonly used 'reference' systems (e.g. ceric, dichromate, alanine).

4. EXPERIMENTAL DETAILS

All irradiations were carried out in the 8 kCi standardized Co-60 facility at Harwell[1]. For the dose rate studies some irradiations were done in a rotating aluminium can mounted close to one of the four source positions, dose rates being altered by changing the number of sources exposed. Uniformity of irradiation within this can was better than ± 1% and dose reproducibility better than ± 0.1%.

Spectrophotometric measurements were made using a Pye-Unicam 8800 instrument with a bandwidth setting of 1 nm. Accuracy of wavelength selection and absorbance readout were monitored with a built in microprocessor-controlled test programme. Absorbance readout accuracy was also checked using a set of standardized neutral density filters calibrated by NPL.

To provide sets of dosimeters with known water concentrations 1000 dosimeters were first dried to 0.1% w/w H_2O by heating to 40°C for 10 days. After removing a sample of 'dry' dosimeters the remaining dosimeters were exposed to humid air (approx. 100% R.H.) at ambient temperatures in a specially made humidity chamber. Samples were removed at intervals and their water concentrations determined by the weighing technique[7]. All samples were stored in sealed aluminium cans to maintain the water concentrations imposed. For the irradiation trials dosimeters were immediately sealed in individual sachets after removing them from the sample cans to avoid changes in water concentration before irradiation.

5. CONCLUSIONS

The dosimetric accuracy attainable with red 4034 dosimeters can be affected by dose rate and irradiation temperature, but for most practical purposes these factors are considered to be unimportant.

Post-irradiation changes in specific absorbance (batch W) fall within ± 2% over 10 days provided that the irradiated dosimeters are stored at normal ambient temperatures (15-25°C).

ACKNOWLEDGEMENT

Dichromate reference dosimeters were supplied and read by Dr. P. G. Sharpe (NPL). His contribution to this work is gratefully acknowledged.

REFERENCES

[1] GLOVER, K. M., KING, M., WATTS, M. F. Paper IAEA-SM-272/6, these Proceedings.
[2] WHITTAKER, B. "Radiation Dosimetry Technique using Commercial Red Perspex", U.K.A.E.A. Rep. AERE R 3360 (1964).
[3] JEFFERSON, S., LEY, F. J., ROGERS, R. "Radiation Sterilisation of Medical Supplies", Nucl. Eng. $\underline{10}$ (1964).
[4] WHITTAKER, B. "Radiation Dosimetry Manual". Eds. HOLM, N.W., BERRY, R. J., MARCEL DEKKER, New York (1970) 363.
[5] SHARPE, P. G., WHITTAKER, B. Private Communications (1983).
[6] RAWCLIFFE, J. Private Communication.
[7] JONES, I. G. "Water Content of Red Perspex Dosimeters" U.K.A.E.A. Rep. AERE M2913 (1977).
[8] OLEJNIK, T. A. "Red 4034 Perspex Dosimeters in Industrial Radiation Sterilisation Process Control", Radiat. Phys. Chem. $\underline{14}$ (1979), 431.
[9] BARRETT, J. H., SHARPE, P. H. G., STUART, I.P. U.K. National Physical Laboratory Rep. RS 52 (1981).
[10] CHU, R. D. H., ANTONIADES, M. T. "Use of Ceric Sulphate and Perspex Dosimeters for the Calibration of Irradiation Facilities". IAEA Report SM-192/14 (1975).
[11] MILLER, A., BJERGBAKKE, E., McLAUGHLIN, W. L., Int. J. Appl. Radiat. Isot. $\underline{26}$ (1975), 611.
[12] MILLER, A., McLAUGHLIN, W. L. "High Dose Measurements in Industrial Radiation Processing", IAEA Technical Rep. No. 205 (1981) 119.
[13] BARRETT, J. H., SHARPE, P. H. G., STUART, I.P., U.K. National Physical Laboratory Rep. RS49 (1980).

IAEA-SM-272/37

DOSIMETRY METHODS APPLIED TO IRRADIATION WITH TESLA-4 MeV LINEAR ELECTRON ACCELERATORS

I. JANOVSKÝ
Nuclear Research Institute,
Řež, Czechoslovakia

Abstract

DOSIMETRY METHODS APPLIED TO IRRADIATION WITH TESLA-4 MeV LINEAR ELECTRON ACCELERATORS.

The calorimetric calibration technique was applied for dose measurements and calibration of routine dosimeters in electron beams of two Tesla-4 MeV linear electron accelerators. A simple graphite differential calorimeter was used in the beam of the 0.1 kW accelerator and a water calorimeter in the beam of the 1 kW accelerator. The experience gained with several dosimeters, both thin films (radiochromic, CTA and PVC) and liquid systems (ethanol chlorobenzene, dichromate and the 'Super Fricke' dosimeter), is reported.

1. INTRODUCTION

The Radiation Chemistry Department of the Nuclear Research Institute, Řež, is equipped with two Tesla-4 MeV linear high-frequency electron accelerators with an output power of 0.1 and 1 kW, respectively.[1] The main characteristics of both accelerators are given in Table I.

The 0.1 kW accelerator was used for pulse radiolysis and irradiation purposes [1], while the 1 kW accelerator is designed solely for irradiation [2]. The irradiated materials comprise mainly polymers, electronic components, textiles, sludges, etc.

For the purpose of routine dosimetry the calorimetric calibration technique was applied and several 'high-level' dosimeters, both thin films and liquid chemical systems, were examined and used.

2. IRRADIATION AND CALIBRATION TECHNIQUE

Calibration of thin film dosimeters in the beam of the 0.1 kW accelerator was performed by using a simple graphite differential calorimeter [3]. It consists of two separate graphite bodies (discs of 30 mm diameter and 4 mm thickness)

---

[1] The accelerators were built in the Tesla Research Institute for Vacuum Electronics, Prague.

TABLE I. MAIN CHARACTERISTICS OF TESLA-4 MeV LINEAR ELECTRON ACCELERATORS

Maximum average output power (kW)	0.1	1
Mean electron energy (MeV)	4	4
Maximum average electron current (μA)	30	300
Maximum pulse electron current (A)	0.2	0.2
Pulse length (μs)	2.5 (also 0.08)	2.5
Pulse repetition frequency (Hz)	50 (also 25, 12.5 and 6.25)	500
Scanner frequency (Hz)	1	2
Maximum scan width (cm)	25	50
Maximum pulse dose rate on the conveyor (Gy/s)	$\sim 10^7$	$\sim 10^7$
Surface dose at maximum scan width, maximum beam current and conveyor speed of 30 cm/min (kGy)	~ 2.5	~ 20

into which thermocouples are embedded, and is thermally insulated by a foam polystyrene. During the passage of the calorimeter on the conveyor through the scanned electron beam only one body is irradiated, while the other is shielded, and the temperature difference time profile is recorded. From the calorimetric curve the 'adiabatic' temperature increase is obtained by the conventional linear extrapolation method. No calibration heaters are used and the average dose absorbed in the graphite body is calculated by using the specific heat of graphite. The calorimeter is convenient for measurement of absorbed doses in the range of about 1 to 7 kGy. The reproducibility estimated from 30 measurements performed under identical conditions at a dose of 3 kGy was ±2%.

Measurement of the depth-dose distribution showed that the local dose in the half-depth of graphite body is practically identical with the average dose within the body. This is the basis of the calibration method when the dosimetric film is irradiated between two layers of the same material as that of the calorimetric body but having half the thickness [4]. The dose absorbed in the film is then calculated directly from the calorimeter dose by applying stopping power correction.

In the beam of the 1 kW accelerator a water calorimeter similar to that developed at the Risø National Laboratory, Roskilde, Denmark [5] was used. The calorimetric body formed by a water-filled polystyrene Petri dish (total thickness corresponding to 1.1 g/cm^2, diameter 9 cm)[2] into which a calibrated

---

[2] The smaller thickness and diameter were chosen with respect to the lower electron energy and the possibility of using a narrower beam scan, respectively.

glass-encapsulated bead thermistor is embedded is insulated by a foam polystyrene envelope. In this case the temperature is measured only before and after irradiation; the absorbed dose is obtained from the temperature difference and calorimeter heat capacity. For the given type, the heat capacity has a value of about $3.4 \text{ J} \cdot \text{g}^{-1} \cdot {}^\circ\text{C}^{-1}$, as calculated from the masses and specific heats of individual components. The calorimeter is used for measurements of absorbed doses of 5 to 70 kGy. Measurements performed with several calorimeters irradiated in close succession were reproducible within 2 to 3%. When the conveyor speed was varied, the measured dose was usually within ±3%, proportional to the reciprocal of the speed. The depth-dose distribution within the calorimetric body was determined using a plastic (nylon) model and thin film dosimeters. Graphic integration of the depth-dose curve yielded the value of the ratio surface dose to average dose within the body of 0.80 to 0.85. This procedure is repeated whenever the thin film dosimeters are calibrated and the actual value is used. The equivalence of the plastic model and calorimetric body was checked by comparing the ratios of surface to bottom dose; within 2% the same ratio value was obtained.

Thin film dosimeters to be calibrated are placed directly on top of the polystyrene Petri dish in the calorimeter and their response is related to the surface dose. Liquid chemical dosimeters are irradiated under conditions closely matching those of the calorimeter, i.e. in a Petri dish inside the foam polystyrene envelope. Both the dosimeter solution and calorimeter are irradiated in close succession and the chemical change in the dosimeter is related to the average dose determined by the calorimeter. In some cases the radiochromic film serving as a 'monitor' is placed on top of both the calorimeter and the dosimeter Petri dish.

The effects of temperature on the thin film dosimeter response were investigated in the beam of the 0.1 kW accelerator using a simple thermostat.[3]

To establish the dose rate effects on dosimeter response, calibrations in a laboratory panoramic ^{60}Co gamma source against the Fricke dosimeter were also carried out.

3. THIN FILM DOSIMETERS

3.1. Radiochromic film

Commercially available radiochromic film FWT-60-20 (from Far West Technology, Inc., Goleta, USA) was used to check the reliability of both calorimetric calibration techniques. The same batch of film was always calibrated simultaneously in the electron beam and the gamma source. Absorbance of the irradiated film was measured at 600 or 604 nm (for electron radiation about 6 hours after irradiation), which was sufficient for full colour development [6]. The results

---

[3] The thermostat was constructed by R. Štětka.

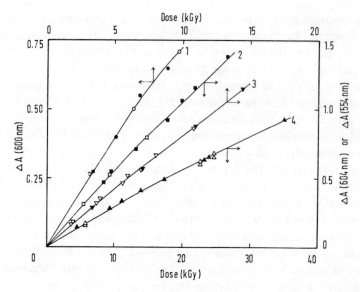

FIG.1. Response curves for radiochromic films: (1) FWT-60: (○) gamma, 0.8 kGy/h; (●) 1 kW accelerator, beam current 50 µA. (2) PVB — pararosaniline cyanide: (□) gamma, 0.7 kGy/h; (■) 1 kW accelerator, beam current 100 µA. (3) FWT-60: (▽) gamma, 1.3 kGy/h; (▼) 0.1 kW accelerator, 3.5 kGy fractions. (4) CTA — rosaniline cyanide: (△) gamma, 1 kGy/h; (▲) 0.1 kW accelerator, 2–3 kGy fractions.

shown in Fig.1 indicate an identical response curve in both cases (scatter within ±3%), which is in accordance with the dose rate independence of this dosimeter.

Two other types of radiochromic films, namely CTA containing rosaniline cyanide[4] and PVB containing pararosaniline cyanide similar to those developed at Risø [5, 7]), were prepared and tested. The radiochromic components were chosen with respect to their relatively low light sensitivity. Both irradiated films showed a similar optical absorption (maximum at ~554 nm); however, their post-irradiation behaviour was different. For the electron-irradiated CTA radiochromic film the final colour developed within a few hours, while for the PVB film both after electron and gamma irradiation some additional absorbance formation continued for several days. To accelerate this process the recommended heat treatment (5 min at 60°C) [5, 6] was applied for PVB films. The response curves for both films are shown in Fig.1; the dose rate independence is again obvious. The dependence of the response of both radiochromic films on temperature is shown in Fig.2. For the CTA film, the temperature coefficient has average values of about ±0.5%/°C and +1%/°C in the regions of 20 to 40°C and 40 to 80°C, respectively. For the PVB film the effect of temperature below 45°C

---

[4] Prepared by S. Kudrna from the Východočeské chemiské závochy, Synthesia np, Pardubice-Semtín, Czechoslovakia.

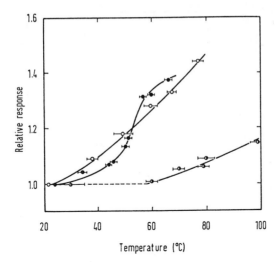

FIG.2. Effect of temperature during electron irradiation on the response of (○) CTA radiochromic film at a dose of ~ 7.5 kGy; (●) PVB radiochromic film at a dose of ~ 5 kGy; (◐) CTA film (Numelec) at a dose of ~ 50 kGy.

is smaller, but becomes more prominent around ~50°C; in these experiments post-irradiation heat treatment (25 min at 60°C) was again performed. The temperature effects observed here are more significant than those reported for PVB and nylon films containing hexa(hydroxyethyl) pararosaniline cyanide [8].

3.2. CTA film

The CTA dosimetric film (supplied by Société Numelec, Versailles, France) was evaluated by measuring the radiation-induced absorbance at 280 nm. Owing to a post-effect (slow additional absorbance increase), the measurement was performed at a fixed time (usually about 4 hours) after electron irradiation. The calibration graphs for gamma and electron radiation are shown in Fig.3. The close agreement of response to gamma and electron irradiations with the 0.1 kW accelerator when doses were cumulated in fractions of 3.3 kGy indicated dose rate independence [9]. However, the response to irradiation with the 1 kW accelerator when doses were delivered in a single pass through the beam is about 18% lower. When the beam current (i.e. pulse dose rate) was changed by a factor of 3, no apparent response change occurred. When repeating the calibration by applying about five different doses, the calibration constant (absorbance per unit dose) was reproducible within 4%. The effect of temperature is shown in Fig.2. The response is influenced markedly only at temperatures above 60°C; in the region of 60 to 100°C it increases by about 0.5%/°C [9].

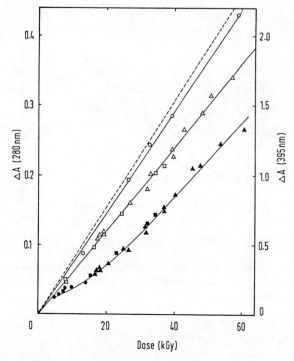

FIG.3. Calibration graphs for CTA dosimetric film (left-hand co-ordinate): (- - -) gamma, 1 kGy/h; (○) 0.1 kW accelerator, 3.3 kGy fractions; (□, △) 1 kW accelerator, 'single' doses at beam currents of 100 and 300 μA, respectively. Calibration graph for PVC film (right-hand co-ordinate): (●, ■, ▲) 1 kW accelerator, 'single' doses at beam currents of 50, 100 and 300 μA, respectively.

A brief comparison with a new CTA film from Japan (Fuji Photo Film Co.) showed that in the dose range of 10 to 70 kGy the response for both materials is, on average, within 3%.

Our observations on post-effect and temperature effect are essentially in accord with recent descriptions of dosimeter properties by French and Japanese authors [10, 11], but the dose rate effect seems to be more complex. The response to electron irradiation is apparently also influenced by dose fractionation, as has been reported earlier [12].

3.3. PVC film

Samples of 0.25 mm thick PVC film (type 37–0015 from the Kunststoffwerke GmbH, Staufen, Federal Republic of Germany) were subjected to conventional heat treatment (30 min at 70°C) immediately after electron irradiation and the optical absorbance was measured at 395 nm. When investigating response

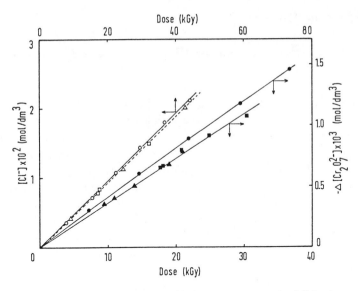

FIG.4. Formation of Cl^- in the ethanol-chlorobenzene dosimeter: (○, full line) gamma, 7 kGy/h; (△, □, dashed line) 1 kW accelerator, beam currents of 100 and 200 µA, respectively. Reduction of dichromate: (●) gamma, 7 kGy/h; (▲, ■) 1 kW accelerator, beam currents of 100 and 200 µA, respectively.

stability some fading was experienced; this was most important during the first day. Its extent decreased with increasing dose; below 10 kGy it amounted to 25 to 30%, in the range of 10 to 20 kGy, on average, to ~15%. For this reason routine readings were performed again at a fixed time (usually about 4 hours) after film heating and reasonable reproducibility was achieved. The calibration graph in Fig.3 shows some non-linearity or an 'induction period'. Similar results were obtained with another material used (Genolon from Kalle, Wiesbaden, Federal Republic of Germany), only the response was somewhat higher.

4. LIQUID DOSIMETERS

4.1. Ethanol chlorobenzene dosimeter

The dosimeter used contained 24% chlorobenzene, 4% water and 0.04% acetone in ethanol. The concentration of chloride after irradiation was determined by mercurimetric titration [13]. Samples were irradiated in the presence of air. The results are shown in Fig.4. The corresponding average yields $G(Cl^-) = 5.33 \pm 0.03$ and $G(Cl^-) = 5.19 \pm 0.10$ for gamma and electron irradiations, respectively, are close (within 3 and 2%) to the values determined by the authors of this dosimeter [14]; they also stated the dose rate independence of the system.

4.2. Dichromate dosimeter

The dosimeter solution contained 2×10^{-3} mol/dm^3 $K_2Cr_2O_7$ and 0.5×10^{-3} mol/dm^3 $Ag_2Cr_2O_7$ in 0.1 mol/dm^3 $HClO_4$ [15]. The radiation-induced dichromate reduction was followed by measuring the optical absorbance at 440 nm. The concentration change was calculated using extinction coefficient $\epsilon_{440}(Cr_2O_7^{2-}) = 46$ m^2/mol (determined with a set of standard dichromate solutions), and correction for absorption of Cr^{3+} ion was made taking $\epsilon_{440}(Cr^{3+}) = 1$ m^2/mol [15]. All solutions were irradiated in the presence of air and the results are plotted in Fig.4. For gamma irradiation the results indicate a decrease in the reduction yield with increasing dose, or an intercept (recommended solution pre-irradiation [15] was not performed). The results obtained for electron irradiation show some scatter but, on average, the reduction yields are about 12% lower than for gamma irradiation. Despite a larger uncertainty in dose determination for electron irradiation, these results seem to indicate dose rate effect. We have previously observed a similar difference with dosimeter solution prepared using $AgNO_3$ instead of $Ag_2Cr_2O_7$ [16].[5]

4.3. 'Super Fricke' dosimeter

The dosimeter solution contained 0.01 mol/dm^3 $Fe(NH_4)_2(SO_4)_2$ in 0.4 mol/dm^3 H_2SO_4 and during or before irradiation was saturated with oxygen. Interpretation of dose was achieved spectrophotometrically taking $G(Fe^{3+}) = 16.1$ [17]. Although no direct comparison of this most useful reference dosimeter with our calorimeter could be made owing to its restricted dose range, some other confrontations were made.

The energy fluence per passage through the scanned beam of the 0.1 kW accelerator was determined by the total absorption of electrons in the dosimeter solution, as well as in a stack of PMMA slabs interleaved with the calibrated CTA dosimetric film. Graphic integration of the depth-dose curve in PMMA gave an energy fluence value which agreed within 4% with that determined using the dosimeter solution [9].

In another experiment the dosimeter solution was used to determine the energy content per pulse E_p from the 0.1 kW accelerator. The dependence of E_p on the average electron current $\bar{I}$ is shown in Fig.5 [18]. From the values $E_p(J)$, $\bar{I}(\mu A)$ and pulse repetition frequency $f(Hz)$, the mean electron energy $\bar{E}(MeV)$ can be calculated according to the equation $\bar{E} = E_p \times \bar{I}^{-1} \times f$. The calculated value decreases from 4.2 MeV at 2.5 μA to 3.8 MeV at 13 μA, which is in agreement with the measured electron energy spectra [1].

---

[5] The G-values in the paper cited are low (by about 5%), since correction for Cr^{3+} absorption was omitted.

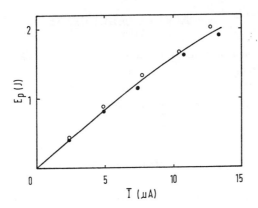

FIG.5. *Dependence of energy content per pulse from the 0.1 kW accelerator on the average beam current at a pulse repetition frequency of 25 Hz (two sets of measurements) (from Ref.[18]).*

5. CONCLUSIONS

It has to be emphasized that the calorimeters used in this work are not considered as standards. No thorough dose intercomparisons were made and the accuracy of measurements was not established. Miller [5] estimated the accuracy of the water calorimeter to be within ±5% and it is probably nearly the same for both the calorimeters used in this study. This is supported by the results obtained with radiochromic films.

From the dosimeters tested, CTA film was used most frequently for routine dose checks in the tens of kilograys range. PVC film served mainly to determine both scanned and straight beam profiles and to check irradiation field homogeneity. For the atter a digital transmission densitometer (Meodenzi TRD 4, Meopta np, Brno (Czechoslovakia), was adopted which enables fast routine measurements to be made. All films were applied to measure depth-dose distributions in various materials, both homogeneous and heterogeneous, and comparisons with some algorithms were attempted. Both radiochromic and PVC films were used to simulate wire insulation when solving the problem of homogeneous irradiation in a radiation vulcanization process; the film readings were performed by using a modified high-resolution scanning densitometer (Schnellphotometer G II, Carl Zeiss, Jena, German Democratic Republic).

Liquid dosimeters are useful to determine an average dose (especially if irradiating liquid samples, e.g. in an ampoule) when, owing to a relatively low electron energy, large dose non-homogeneity occurs and transfer from the calorimeter or thin film dosimeter is problematical.

ACKNOWLEDGEMENTS

The author wishes to thank his colleagues of the Radiation Chemistry Department for their collaboration throughout this work, S. Kudrna for preparing one of the radiochromic films and A. Miller from the Risø National Laboratory, Roskilde, Denmark, for his support.

REFERENCES

[1] TEPLÝ, J., SOJKA, B., JANOVSKÝ, I., VOCÍLKA, J., HLUBUČEK, V., VÁŇA, V., FOJTÍK, A., Nukleon 1 (1977) 3.
[2] HALOUSEK, M., TEPLÝ, J., Nukleonika 26 7–8 (1981) 797.
[3] JANOVSKÝ, I., ŠTĚTKA, R., Jad. Energ. 26 12 (1980) 444.
[4] EISEN, H., ROSENSTEIN, M., SILVERMAN, J., Int. J. Appl. Radiat. Isot. 23 3 (1972) 97.
[5] MILLER, A., "Dosimetry for Electron Beam Applications", Risø National Laboratory, Rep. Risø-M-2401 (1983).
[6] CHAPPAS, W.J., Radiat. Phys. Chem. 18 5–6 (1981) 1017.
[7] McLAUGHLIN, W.L., MILLER, A., FIDAN, S., PEJTERSEN, K., BATSBERG PEDERSEN, W., Radiat. Phys. Chem. 10 2 (1977) 119.
[8] MILLER, A., McLAUGHLIN, W.L., "Evaluation of radiochromic dye films and other plastic dose meters under radiation processing conditions", High-Dose Measurement in Industrial Radiation Processing, Technical Reports Series No.205, IAEA, Vienna (1981) 119–138.
[9] JANOVSKÝ, I., Radiochem. Radioanal. Lett. 47 4 (1981) 251.
[10] LAIZIER, J., "Techniques utilisées au Centre d'applications et de promotion des rayonnements ionisants pour la dosimétrie des irradiations gamma et sons faisceaux d'électrons", High-Dose Measurements in Industrial Radiation Processing, Technical Reports Series No.205, IAEA, Vienna (1981) 57–68.
[11] TAMURA, N., TANAKA, R., MITOMO, S., MATSUDA, K., NAGAI, S., Radiat. Phys. Chem. 18 5–6 (1981) 947.
[12] MILLER, A., BJERGBAKKE, E., McLAUGHLIN, W.L., Int. J. Appl. Radiat. Isot. 26 10 (1975) 611.
[13] DVORNIK, I., in Manual on Radiation Dosimetry (HOLM, N.W., BERRY, R.J., Eds), Marcel Dekker, New York (1970) 345.
[14] DVORNIK, I., RAŽEM, D., BARIĆ, M., "Application of the ethanol-chlorobenzene dosimeter to electron beam dosimetry: Pulsed 10 MeV electrons", Large Radiation Sources for Industrial Processes (Proc. Symp. Munich, 1969), IAEA, Vienna (1969) 613–622.
[15] SHARPE, P.H.G., BARRETT, J.H., BERKLEY, A.M., Dichromate Solution as a Reference Dosemeter for Use in Industrial Irradiation Plants, National Physical Laboratory, Rep. RS(EXT) 60 (1982).
[16] JANOVSKÝ, I., Radiochem. Radioanal. Lett. 57 4 (1983) 197.
[17] SEHESTED, K., in Manual on Radiation Dosimetry (HOLM, N.W., BERRY, R.J., Eds), Marcel Dekker, New York (1970) 313.
[18] JANOVSKÝ, I., Jad. Energ. 28 7 (1982) 249.

IAEA-SM-272/18

METHODS FOR MEASURING DOSE AND BEAM PROFILES OF PROCESSING ELECTRON ACCELERATORS

R. TANAKA, H. SUNAGA, T. AGEMATSU
Takasaki Radiation Chemistry
  Research Establishment,
Japan Atomic Energy Research Institute,
Takasaki, Gunma-ken, Japan

Abstract

METHODS FOR MEASURING DOSE AND BEAM PROFILES OF PROCESSING ELECTRON ACCELERATORS.
    Practical methods for measuring dose and beam profiles of processing electron accelerators are presented and discussed on the basis of dosimetric methods using the CTA film dosimeter and the electron current densitometer. In electron beam dosimetry, film dosimeters are the most effective means of directly obtaining information on dose profiles in irradiated material, but they are not very effective in obtaining information on the radiation field such as electric charge and energy fluence, intensity and their change with time. Measurement of electron current density provides such information. The paper is chiefly concerned with the application of both measurement systems for various purposes in electron radiation processing studies. A readout system for the CTA dosimeter providing a hollow cathode lamp which emits monochromatic light was developed for precise and simple measurement. This system simplifies the procedure of obtaining depth-dose and lateral dose profiles by automatic traces along the CTA tape. A double-scanning microspectrometer was also designed and used on a trial basis to measure two-dimensional dose distributions with the CTA dosimeter and other plastic or dye-plastic dosimeters. The apparatus allows local distributions of dose and beam intensity in small areas to be obtained, as well as dose maps in relatively large areas to ensure dose uniformity. Electron current densitometers are widely applied for measurements of quantities associated with the radiation field, dose evaluation in the absorber, calibration of dosimeters and time monitoring of electron charge fluence and electron current density in order to obtain accurate dose dependence and various dosimetric characteristics. The influence of ionic charge is described as the main source of uncertainty or error in determining electron current density. Limitation in applications of the CTA dosimeter and the electron current densitometer to an electron energy lower than 500 keV is also discussed.

1. INTRODUCTION

The recent extensive advances made in electron accelerators for processing have led to the expansion of application fields in radiation processing. The requirements in radiation measurement and monitoring have been increased and diversified in electron radiation processing, especially in research and developing stages, and radiation resistance tests. Simple handling, accuracy and wide applicability are basically required in such fields.

In processing electron accelerators, film dosimeters are the most effective means of directly obtaining information on various dose profiles in irradiated material. We have applied the CTA dosimeter [1, 2] for various requirements in the electron irradiation technique. However, film dosimeters are not very effective in obtaining information on the radiation field such as electric charge and energy fluence, intensity and their change with time. Measurement of electron current density [3, 4] provides such information and makes up for the defects of film dosimeters in electron beam dosimetry [5]. We have developed the CTA dosimeter and the electron current densitometer in the Japan Atomic Energy Research Institute (JAERI). This paper is chiefly concerned with the application of both measurement systems for various dosimetric purposes.

2. APPLICATIONS OF THE CTA DOSIMETER

2.1. Measurement of one-dimensional dose profiles [6]

The CTA dosimeter has a linear response with a dose of up to 150 kGy and CTA film is in the form of a long tape (thickness: 0.125 mm, width: 8 mm). These advantages enable the dose distribution to be traced automatically and directly along the CTA tape. The optical density change per unit dose does not depend on temperature and humidity during irradiation at dose rates higher than 1 MGy/h, which are typical for electron radiation processing.

UV spectrophotometers are commonly used as the readout system for plastic and liquid chemical dosimeters because they are easy to obtain and the readout wavelength is variable. However, they are not as simple or as convenient as the reader of a single dosimeter, since for optical measurement they have many functions which can be omitted for the dosimetry procedure, and also because they are generally not portable nor handy to use. UV spectrophotometers also have another problem when applied to the CTA dosimeter: the setting error of the 280 nm readout wavelength results in appreciable error in the dosimetry, since the radiation-induced optical density considerably changes with the wavelength around the readout wavelength.

These problems can be eliminated by using a light source which emits the monochromatic light of the readout wavelength without a monochromator. As a result of the above considerations, we adopted a hollow cathode lamp (Hamamatsu Photonics, L233-25NU) as the light source.

The hollow cathode lamp emits light with a line spectrum, depending on the cathodic material and the construction. The line spectrum usually consists of many monochromatic lights of different wavelengths and intensities. The monochromatic light of the readout wavelength for a specific dosimeter can be

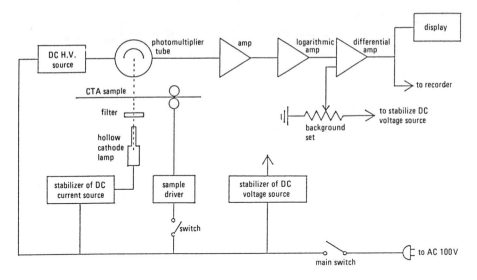

FIG.1. Diagram of the readout system for the CTA dosimeter using a manganese hollow cathode lamp.

obtained by selecting a suitable cathodic material and by filtrating extraneous light using interference filters. When manganese is used as the cathodic material, the lamp emits two strong monochromatic lights of 279.48 and 403.08 nm, among a number of different monochromatic lights. The former light is best suited for the CTA dosimeter; another light may be used for the cobalt-glass dosimeter [7].

The hollow cathode lamp requires a small power supply with a voltage of 170 V and a maximum current of 20 mA. It usually measures 39 mm in diameter (maximum) and 165 mm in length (maximum), and requires only a small storage space. The nominal lifetime of the lamp is about 500 h under maximum current.

Figure 1 shows a diagram of the compact readout system for the CTA dosimeter using the manganese hollow cathode lamp. The system was developed at JAERI and is now commercially available (Nisshin film dose reader). The light-transmitting CTA tape is detected by a photomultiplier and the electric signal from the photomultiplier is converted into optical density with a logarithmic amplifier. The readout system also provides an automatic tape feed mechanism which is capable of reading and recording the dose profile along a long strip of the CTA tape; the dose can also be read individually by using an adapter for short strips. Since the system only weighs 7.5 kg, it is handy to use in radiation processing.

The readout system for the CTA dosimeter simplifies the procedure of obtaining depth-dose distribution and other dose profiles in the lateral direction of the absorber. Depth-dose profiles in the aluminium absorber which passed through the electron radiation field under the beam window are shown in Fig. 2

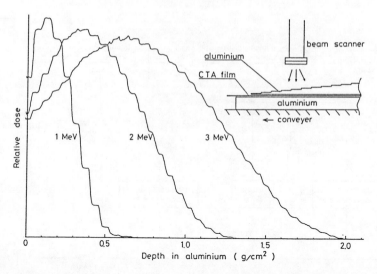

FIG.2. Depth-dose profiles in the aluminium absorber which passed through the electron radiation field under the beam window and the irradiation method using a wedge-shaped aluminium absorber graded stepwise.

as a typical example of applications of the system for electron beam dosimetry. A long strip of CTA film is inserted between a wedge-shaped aluminium absorber graded stepwise, with a very small gradient and a thick aluminium plate, as shown in Fig. 2. The thickness of each step was 0.2 mm. This method gives definite dose values discontinuously along the CTA tape instead of a smooth depth-dose curve [8]. It has two advantages in comparison with the linear wedge method: the precise relation between dose and depth can be redrawn from the original uneven trace (as shown in Fig. 2) even for a relatively low energy electron beam, and it is generally easy to prepare the stepped wedge with a very small gradient for various kinds of materials in the form of film or foil.

The fluctuation of optical density along the unirradiated CTA tape is usually less than ±0.002 and increases with dose. Increment of the fluctuation is chiefly due to the fluctuation of thickness of the CTA tape. It can be reduced by correcting the radiation-induced optical density for thickness.

The readout system for the CTA dosimeter is also useful for dosimetry in a large dimensional area. Figure 3 shows an example of isodose rate charts in a large electron radiation field obtained with very long strips of CTA tape. The charts show air dose rate distributions in the irradiation room for a horizontal beam from the 2 MeV dual-beam electron accelerator at JAERI [6, 9]. The height of the horizontal beam axis above the floor is 1.75 m and the electron beam is scanned horizontally, with a scanning width of 60 cm at the beam window. Figure 3(a) and (b) shows the isodose rate charts with the beam axis in horizontal and vertical

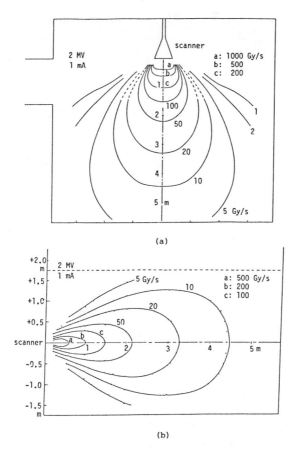

FIG.3. *Isodose rate charts in the irradiation room for a horizontal beam from the 2 MeV dual-beam electron accelerator measured with the CTA dosimeter: (a) and (b) show charts with the beam axis in horizontal and vertical planes, respectively.*

planes, respectively. The air dose rate is approximately given by the product of dose rate in the CTA and the mass collision stopping power ratio of air to CTA.

The isodose rate chart in the electron radiation field is also calculated on the basis of the approximation formula for electron current density distribution [4, 10]. In the vertical plane, the calculated chart agrees well with the chart measured at a relatively short distance from the beam window but, in the horizontal plane, the agreement is not good in the area far from the beam axis, since the formula does not give good approximation of oblique incidence to the beam window.

2.2. Measurement of two-dimensional dose profiles

The CTA dosimeter is also manufactured in sheet form, which enables measurement of two-dimensional dose profiles. We designed and used on a trial

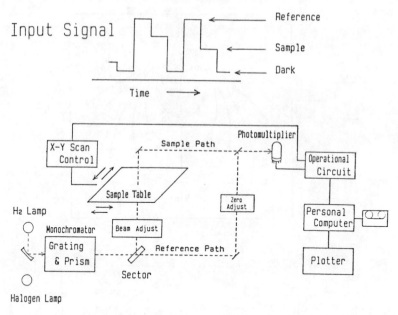

FIG.4. Diagram of double-scanning microspectrophotometer to measure two-dimensional dose distribution (CTA dosimeter in sheet form).

basis a double-scanning microspectrophotometer for the CTA dosimeter and other plastic or dye-plastic dosimeters. In conventional microspectrophotometers the available wavelength range is usually limited down to about 400 nm in visible regions. In the new apparatus the available range was extended to the near UV region, i.e. down to 250 nm; a sample size of 300 × 240 mm maximum can also be scanned with a light beam of 0.05 to 5 mm diameter.

Figure 4 shows a diagram of the microspectrophotometer. The optical path is divided into double paths, i.e. reference and sample. The monochromatic light is divided periodically into three patterns by rotation of the sector, namely reference, sample and dark, as shown in the figure, and is applied to the photomultiplier in turn as the input signal. In the geometrical arrangement of the sample scanning system, extraneous light is unavoidably mixed with the signal light, but this influence can be compensated for by subtracting the dark signal from the sample and reference signals in the operational circuit. The isodose rate chart is finally displayed by a plotter using a personal computer. The double-scanning microspectrophotometer and peripheral equipment are shown in Fig. 5.

The apparatus allows local distributions of dose and beam intensity in small areas to be obtained as well as dose maps in relatively large areas to ensure dose uniformity. An example of isodose charts measured with this apparatus is shown in Fig. 6, giving the dose profile in a cross-section of the wire insulation of a 25 mm diameter power cable irradiated from one side with 2 MeV electrons.

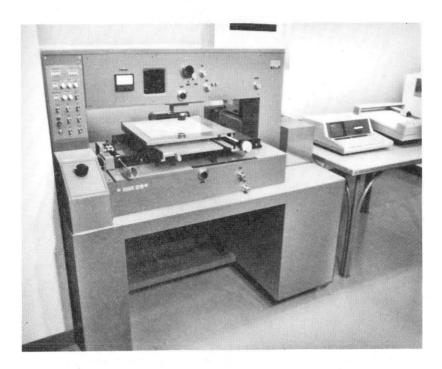

FIG.5. Double-scanning microspectrophotometer and peripheral equipment for the CTA dosimeter.

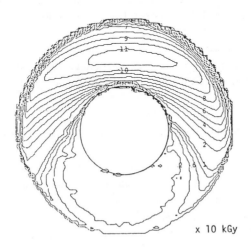

FIG.6. Isodose chart showing dose distribution in a cross-section of the wire insulation of 25 mm diameter power cable irradiated from one side with 2 MeV electrons.

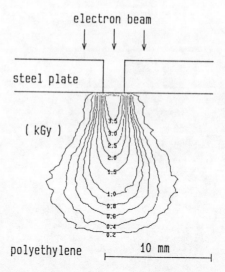

FIG.7. Isodose chart showing the local dose distribution of polyethylene bulk exposed to a 3 MeV electron beam which passed through a narrow slit (width: 2 mm).

A sheet of CTA film cut out in a special form was inserted between two sections of the wire insulation, and was then irradiated. The isodose chart in unevenly distributed absorbers is obtained clearly and simply with this method which can be also applied for thinner cables; however, when the thickness of the insulation is of the order of 1 mm, this method may be replaced with a simulation method by winding a dosimeter film strip many times around a copper wire core [11].

Figure 7 shows another example of an isodose chart measured with the apparatus, giving the local dose distribution of polyethylene bulk exposed to a 3 MeV electron beam which passed through a narrow slit (width: 2 mm). For irradiation, the radiochromic dye film dosimeter in sheet form was inserted between two polyethylene blocks. The isodose chart enables understanding of the lateral diffusion effect of an electron beam in the medium.

3. APPLICATIONS OF ELECTRON CURRENT DENSITOMETER

A calorimeter is the most commonly used dose calibration system for processing electron accelerators, but it has some disadvantages, namely (1) handling is generally complicated for accurate measurement; (2) it does not have wide effective measuring ranges of dose and dose rate; (3) it does not give dose and dose rate simultaneously, or other information on the radiation field.

Electron current density measurement is not as direct a dosimetric method as thin foil calorimetry [12], but it has wide applications for radiation measure-

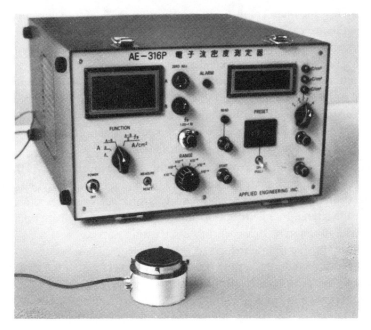

FIG.8. *Detector and readout system of electron current densitometer.*

ments including dose calibration at electron energies higher than 500 keV. The advantages of this system are:

(1) The simplicity of electric charge measurement is suited to radiation processing, since it does not require vacuum conditions
(2) Energy fluence and absorbed energy per unit area of light atomic number materials can be accurately evaluated if the electron energy in the acceleration voltage is calibrated
(3) It can cover wide ranges of quantities associated with the radiation itself [13], electron current density (fluence rate), electron charge fluence (particle fluence), energy fluence rate and energy fluence, since it has good heat-resistance properties
(4) The system can also be used as a time monitor and a beam profile detector.

3.1. Applications to radiometry [13] and dose evaluation

Electron current density depends on electron energy, beam current and other irradiation parameters, but does not depend on irradiated materials. In this sense, the role of electron current density in electron beam dosimetry is analogous to that of the exposure rate in gamma ray dosimetry.

Figure 8 shows the detector and readout system of the electron current densitometer [3]. The detector measures 58 mm in diameter and 42 mm in height.

The diameter of the main graphite absorber is 20 mm. Electron current density and charge fluence can be simultaneously monitored and recorded with the readout system which is capable of measuring over wide ranges.

At the Takasaki Radiation Chemistry Research Establishment, JAERI, the dose evaluation method used in usual electron irradiation is based on the following calibration or measurements [14]: (1) electron energy in the acceleration tube; (2) lateral distributions of electron current density in the radiation field under the beam window, measured with the electron current densitometer; (3) depth-dose distribution measured with the CTA dosimeter.

The results of item (2) provide information on the mean value of the radiation intensity (energy fluence rate) and the dose rate, and on their lateral uniformity in the radiation field. Relative beam intensity profiles can also be obtained from measuring the lateral dose distribution along the surface of irradiated material with the CTA dosimeter. However, this distrbiution does not give the same relative beam intensity profile as that measured with the electron current densitometer because the absorbed dose is a spherical intensity [13, 15][1] and the surface dose is affected by backscattered electrons from the irradiated material.

Direct conversion from electron current density to surface dose rate is often required in radiation processing. A rough evaluation of surface dose rate is given by the product of electron current density and mass collision stopping power for electron energies higher than a few hundred keV and for low atomic number materials. For accurate estimation, we must consider detailed information on the radiation field and the interaction of electrons with thick matter. To irradiate a slab layer with a continuously moving conveyor, a more practical method for the direct conversion from electron charge fluence to dose as a function of depth of the absorber can be given by the calculation code EDMULT [16] (for evaluating electron depth-dose distribution in multilayer slab absorbers).

When a slab absorber is moved through an electron beam in a direction that is perpendicular to the direction of the beam scanning, the depth-dose distribution in the uniformly irradiated slab absorber in the neighbourhood of the beam axis is ascribed to the depth-dose distribution for the plane-parallel electron beam normally incident on a triple-layer slab absorber (beam window–air gap–slab layer) by a geometrical analogy of the irradiation model. The depth-dose profile $D(x)$ in the slab layer is evaluated by the expression

$$D(x) = \frac{K\, d(x,E)}{\eta_t} \int j\, dt \qquad (1)$$

---

[1] The electron current density (A/cm^2) is not exactly the same as the electron fluence rate ($cm^{-2} \cdot s^{-1}$) since the former is a plane intensity which refers to a given plane surface, and the latter is a spherical intensity which refers to a cross-sectional area of a sphere.

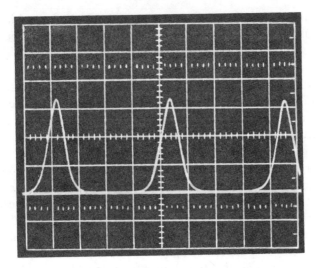

FIG.9. Instantaneous current density profile of a 1 MeV scanned electron beam which was monitored with the electron current densitometer at a distance of 20 cm from the beam window along the beam axis (scanning width: 60 cm).

where K is a constant, $\int j dt$ is the electron charge fluence obtained by integration of the measured electron current density, η_t is the electron number transmission coefficient of the beam window and the air gap given by an empirical formula [17], and d(x,E) is the dose at depth x in the third (slab) layer per unit fluence of the plane-parallel electron beam of incident energy E normally impinging on the surface of the first layer. The value of d(x,E) is given by the EDMULT code.

For irradiation with beam scanning, the electron current density measured with an ammeter gives the value obtained by averaging periodical changes due to beam scanning [4]. Instantaneous electron current density can be obtained by monitoring the signal current from the electron current densitometer with an oscilloscope. Monitoring allows measurement of the instantaneous dose rate under irradiation. Figure 9 shows the instantaneous current density profile of a 1 MeV scanned electron beam which was monitored with the electron current densitometer at a distance of 20 cm from the beam window along the beam axis (scanning width: 60 cm). The measured profile agreed with the calculated profile obtained by the approximation formula [4, 10] given for a typical irradiation condition.

3.2. Calibration of dosimeters

The relation between the radiation-induced optical density of the CTA dosimeter and the depth in CTA obtained by irradiating a stack of CTA films gives the relative dose profile of CTA itself. Calibration of the CTA dosimeter is carried out by normalizing the integrated area of the profile by the total absorbed energy,

W, per unit area of the CTA stack. The value of W is measured with the total absorption calorimeter; it can also be measured with the electron current densitometer if an accurate value of the electron energy is given.

In comparison to the total absorption calorimeter, the electron current densitometer is simple to handle and covers a wide effective measuring range from 0.01 to 100 $\mu A/cm^2$, roughly corresponding to a dose rate range from 20 Gy/s to 200 kGy/s. It also covers a much wider dose range by integration of the electron current density. Furthermore, the lower limit can be reduced with careful correction of the effect of the ionic charge produced in air.

Another advantage of this method is that measurement of the electron charge fluence and monitoring of the electron current density can be carried out simultaneously with irradiation of the dosimeters in order to obtain the calibration curve and various dosimetric characteristics with electron beams. Uncertainties in dosimetry due to the fluctuation of irradiation parameters such as the beam current and conveyor speed can be avoided with such simultaneous irradiation. We obtained good reproducibility (within ±1%) using this method to determine both the calibration constant [8] and the dose rate dependence of the CTA dosimeter over the wide range from tens of kGy/h to tens of MGy/h. It is also useful and convenient for accurate determination of the dose calibration curve and various radiation effects, especially fluence dependence on the radiation effects of semiconductor devices.

3.3. Error sources in electron current density measurement [3]

The main source of uncertainty in determining the electron current density is the influence of ionic charge produced in air around the main absorber. This results from collection of ionic charge due to the electric field formed in the surrounding air. It is believed that three factors contribute to the formation of electric field: (1) contact potential between the main absorber and the guard ring; (2) negative charge deposition in the surrounding air due to energy degradation and thermalization of electrons; (3) charging of neighbouring bare insulators.

The contact potential may depend on good distinction between both surface states. It is usually negligibly small for graphite compared with metals such as aluminium and copper but, in the extreme case, it becomes more than a few tens of millivolts even for graphite. The influence of contact potential can be checked by sealing the detector in a vacuum vessel with a thin metal foil beam window. The ion recombination rate increases with the dose rate in air. The effect of the contact potential can almost be neglected at electron current densities higher than 0.1 $\mu A/cm^2$.

The influence of the negative space charge in the surrounding air is usually negligible in irradiating conditions with electron energies higher than 500 keV, but it becomes appreciable for lower electron energy and at long distances from

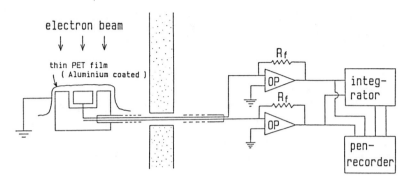

FIG.10. *Method of measuring electron current density to avoid the influence of the extraneous electric field in the surrounding air and to compensate for radiation-induced current from the signal cable (R_f is feedback resistance; OP is operational amplifier).*

the beam window to the main absorber of the detector, even for higher energy. Electron backscattering from heavy metal around the detector also enhances the space charge density in the surrounding air.

It is necessary that large bare insulators be kept away from the detector in order to avoid the formation of extraneous electric field around the main absorber. This effect can be experimentally checked.

When the space charge or large bare insulators are unavoidable during irradiation, it is recommended that the detector surface be covered with very thin polyethylene terephthalate (PET) film (0.01 mm or less) coated with aluminium on the outside surface, as shown in Fig.10. The influence of the extraneous electric field in the surrounding air on the detected current is effectively removed by the thin film.

Figure 10 also illustrates a method for compensating for radiation-induced current from the signal cable using a double-core coaxial cable. The radiation-induced current is compensated for by subtracting the dummy current from the signal current using operational amplifiers. The potential decrease at the absorber and the resultant absorption of positive ionic charge is avoided by inverting gain-controlled operational amplifiers.

When converting from charge fluence to energy fluence, another uncertainty may originate from head-on collision in the beam window and the air gap, producing relatively low energy electrons (delta rays). This uncertainty has not yet been estimated, but it is presumably negligible because calculation by the ETRAN code gives a small production rate of delta rays in the beam window and air, and the detection rate of delta rays emitted from the beam window is considered to be very small compared with primary electrons, since they emit at relatively large angles.

4. LIMITATION IN APPLICATIONS TO LOW ENERGY ELECTRON BEAM DOSIMETRY

Application of film dosimeters to low energy electron beam dosimetry is limited in the thickness of dosimeters, since the thickness must allow enough spatial resolution to draw a depth-dose profile in the form of a histogram. The lower limit of electron energy for which the CTA dosimeter (0.125 mm thickness) can be applied is about 300 keV. However, the dosimeter is expected to be useful for drawing lateral profiles of relative beam intensity of lower electron energy, since it is in the form of a long tape. The CTA film (0.038 mm thickness), which is commercially available for general use, may be applied for a dosimeter for low electron energy. However, the response is insensitive to dose and the optical density of the unirradiated film is high. Improvement of the film is necessary for dosimetry use. At present, the radiochromic dye film dosimeter is considered to be the most useful for energy electron beams lower than 300 keV.

The applicability of electron current density measurement to an energy electron beam lower than 500 keV has not yet been examined. It is assumed that the overall uncertainty in determination of current density increases with decreasing electron energy for three reasons: (1) increase of oblique incidence to the main absorber, which results in an increase of backscattering correction for oblique incidence; (2) increase of backscattering correction for normal incidence; (3) increase of negative space charge density in the surrounding air described in subsection 3.3.

The overall uncertainty increases further if the angular distribution of the incident electrons is not accurately given. The uncertainty in determination of energy fluence is caused by an increase in the uncertainty of the calculated mean energy of incident electrons with a decrease in electron energy.

Compared with the middle energy region (0.5 to 3.0 MeV), low energy electron beam dosimetry involves basic limitations such as sensitivity to the irradiation parameters, the limited spatial resolution of film dosimeters, the difficulty in conversion from dose in the dosimeter itself to that in other materials, etc. However, most irradiating conditions for low energy electron beams are simple, since they are limited to single-direction, single-pass irradiators. If there is a peak in the depth-dose profile under the irradiating condition, the peak absorbed dose may be approximately calculated from a simple formula as a function of several irradiation parameters: electron energy, beam current, beam width, conveyor speed, number of transmission coefficients, etc., because the peak dose depends slightly on the thickness of the beam window and the air-gap distance. Considering the limitations in the dosimetry described, the approximation formula will play an important role in low energy electron beam dosimetry and in comparisons of radiation effects observed under different irradiating conditions.

REFERENCES

[1] TAMURA, N., TANAKA, R., MITOMO, S., MATSUDA, K., NAGAI, S., Radiat. Phys. Chem. **18** (1981) 947.
[2] TANAKA, R., MITOMO, S., TAMURA, N., Int. J. Appl. Radiat. Isot. **35** (1984) 895.
[3] TANAKA, R., MIZUHASHI, K., SUNAGA, H., TAMURA, N., Nucl. Instrum. Methods **174** (1980) 201.
[4] TANAKA, R., SUNAGA, H., YOTSUMOTO, K., MIZUHASHI, K., TAMURA, N., Radiat. Phys. Chem. **18** (1981) 927.
[5] TANAKA, R., Proc. Symp. Accelerators for Processing, Lake Kawaguchi, Yamanashi, Japan, Japan Atomic Industrial Forum Inc., Tokyo (1984) 115.
[6] TAMURA, N., Radiat. Phys. Chem. **22** (1983) 81.
[7] NAKAMURA, Y., TANAKA, R., Japan Atomic Energy Research Institute, Tokyo, Rep. JAERI-M 5976 (1975) (in Japanese).
[8] TANAKA, R., SUNAGA, H., MITOMO, S. (to be published in Int. J. Appl. Radiat. Isot.).
[9] YOTSUMOTO, K., Japan Atomic Energy Research Institute, Tokyo, Rep. JAERI-M 84-032 (1984) (in Japanese).
[10] TANAKA, R., Oyo Butsuri **48** (1979) 432 (in Japanese).
[11] McLAUGHLIN, W.L., MILLER, A., PEJTERSEN, K., PEDERSEN, W.B., Radiat. Phys. Chem. **11** (1978) 39.
[12] CHAPPEL, S.E., HUMPHREYS, J.C., IEEE Trans. Nucl. Sci. **19** (1972) 175.
[13] INTERNATIONAL COMMISSION ON RADIATION UNITS AND MEASUREMENTS, Report 33, Radiation Quantity and Units, ICRU, Washington, DC (1980).
[14] SUNAGA, H., MIZUHASHI, K., YOTSUMOTO, K., TANAKA, R., TAMURA, N., Japan, Atomic Energy Research Institute, Tokyo, Rep. JAERI-M 82-142 (1982) (in Japanese).
[15] WHYTE, G.N., Principles of Radiation Dosimetry, John Wiley & Sons, Inc., New York (1959).
[16] TABATA, T., ITO, R., Radiation Center Osaka Prefecture, Technical Report 1 (1981).
[17] TABATA, T., ITO, R., Nucl. Instrum. Methods **136** (1976) 553.

IAEA-SM-272/38

OXYGEN EFFECTS IN CELLULOSE TRIACETATE DOSIMETRY*

P. GEHRINGER, E. PROKSCH,
H. ESCHWEILER
Department of Chemistry,
Austrian Research Centre Seibersdorf,
Seibersdorf, Austria

Abstract

OXYGEN EFFECTS IN CELLULOSE TRIACETATE DOSIMETRY.
  Gamma irradiation of CTA films in air results in a coloration in the UV region which is higher than that from electron irradiation and which is (contrary to the latter) dependent on the humidity of the air with which the films are in equilibrium. Replacement of the air with argon removes the humidity dependence and lowers the gamma irradiation response to a level close to that from electron irradiation. Thus, it may be concluded that the dose rate and the humidity effect are in fact oxygen effects caused by reactions with some reactive intermediates (most likely long-living radicals) of oxygen diffusing into the films during irradiation. Post-irradiation coloration occurs after irradiation in argon and in air. It is caused mainly by the reaction with the above-mentioned intermediates of oxygen diffusing into the films after irradiation.

1. INTRODUCTION

In radiation processing plants certain plastic or dyed plastic films are commonly used as routine dosimeters for quality control. In spite of some weak points (e.g. possible variation of response with dose rate and/or environmental factors; batch-to-batch variation in sensitivity) these dosimeters enjoy wide popularity because of their low cost, ease of handling and readout, ruggedness, availability in large batches, and compatibility with various geometrical configurations. The CTA film dosimeter is a typical example of such a routine dosimeter showing almost all the above-mentioned advantages and disadvantages. Use of this system has been described in the literature since 1974 and many different research groups have made rather comprehensive investigations; nevertheless, some of the data given are not fully consistent.
  Puig et al. [1] reported an excellent sensitivity and a linearity between dose and absorbance when the measurements are performed at 280 nm. This has been

---

  * Research carried out with the support of the IAEA under Research Contract No. 2333/RB.

confirmed by other authors [2–6] and 280 nm is now the generally accepted standard wavelength.

Most researchers [1, 5–9] found that a dose rate effect existed, although the reported magnitudes of this effect and the dose ranges vary somewhat. Some authors [3, 4] did not observe any rate effect at all. More recent investigations [5–7], however, agree that there is essentially no dose rate effect above about 10^6 Gy/h, a pronounced effect between about 10^4 and 10^6 Gy/h and a somewhat smaller effect below 10^4 Gy/h. A humidity effect was found for gamma irradiations (although the detailed results are not in mutual agreement) but not for electron irradiations [6, 10].

There is no agreement in the literature with respect to a post-irradiation coloration effect. Whereas some investigators [1, 11, 12] found a stable response during post-irradiation storage up to 7 days, others [4–8] found a pronounced increase.

The present work was aimed at investigating some of the controversial points and trying to remove the possible sources of errors. Special attention has been paid to the influence of dose rate, film moisture and presence or absence of oxygen during irradiation, the response of CTA films and their post-irradiation behaviour.

2. EXPERIMENTAL PROCEDURES

2.1. Dosimetric system

The CTA dosimeter films used were obtained commercially from Société Numelec, France. They were supplied in long rolls (8 mm in width and 0.125 mm in thickness) and contained 15 wt% triphenyl phosphate.

Unfortunately, these films are no longer produced by Numelec, but the Fuji Photo Film Co., Japan, now manufactures films with analogous specifications. The preliminary tests performed seem to confirm that both products behave very similarly upon irradiation.

2.2. Preconditioning of films

Before irradiation all dosimetric films were conditioned carefully to a well-defined moisture content. For this purpose the films were stored for at least 1 week in a glove box flushed with gas (air or 99.999% argon) of an exactly defined relative humidity. The latter was achieved by passing the gas through a water scrubber held at an appropriate temperature. The glove box was installed in a room thermostated to 25°C. The temperature and humidity of the gas stream as well as the inside of the glove box were measured permanently by a Temperature/Relative Humidity Indicator Model H-10 (Weather Measure Corporation, Sacramento, USA).

2.3. ^{60}Co γ-irradiation

An AECL Gammacell-220 with a dose rate of 2.0 Gy/s was used. Dose calibration was performed by Fricke dosimetry. All the dose values given are expressed in terms of dose in water. To maintain the CTA dosimeters in electronic equilibrium during irradiation, packages were prepared by stacking five films between 5 mm PMMA front and back platelets, the PMMA having been preconditioned analogous to the dosimetric films. During irradiation, the packages were thermostated to 25°C and permanently flushed with the same gas as that used during preconditioning.

2.4. Electron beam irradiation

The electron processing facility at Seibersdorf was used as the electron beam source. It is equipped with a 500 keV/25 mA ICT-500 electron accelerator (High Voltage Engineering, Burlington, USA). All irradiations were performed 20 cm below the exit window. The dose rate at the maximum beam current (25 mA) was about 2.7×10^4 Gy/s. All doses were determined by the energy balance method described in Ref. [13].

For irradiation five CTA film pieces were fixed side by side to a PMMA base plate; this plate was then covered by a 15-μm thick polyester tube which was finally sealed on both sides. This procedure was performed in the glove box used for preconditioning; thus, defined humidity conditions could be maintained even during irradiation. During irradiation the samples were heated (depending on the dose) to a maximum of about 75°C. No corrections were made for the varying temperatures, since according to Ref. [5] there is no temperature effect at dose rates above about 3×10^2 Gy/s.

2.5. Post-irradiation storage

After irradiation all the samples were stored at ambient conditions (i.e. in air at about 22°C and about 40 to 60% relative humidity) in darkness.

2.6. Evaluation of irradiated films

All the films were read out at the 280 nm wavelength using a CARY model 17 spectrophotometer. The absorbance of each film was measured before and, except when stated otherwise, 1 hour after irradiation. All results stated here are expressed as absorbance changes ΔA. The thicknesses of the film pieces were checked and found to be sufficiently constant, therefore no thickness corrections had to be made. Only the mean ΔA values averaged over the five samples irradiated together are reported in the following section.

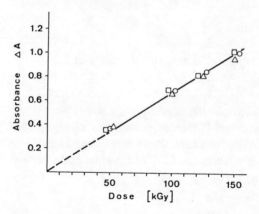

FIG.1. Optical absorbance change at 280 nm versus dose of 125-μm thick CTA films irradiated in air with 500 keV electrons at a dose rate of 1.1×10^4 Gy/s at different relative humidities (○ 10%; □ 32%; △ 86%).

3. RESULTS AND DISCUSSION

3.1. Dependence on dose rate and irradiation atmosphere

When irradiations are performed in air (Figs 1 and 2) there is a marked influence of irradiation conditions. Over the whole dose range covered, gamma irradiations result in a much higher absorbance change than electron irradiations. The difference is, on average, about 12.5%. This is in fair to good agreement with the dose rate effects reported in the literature [1, 5–7]. There is no discernible influence of the water content of films for electron irradiations (Fig. 1). However, for gamma irradiations (Fig. 2) such an influence clearly exists; the absorbance changes increase by about 23% when the relative humidity of the surrounding atmosphere (before and during irradiation) is increased from 10 to 86%. This is in good agreement with Ref. [6].

Replacing air with argon results in only minor and hardly significant changes for electron irradiations (Fig. 3). However, for gamma irradiations exclusion of oxygen completely removes any humidity dependence (Fig. 4). To the authors' best knowledge, such an effect has not been reported previously. Additionally, the response to gamma irradiation in argon now becomes almost identical to that found for electron irradiation. This can be seen even more clearly in Table I, which summarizes the slope values k in Figs 1 to 4. The remaining small differences (about ±3.5%) are most likely due to dose inaccuracies (especially in the case of electron irradiations), to the influence of traces of oxygen still present in the argon used, and/or to the different post-irradiation behaviour of gamma- and electron-irradiated samples (see below).

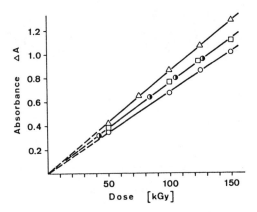

FIG.2. *Optical absorbance change at 280 nm versus dose of 125-μm thick CTA films irradiated in air with ^{60}Co γ-rays at 25°C at a dose rate of 2.0 Gy/s at different relative humidities (○ 10%; □ 32%; ● 45%; △ 86%).*

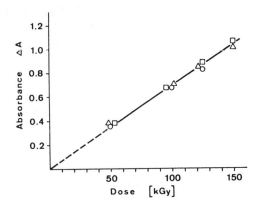

FIG.3. *Optical absorbance change at 280 nm versus dose of 125-μm thick CTA films irradiated in argon with 500 keV electrons at a dose rate of 1.1×10^4 Gy/s at different relative humidities (for explanation of symbols see Fig. 1).*

These findings are of high practical importance if CTA films are to be used as gamma dosimeters. Only when using argon can a calibration be made which may be used subsequently for any dose rate and any humidity. Possibly, this would also eliminate the temperature effect which is otherwise present (i.e. in air) at dose rates typical for gamma irradiations.

From these findings, some very interesting conclusions may also be drawn with respect to the mechanism of the radiation coloration of CTA films. The results given above may best be interpreted by the assumption that the coloration is made up from two components. The main component is due to some direct

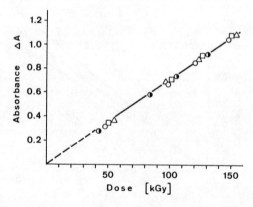

FIG.4. Optical absorbance change at 280 nm versus dose of 125-µm thick CTA films irradiated in argon with ^{60}Co γ-rays at 25°C at a dose rate of 2.0 Gy/s at different relative humidities (for explanation of symbols see Fig. 2).

TABLE I. INFLUENCE OF DOSE RATE, HUMIDITY AND TYPE OF IRRADIATION ATMOSPHERE ON THE RESPONSE AT 280 nm OF 125 µm CTA FILMS (k-value: slope of the respective absorbance change versus dose curves)

	Radiation source	Dose rate (Gy/s)	Relative humidity (%)	Irradiation atmosphere	k-value (ΔA/kGy)
1			10		0.0068
2	^{60}Co		32; 45	air	0.0075
3		2.0	86		0.0086
4			10; 32; 45; 86	argon	0.0071
5	500 keV	1.1×10^4	10; 32; 86	air	0.0067
6	Electron beam			argon	0.0069

radiation-chemical reaction, involving only the CTA film itself. The second, smaller component involves a reaction of some irradiation product of the CTA film with oxygen diffusing into the film during irradiation. At high dose rates (i.e. under electron irradiation) this component can be neglected because oxygen diffusion is too slow and thus inefficient. At low dose rates (i.e. under gamma irradiation) oxygen diffusion is efficient enough to make a substantial contri-

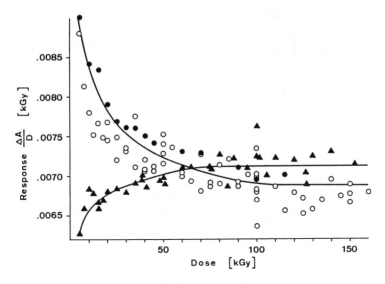

FIG.5. Influence of dose and dose rate on the response at 280 nm of 125-μm thick CTA films irradiated with 500 keV electrons at a dose rate of 1.1×10^4 Gy/s in air (○) and argon (●), and ^{60}Co γ-rays at a dose rate of 2.0 Gy/s in argon (▲).

bution to the overall response. In this region, the response should increase with decreasing dose rate, as has been found in the literature. Water, which acts as a plasticizer to CTA, is expected to increase the diffusivity of oxygen; an increase of the water content should therefore increase the response of the film and this is again in agreement with experimental results. Dose rate and humidity effects therefore are in fact oxygen effects; they disappear if oxygen is not present or when the dose is administered in such a short time that oxygen diffusion can be neglected. This model can also explain why temperature effects, according to the literature [5–7], are present only over the low dose rate range; temperature only affects (via the oxygen diffusion coefficient and the second-order rate constant) the oxygen-dependent component and not the direct radiation-chemical component of CTA coloration.

3.2. Dose dependence

During the investigations described, it was tacitly assumed that the absorbance versus dose relation is indeed linear, as has been stated in the literature [1–7, 12]. However, a more detailed study covering a broader dose range shows that this linearity is only an approximate one. To show this more clearly, absorbance values per unit dose (response values) have been plotted versus dose (Fig. 5).

A truly linear relation between absorbance and dose exists only at doses above about 100 kGy. At lower doses deviations exist which increase with

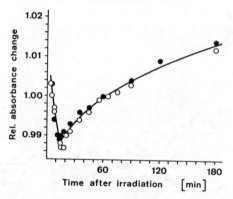

FIG.6. Relative optical absorbance change at 280 nm (normalized to 1 hour's storage) as a function of storage time at room temperature. Irradiation dose was 100 kGy and irradiations were performed with 500 keV electrons at a dose rate of 1.1×10^4 Gy/s in air (○) and argon (●).

decreasing dose. If such deviations should be limited to, for example, $<\pm 2\%$ then a lower limit to the useful dose range of about 75 kGy results for electron irradiations and about 50 kGy for gamma irradiations in argon.

On the high-dose side, the useful range is limited by two effects. First, films irradiated at >250 kGy become markedly brittle. Such films are difficult to handle and often fall apart despite great care. Additionally, the absorbance values at 250 kGy (film thickness 0.125 mm) are already close to 2, which is usually a limit for accurate operation of many of the more simple spectrophotometers; therefore 250 kGy can be considered as an upper limit to the useful dose range.

Deviations from linearity at low doses are opposite for gamma and electron irradiations. The reason for this is not yet completely clear. Possibly, a part of the positive deviations in the case of electrons is a consequence of oxygen traces dissolved in the CTA films and subsequently used-up during irradiation. The negative deviations found for gamma irradiations could possibly be due to oxygen traces in the argon used.

3.3. Post-irradiation behaviour

The radiation-induced absorbance changes at 280 nm were found to be subject to rather complex variations during post-irradiation storage under ambient conditions. In accordance with most of the literature [4–8] there is a general trend for a long-lasting increase with time, which additionally depends on dose, dose rate and irradiation atmosphere. It will be shown that all these effects are in accordance with the two-component model described in subsection 3.1.

After electron irradiations, irrespective of the irradiation atmosphere, absorbance decreases rapidly, reaches a sharp minimum after about 15 min and slowly increases again (Fig.6). This is in general agreement with the literature

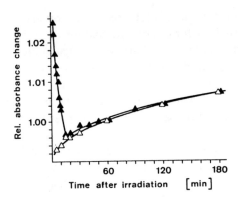

FIG.7. *Relative optical absorbance change at 280 nm (normalized to 1 hour's storage) as a function of storage time at room temperature. Irradiation dose was 100 kGy and irradiations were performed with ^{60}Co γ-rays at 25°C at a dose rate of 2.0 Gy/s in air (△) and argon (▲).*

for electron irradiation in air [5, 6]. Immediately after irradiation a high concentration of the above-mentioned 'irradiation product' must have been present which up to that moment had had no opportunity of reacting with oxygen (nor in any other way). Exposure to air results in its rapid reaction with oxygen (downward branch of the experimental curve) and the formation of a stable product (upward branch), obviously via some slowly decaying intermediate(s). Existence of the latter must be postulated in order to account for the obvious delay found experimentally; the intermediate(s) may be identified with various peroxidic compounds. This explanation agrees with that put forward by Tamura et al. [5] only for the initial absorbance decrease, which they explained by the reaction of one of the tertiary CTA radicals with oxygen. This radical may well be the 'irradiation product' of our model. The subsequent increase in absorbance is explained by Tamura et al. by a reaction between CTA molecules and NO_2 produced by irradiation and adsorbed on to the CTA films. This explanation, however, is ruled out by the fact that irradiation in argon results in the same post-coloration behaviour as that in air.

After gamma irradiation in argon, a similar curve to that obtained after electron irradiation may be expected. Figure 7 shows that this is indeed the case. The lower slope of the upward branch may be explained by the decay of some of the CTA radicals to some uncoloured products already during irradiation. After gamma irradiation in air, no radicals should remain and no downward branch should be observable. As can be seen from Fig.7, this is again the case. The slope of the upward branch is again low; a large fraction of the radicals has already completed its transformation to the final product during irradiation. This behaviour is again in general agreement with the literature [5, 6].

From Figs 6 and 7 it may be deduced (either directly or by back-extrapolation of the upward branches) that for the 1 hour standard storage time the post-irradiation contribution to total response is around 0.8% for gamma irradiation and around 3% for electron irradiation.

That the downward branches in these two figures are due to the reaction with oxygen is ascertained by keeping some samples after irradiation for 20 min under an inert atmosphere (Ar or CO_2) and exposing them to air only thereafter; the resulting absorbance changes (starting from that point) are identical to those which had been measured for samples already exposed to air at the end of irradiation.

Post-coloration continues for a very long time. After 42 days at room temperature it has reached levels around +10% compared with 1 h after irradiation and in most cases has not yet ended. It is still somewhat dependent on irradiation conditions.

The higher the humidity during irradiation the more intensive the post-coloration after electron irradiation (in air or argon). This indicates that effects measured after weeks still have their origin in processes that started during or shortly after irradiation.

For gamma irradiation a moisture effect is also present, but in this case the presence or absence of oxygen is the main factor. After irradiation in air the signal becomes stable after about 2 weeks but continues to increase even after 6 weeks when irradiation is performed in argon. Obviously, in the former case some of the intermediates formed from the CTA radicals and oxygen (most likely peroxy radicals and peroxides) have already been destroyed during irradiation.

Several attempts have been made to find a procedure for stabilizing the post-irradiation effect. Application of higher temperatures during storage in air failed completely; coloration was accelerated but did not reach a stable state. Attempts to destroy the unstable intermediates by exposure to chemical reactants also failed. Heat treatments in argon (around 1 h at 100 to 120°C) effectively stop any further post-coloration. However, the original absorbance changes are increased by about 50 to 100% by this procedure, the final values being so sensitive to annealing temperature and annealing time that such a procedure does not seem to be feasible except for special cases.

3.4. Precision and accuracy

The standard deviation determined from absorbance measurements of five films irradiated together is, on average, about ±1%. Compared with other film dosimeters (e.g. radiochromic films [14]) this is a rather high precision. However, to fully utilize this advantage during practical application of the system, i.e. to obtain high accuracy as well, due care has to be taken of the possible influences of dose, dose rate, temperature and humidity. The possible influences of post-irradiation effects should be minimized by using short storage times, around 1 hour. Additionally, use of a spectrophotometer with high wavelength precision is a necessary prerequisite.

4. CONCLUSIONS

Regarding present investigations performed in air as the irradiation atmosphere, results already present in the literature have, in general, been confirmed. Irradiations in argon, however, have led to completely new results. These are of direct practical importance (e.g. the possibility of excluding dose rate, temperature and moisture effects during gamma irradiation by using an oxygen-free gas atmosphere, e.g. argon) and they cast additional light on the complicated mechanism of the radiation-coloration of CTA. Investigations into post-irradiation coloration have shown that this phenomenon is caused by oxygen reacting with radicals present after irradiations.

ACKNOWLEDGEMENTS

The authors wish to express their thanks to the Jubiläumsfonds der österreichischen Nationalbank and to the Fonds zur Förderung der wissenschaftlichen Forschung who generously supported this work by making available the electron accelerator, the ^{60}Co source and the spectrophotometer.

REFERENCES

[1] PUIG, J.R., LAIZIER, J., SUNDARI, F., Le film TAC, dosimètre plastique pour la mesure pratique des doses d'irradiation reçues en stérilisation, Radiosterilization of Medical Products (Proc. Conf. Bombay, 1974), IAEA, Vienna (1975) 113.
[2] NAKAI, Y., MATSUDA, K., TAKAGAKI, T., Dosimetry of Electron Beams and γ-rays by Cellulose Triacetate (CTA) Film, Japan Atomic Energy Research Institute, Rep. JAERI 5029 (1974) 153.
[3] VERKHGRADSKII, O.P., Radiation physics and chemistry of polymers (MAKHLIS, F.A., Ed.), Keter Publishing House Jerusalem Ltd., Jerusalem (1975) 88.
[4] JANOVSKÝ, I., Radiochem. Radioanal. Lett. 47 (1981) 251.
[5] TAMURA, N., TANAKA, R., MITOMO, S., MATSUDA, K., NAGAI, S., Radiat. Phys. Chem. 18 5-6 (1981) 947.
[6] TANAKA, R., MITOMO, S., SUNAGA, H., MATSUDA, K., TAMURA, N., Manual of CTA Dose Meter, Japan Atomic Energy Research Institute, Rep. JAERI-M-82-033 (1982).
[7] LAIZIER, J., "Techniques utilisées au Centre d'applications et de promotion des rayonnements ionisants pour la dosimétrie des irradiations gamma et sous faisceaux d'électrons", High-Dose Measurements in Industrial Radiation Processing, Technical Reports Series No. 205, IAEA, Vienna (1981) 57.
[8] McLAUGHLIN, W.L., HUMPHREYS, J.C., RADAK, B.B., MILLER, A., OLEJNIK, T.A., Radiat. Phys. Chem. 14 (1979) 535.
[9] TANAKA, R., YOTUMOTO, K., TAJIMA, S., KAWAI, M., MIZUHASHI, K., Japan Atomic Energy Research Institute, Rep. JAERI-M-5608 (1974).
[10] LEVINE, H., McLAUGHLIN, W.L., MILLER, A., Radiat. Phys. Chem. 14 (1979) 551.

[11] SUNDARI, F., Majalah Batan **9** 2 (1976) 2.
[12] DAFFERNER, J.M., Dosimetry in the Region of 0.25 Mrad and 25 Mrad Using Cellulose Triacetate Films (CTA), Instituto de Energia Atomica, São Paulo, Rep. IEA-DT-088 (1978).
[13] PROKSCH, E., GEHRINGER, P., ESCHWEILER, H., Int. J. Appl. Radiat. Isot. **30** (1979) 279.
[14] GEHRINGER, P., ESCHWEILER, H., PROKSCH, E., Int. J. Appl. Radiat. Isot. **31** (1980) 595.

RESPONSE OF RADIOCHROMIC FILM DOSIMETERS TO ELECTRON BEAMS IN DIFFERENT ATMOSPHERES

Wenxiu CHEN, Haishen JIA
Department of Chemistry,
Beijing Normal University,
Beijing, China

W.L. McLAUGHLIN
Center for Radiation Research,
National Bureau of Standards,
Gaithersburg, Maryland,
United States of America

Abstract

RESPONSE OF RADIOCHROMIC FILM DOSIMETERS TO ELECTRON BEAMS IN DIFFERENT ATMOSPHERES.

Electron beam responses (0.01–600 kGy) of radiochromic film dosimeters, with ten kinds of plastic matrices (polychlorostyrene containing 1 or 25% Cl, polybromostyrene containing 2 or 43% Br, nylon, polyvinyl butyral, polyvinyl pyrollidone, polyvinyl chloride, cellulose triacetate and an aromatic polyamide), were examined when irradiated in vacuum and under different atmospheres (air, oxygen, nitrous oxide and nitrogen). In addition, the storage stability of the films was studied for periods of up to one month after irradiation under given conditions. The response of polyhalostyrene films generally did not show much difference in different atmospheres, especially the film with a larger amount of bromine (43%) showing less difference than the other polyhalostyrene films. These results resemble those from gamma ray irradiation. Nylon films also show little difference in response in different atmospheres, which is similar to polyhalostyrene films. However, for polyvinyl butyral (PVB), polyvinyl pyrollidone (PVD) and cellulose triacetate (CTA) films, differences are much greater. Especially in O_2 atmosphere, the $\Delta OD/mm$ values were diminished. CTA films vary less in different atmospheres than those of PVB and PVP. The $\Delta OD/mm$ values of polyvinyl chloride (PVC) and trogamid (TGM) films with low dye precursor concentrations are suitable for very high dose measurement. The response curves of PVC and polyethylene (PE) films are very similar, i.e. the $\Delta OD/mm$ values ascend in the same O_2, air, N_2O, vacuum and N_2 order. The storage stability of these films is related to their response in different atmospheres, and films irradiated by electron beams are more stable than those that are gamma ray irradiated. From our electron beam results, it appears to be important to control the matrix material and the dye composition according to definite irradiation conditions which may cause marked differences in the ultimate response factors. Accurate dose measurements with radiochromic materials depend on careful planning of irradiation and storage conditions.

1. INTRODUCTION

Radiochromic dosimeter films are particularly successful for routine dosimetry and have been used extensively in industrial radiation processing for more than ten years because they provide a relatively convenient method of analysis (spectrophotometry), are easily calibrated, and are usually commercially available in large reproducible batches [1]. Since 1977, radiochromic dosimetry has been introduced into the low-, medium- and high-dose intercomparisons organized by the IAEA [2]; thus it is able to cover a wide dose range (0.01–600 kGy) of interest with reliable accuracy. To guarantee that values of dose interpretation under various radiation processing conditions are precise, changes in dosimeter response to changes in environment conditions (temperature, light, relative humidity, the presence of gases, etc.) must be carefully studied during both irradiation and storage.

The present study was carried out in order to examine the variations in response[1] of some radiochromic dosimeter films irradiated by electron beams in vacuum and in the atmospheres of different gases that might occur during radiation processing, and to determine the constancy of optical density after irradiation. Results are given for particular lots and batches of commercial radiochromic dosimeter films; however, allowance must be made for the fact that slight differences in effect may exist.

2. EXPERIMENTAL MATERIAL AND PROCEDURES

2.1. Radiochromic dye dosimeter films

Ten varieties of the dosimeter films used are listed in Table I; the polyethylene (PE) film used has been described in a previous paper [3].

2.2. Irradiation and gas-filling procedure

The dosimeter films were irradiated with electron beams using the Beijing Radiation Centre BJ-5 linear accelerator (beam current 100 μA, energy 3 MeV).

The absorbed dose was checked by both Fricke dosimetry and ceric-cerous sulphate dosimetry. A PE sample frame (8 × 8 mm grates, 1 mm thickness) was used for holding the square film sample on one side (see Fig. 1). It was then wrapped loosely with black PE films for protection from light and placed in a glass tube evacuated to ~ 1×10^{-5} or 5×10^{-6} torr;[2] the tube could then be either

---

[1] 'Response' here is defined as the increase in optical density per unit thickness (ΔOD/mm) as a function of absorbed dose.

[2] 1 torr = 1.333×10^2 Pa.

TABLE I. RADIOCHROMIC DYE DOSIMETER FILMS

Film No.	Description	Analysis wavelengths (nm)	Approximate thickness (mm)	NBS batch No.	Dosimeter supplier
1	Malachite green methoxide in polychlorostyrene ($MG-OCH_3$ in PS-Cl-1) (containing 1% Cl)	630, 430	0.120	A	FWT[a]
2	Malachite green methoxide in polychlorostyrene ($MG-OCH_3$ in PS-Cl-25) (containing 25% Cl)	630, 430	0.083	E	FWT[a]
3	Malachite green methoxide in polybromostyrene ($MG-OCH_3$ in PS-Br-2) (containing 2% Br)	630, 430	0.077	C	FWT[a]
4	Malachite green methoxide in polybromostyrene ($MG-OCH_3$ in PS-Br-43) (containing 43% Br)	630, 430	0.090	D	FWT[a]
5	Hexa(hydroxyethyl) pararosaniline cyanide in nylon film (HPR-CN in nylon)	605	0.053	118	FWT[a]
6	Hexa(hydroxyethyl) pararosaniline cyanide in polyvinyl butyral (HPR-CN in PVB)	600	0.059	37	FWT[a]
7	Hexa(hydroxyethyl) pararosaniline cyanide in polyvinyl pyrollidone (HPR-CN in PVP)	598	0.134	109	FWT[a]
8	Hexa(hydroxyethyl) pararosaniline cyanide in cellulose triacetate (HPR-CN in CTA)	595	0.113	3CR2 HQ III	Risø[b]
9	New fuchsin cyanide in polyvinyl chloride (NF-CN in PVC)	565	0.041	108	FWT[a]
10	New fuchsin cyanide in trogamid (aromatic nylon) (NF-CN in TGM)	563	0.053	Y	FWT[a]

[a] FWT films from Far West Technology, Inc., Goleta, California, United States of America.
[b] Risø films from Risø National Laboratory, Roskilde, Denmark.

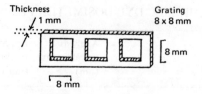

FIG.1. *Diagram of polyethylene sample frame.*

left evacuated or filled with O_2, N_2O or N_2 and finally sealed. Conditioning was completed about one hour before irradiation. The frames with their sample holding side were oriented perpendicularly to the electron beam so that each film sample absorbed the same dose uniformly. After irradiation the films were removed for spectrophotometric readout and were stored for various periods in the dark in air at room temperature and at 50 to 60% relative humidity. Conditionings in different atmospheres before and during irradiation are not meant to achieve saturation conditions but rather to simulate extreme atmospheric conditions that might occur during radiation processing.

2.3. Optical density and thickness measurement

Before and at various times after irradiation, each film was analysed using a Hitachi Model 200-20 double-beam spectrophotometer set at selected wavelengths (shown in Table I). The absorption spectra of the irradiated dosimeter films 1 to 10 were the same as previously obtained [4].

The thickness of the film was measured using an Elecont micrometer made by Mitutoyo, with a reproducibility of ± 0.2 µm.

3. RESULTS AND DISCUSSION

3.1. Response curves in different atmospheres

Three kinds of radiochromic dye precursor (malachite green methoxide, hexa(hydroxyethyl) pararosaniline cyanide and new fuchsin cyanide) were fabricated in ten types of plastic matrices, as indicated in Table I. The effects of various atmospheres during irradiation are shown in the following subsections in terms of response curves, i.e. increases in optical density per unit thickness (ΔOD/mm) at the indicated wavelengths of analysis as a function of absorbed dose.

(a) Polyhalostyrene films

The radiochromic dye films 1 to 4 are polyhalostyrene films (containing malachite green methoxide) used under analysis at two optical wavelengths

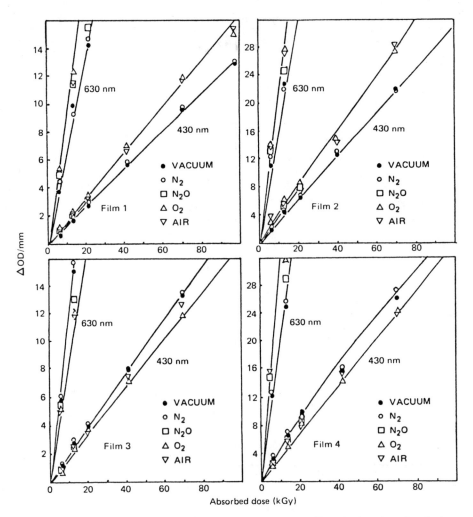

FIG.2. *Response curves (ΔOD/mm versus dose) for dosimeter films 1–4 irradiated with electron beams in different atmospheres and analysed at 630 and 430 nm wavelengths.*

(630 and 430 nm). The response of films 1 to 4 generally did not show much difference in response in different atmospheres (as shown in Fig. 2(a–d)); even in the high dose range (more than 70 kGy) the maximum variation of ΔOD/mm values was around 20% for films 1 and 2 and 13% for films 3 and 4 in different atmospheres. For a given absorbed dose the ΔOD/mm values of films 1 and 2 in N_2O atmosphere were lower than those in O_2 or air but higher than those in N_2 or vacuum, as shown in Fig. 2(a, b) (chlorinated); however, ΔOD/mm values of film 3 (brominate) in N_2O were higher than those in O_2 or air. Film 4, with a greater amount of bromine (see Fig. 2(d)), showed less difference due to different

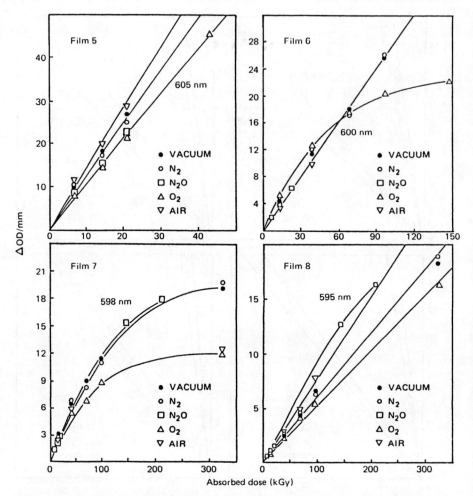

FIG.3. *Response curves (ΔOD/mm versus dose) for dosimeter films 5–8 irradiated with electron beams in different atmospheres and analysed at 605, 600, 598 and 595 nm wavelengths, respectively.*

atmospheres than the effects of films 1 to 3 (Cl-1, Cl-25 and Br 2%) and the ΔOD/mm values of film 4 in N_2O were close to those in O_2 or air. These effects from electron beam irradiation were similar to those from gamma ray irradiation [4].

(b) Nylon, polyvinyl butyral (PVB), polyvinyl pyrollidone (PVP) and cellulose triacetate (CTA) films

The radiochromic dye precursor hexa(hydroxyethyl) pararosaniline cyanide was dissolved and cast in four different matrix materials (nylon, PVB, PVP and CTA

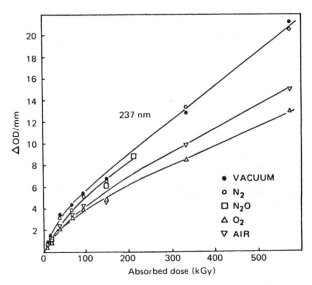

FIG.4. *Response curves ($\Delta OD/mm$ versus dose) for dosimeter PE film irradiated with electron beams in different atmospheres and analysed at 237 nm.*

bases, films 5 to 8, respectively) (Fig.3). Film 5 did not show a greater difference in response in different atmospheres than film 4. However, there were much greater differences in response between different atmospheres for PVB, PVP and CTA, i.e. hydrocarbon polymers base films. Especially in O_2, the $\Delta OD/mm$ values were diminished, and with high absorbed dosage (more than 100 kGy) the $\Delta OD/mm$ values were constant, and the shape of response curves of films 6 and 7 changed form, indicating the significant effect of O_2 lowering of the radiation chemical yield of the radiochromic dye in them as in polar organic solutions [5], i.e. in O_2 atmosphere the precursor of radiochromic dye reacted with O_2 quickly so in this case O_2 was insufficient in the high dose range. Film 8 showed less variation under different atmospheres than films 6 and 7. In films 5 to 8 the response curves in O_2 are the lowest among the atmospheres, i.e. these dosimetry systems give dose value readings lower than those under other conditions. In electron beam irradiation these phenomena are more apparent due to the high dose rate and inadequate O_2 diffusion in the film.

Considering PE film, the response curve involving analysis at 237 nm was related to the formation of the double bonds [3]; the film is affected by oxidation in O_2 atmosphere (Fig.4). This means that the response value ($\Delta OD/mm$) depends on the film varying in different atmospheres, changing upward and following the order of O_2, air, N_2O, vacuum and N_2, as indicated in Fig.4. Hence the matrix material of dosimeter film plays an important role concerning the $\Delta OD/mm$ values in these systems.

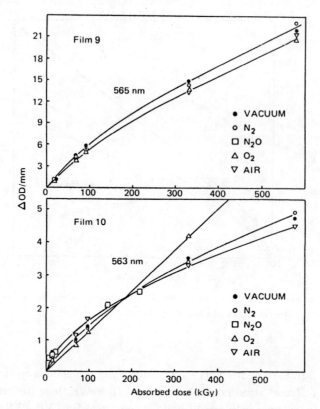

FIG.5. *Response curves ($\Delta OD/mm$ versus dose) for dosimeter films 9 and 10 irradiated with electron beams in different atmospheres and analysed at 565 and 563 nm, respectively.*

(c) PVC and trogamid (TGM) films

The $\Delta OD/mm$ values of films 9 and 10 (PVC base and aromatic nylon (TGM), respectively), where the dye precursor concentrations are low, are suitable for very high-dose measurement.

It is interesting that the response curves of film 9 are very similar to those of the PE film (compare Figs 4 and 5(a)). The $\Delta OD/mm$ values also ascend in the order of O_2, air, N_2O, vacuum and N_2, which are similar to those of PE film. The variation of values ($\Delta OD/mm$) in different atmospheres is under the influence of matrix material and is not directly dependent on O_2 or radiation. The value ($\Delta OD/mm$) of film 10 (TGM) was highest in O_2. The TGM film retained its flexibility even up to 600 kGy, i.e. in O_2, and there is less radiation damage effect on film 10 than on the PVC film.

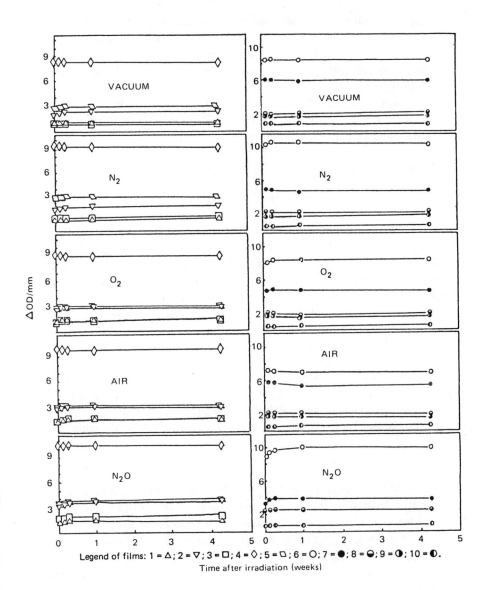

FIG.6. ΔOD/mm of dosimeter films 1 to 10 for different storage times after electron beam irradiation at given high-dose ranges in different atmospheres using the optical wavelengths indicated in Table I for each kind of film.

3.2. Stability of films irradiated in different atmospheres and stored in air

In general, radiochromic films stored in the dark in air after irradiation show fairly good long-term (several weeks) stability, as long as temperatures are below 40°C [6]. The stability of the optical density under different atmospheric conditions with different types of dosimeter films after electron beam irradiation is shown in Fig.6.

There was some instability in optical density readings during the first day after irradiation, particularly with PVB and PVP films, in oxygen, air or N_2O atmospheres and with a high absorbed dose.

(a) Polyhalostyrene films

Figure 6 shows that film 1 (MG-OCH$_3$ in PS-Cl-1) is essentially stable for different storage times up to a month after irradiation in different atmospheres. Film 2 (PS-Cl-25) is somewhat unstable when analysed at 430 nm after irradiation in N_2O atmosphere; the OD/mm values decrease slightly with storage times, as indicated in Fig.6. Film 3 (PS-Br-2) is stable at 430 nm in N_2 and vacuum and increases slightly in air, O_2 and N_2O. The OD/mm values of film 4 (PS-BR-43) are stable with storage times at 430 nm. These results from electron beam irradiation were somewhat different from those of gamma ray irradiation [4].

(b) Nylon film

Film 5 (nylon base) shows excellent, long-term stability, even during the first day after electron beam irradiation; because the temperature under a high dose rate of electron beams was raised as high as ~ 45°C to accelerate slower components of dye formation [7], good stability resulted from the start.

(c) PVB, PVP and CTA films

The $\Delta OD/mm$ values of film 8 (CTA) were the most stable during the first day after irradiation in different atmospheres. The $\Delta OD/mm$ values of film 7 (PVP) with a high absorbed dose decreased slowly after electron beam irradiation under all conditions. Perhaps it is sensitive to humidity during storage; with gamma ray irradiation it showed the opposite effect. The $\Delta OD/mm$ values of film 6 were decreasing slightly after high dosage in air and were stable in other atmospheres.

(d) PVC and TGM films

The $\Delta OD/mm$ of film 9 (PVC) increased slightly with storage time in different atmospheres. For film 10 (TGM) it was stable in oxygen, air and N_2O from the start and increased slightly after irradiation in vacuum and N_2.

In general, the stability of these films for storage time immediately after irradiation with electron beams is better than that with gamma ray irradiation, since the high dose rate of the former warms up the film and accelerates the initial change of dye [7] to a steady situation, and the O_2 effect is limited due to the slower O_2 diffusion than its consumption by the film material.

4. CONCLUSIONS

The electron beam responses of dosimeter films 1 to 4 did not show much difference when irradiated in different atmospheres. Such effects are apparently affected by the gas diffusion properties of the matrix materials of the dosimeter film. Film 8 (CTA) is the type least affected by atmospheric differences. Nylon film 5 and TGM film 10 are fairly resistant to the effects of the different atmospheres; film 10 is more suitable for very high absorbed doses ($\sim 10^6$ Gy), especially in the presence of air or O_2.

The response curve of film 9 is very similar to that of PE film, which implies that the diffusion properties of the matrix material play an important role in the response of the dosimeter system in different atmospheres, and that the plastic dosimeter film must be affected to some degree when using different conditions during irradiation.

The storage stability of these films is related to their responses in different atmospheres. The $\Delta OD/mm$ values of polyhalostyrene, nylon and CTA base are quite stable for long storage times; film 5 (nylon) is stable even during the first day after electron irradiation. Films 6, 7 and 9 are somewhat unstable during the first day or two after irradiation. Electron beams of a high dose rate raise the temperature as high as $\sim 45°C$ and accelerate the change of dye to steadiness. Thus, the stability of these films after electron beam irradiation is greater than that after gamma ray irradiation.

From the above results with electron beams when choosing a radiochromic dosimeter film, it is important to consider the matrix material and also the dye composition, depending on the given atmosphere conditions of irradiation. Accurate dose measurement with film dosimeters thus depends on careful planning of irradiation and storage circumstances.

REFERENCES

[1] KANTZ, A.D., and HUMPHREYS, K.C., Radiat. Phys. Chem. **9** (1977) 737.
[2] NAM, J.W., "The High-Dose Standardization and Intercomparison for Industrial Radiation Processing Programme of the International Atomic Energy Agency", XIII Annual Meeting of the European Society of Nuclear Methods in Agriculture, 6–11 September 1982, Brno, Czechoslovakia (1982).

[3] CHEN, Wenxiu, JIA, Haishen, LU, Xiangdi, LUI, Dongyuan, BAO, Huaying, Radiat. Phys. Chem. **16** (1980) 195.
[4] CHEN, Wenxiu, HUMPHREYS, J.C., McLAUGHLIN, W.L., Reponse of radiochromic dosimeter films to gamma rays in different atmospheres (accepted for publication in Radiat. Phys. Chem.).
[5] McLAUGHLIN, W.L., KOSANIC, M.M., Int. J. Appl. Radiat. Isot. **25** (1974) 249.
[6] LEVINE, H., McLAUGHLIN, W.L., MILLER, A., Radiat. Phys. Chem. **14** (1979) 571.
[7] CHAPPAS, W.J., Radiat. Phys. Chem. **18** (1981) 1017.

IAEA-SM-272/44

Invited Paper

STANDARDIZATION OF HIGH-DOSE MEASUREMENT OF ELECTRON AND GAMMA RAY ABSORBED DOSES AND DOSE RATES

W.L. McLAUGHLIN
Center for Radiation Research,
National Bureau of Standards,
Gaithersburg, Maryland,
United States of America

Abstract

STANDARDIZATION OF HIGH-DOSE MEASUREMENT OF ELECTRON AND GAMMA RAY ABSORBED DOSES AND DOSE RATES.
    Intense electron beams and gamma radiation fields are used for sterilizing medical devices, treating municipal wastes, processing industrial goods, controlling parasites and pathogens, and extending the shelf-life of foods. Quality control of such radiation processes depends largely on maintaining measurement quality assurance through sound dosimetry procedures in the research leading to each process, in the commissioning of that process, and in the routine dose monitoring practices. This affords documentation as to whether satisfactory dose uniformity is maintained throughout the product and throughout the process. Therefore, dosimetry at high doses and dose rates must in many radiation processes be standardized carefully, so that 'dosimetry release' of a product is verified. This standardization is initiated through preliminary dosimetry intercomparison studies such as those sponsored recently by the IAEA. This is followed by establishing periodic exercises in traceability to national or international standards of absorbed dose and dose rate. Traceability is achieved by careful selection of dosimetry methods and proven reference dosimeters capable of giving sufficiently accurate and precise 'transfer' dose assessments: (1) they must be calibrated or have well-established radiation-yield indices; (2) their radiation response characteristics must be reproducible and cover the dose range of interest; (3) they must withstand the rigours of back-and-forth mailing between a central standardizing laboratory and radiation processing facilities, without excessive errors arising due to instabilities, dosimeter batch non-uniformities, and environmental and handling stresses.

1. INTRODUCTION

Electron beams ($\sim$ 1 to 10 MeV), X-rays ($\sim$ 1 to 5 MeV) and radionuclide sources of gamma radiation (^{137}Cs, 0.662 MeV; ^{60}Co, 1.17, 1.33 MeV) are used for radiation treatments of foods, medical and surgical devices, municipal wastes and for selective modification of a wide variety of industrial goods. For such radiation processing, the absorbed doses need to be large, reaching from about

10 Gy to greater than 10^5 Gy. Absorbed rates range from about 0.01 to 10 Gy·s^{-1} for gamma rays and up to instantaneous dose rates as high as 10^{10} Gy·s^{-1} for electron beams. The doses also need to be fairly uniform over large volumes of product, and they need to be delivered in an industrial production-line environment with sufficient accuracy and precision to effect the process without exceeding prescribed uncertainty limits. The method of choice for such quality control is standardized radiation dosimetry, particularly where limits of uncertainty have to be relatively small for the sake of public health and safety. For example, in the irradiation of a large quantity of a given food supply, doses falling below the mean minimum dose designed for a given food preservation process (less than acceptable limits) would result in unacceptable relaxation of specification and ultimate failure of the process. Doses exceeding the design maximum mean dose for each foodstuff must also be controlled to within agreed upon statistical limits. Similar care is required in many other radiation processes.

Routine dosimetry for the measurement of large doses and dose rates, therefore, should be standardized. This standardization generally consists of timely calibration (often at central standards laboratories) of the practical dosimeters chosen for quality control in each dose range of interest. The standardization procedure is often somewhat different from one radiation source to another and from one dose range to another. Typical procedures for standardizing both electron and photon quality control dosimetry are described. They include traceability to primary measurement systems such as calorimeters and ionization chambers, and the use of certain acceptable reference dosimeters such as the Fricke and potassium dichromate aqueous chemical systems, as well as dose intercomparison studies with these reference systems or others showing sufficient stability and reproducibility to allow mailing over large distances (e.g. alanine, ethanol chlorobenzene or radiochromic dosimeters). The differences in dosimetry during routine practice and at the time of commissioning of a radiation process must be emphasized. Dosimeter selection criteria and sources of routine dosimetry uncertainty are also critical factors affecting quality control in radiation processing.

Standardizing electron and photon measurement of absorbed doses and dose rates in practical radiation applications (e.g. in radiation processing, sterilization, and food preservation) has become an acceptable means of quality control [1–3]. By periodic traceability of transfer reference or routine dosimeter calibrations to primary measurements systems, confidence is reached that reasonable, accurate and precise routine dosimetry can be maintained from day to day, as well as over long-term use in a radiation plant [4–9]. Another advantage of such standardization is that correction of problems that arise in dosimetry and quality control is simplified. For example, routine dosimeter readings may stray from the norm, even while parameters for a given commissioned process apparently remain under tight control.

Immediate recalibration of the routine system against a standard may then indicate a random or systematic and a sudden or gradual change in dosimeter response or in some of the process parameters themselves. In addition, a back-up or alternate routine dosimeter may have to be calibrated for specialized applications as, for example, mapping detailed dose distributions in heterogeneous products, correcting anomalous dose readings, or covering a dose range or dose-rate range outside of that specified for the main dosimeter system.

Codes of practice have been established for maintaining quality control by means of high-dose dosimetry [10–14]. The IAEA continues sponsorship of a programme of high-dose intercomparison and standardization of dosimetry for industrial radiation processing by both ionizing photons and electrons [15–18]. Industrial trade associations have developed standardized dosimetry protocols for radiation sterilization applications [19–21]. Food irradiation dosimetry guidelines and standardization procedures are also found in the literature [3, 4, 9, 22].

2. RADIATION MEASUREMENT STEPS FOR MAINTAINING DOSE STANDARDIZATION

There are certain steps that should be taken to minimize dosimetry errors that are apt to occur in processing and testing applications of large radiation sources. These may be listed as follows:

(1) Periodic calibrations of reference and routine dosimeters by a central calibration laboratory (traceability to primary standards)
(2) Demonstration of the capability of making accurate and precise determination of the absorbed dose in a reference material (e.g. water) with a calibrated dosimeter
(3) Correct representation of dose assessments by applying calibration data and appropriate correction factors as, for example, temperature dependence of dosimeter response
(4) Conversion of dose readings in the reference material dose in the product or test material
(5) Determinations of dose distribution patterns, positions and values of minimum and maximum doses for a given irradiation geometry
(6) Maintaining checks on dosimeter batches, dosimeter shelf-life and stability, changes in geometrical factors (size, shape, wall materials, etc.) and environmental effects so that appropriate corrections can be made
(7) Specifying and adjusting process and test parameters, i.e. relating such parameters as conveyor speed, source decay, product geometry, beam current, bulk density, dwell time, etc. to minimum and maximum values of absorbed dose, as in the commissioning of a radiation process or test procedure
(8) Correction of dosimetry anomalies whenever routine dosimetry results show obvious errors, as when routine dosimeter quality control itself may fail or when unexpected environmental or geometrical discontinuities occur

(9) Testing of the properties of the product or test material itself such as microbiological assays in the case of radiation sterilization or standard mechanical, electrical or thermal tests of materials, as a means of monitoring the established relationships between routine dosimetry results and prior dose-setting specifications

(10) Thorough documentation, in the form of dosimetry record keeping, statistics procedures and accurate communications between technical and administrative levels.

3. ROLES OF THE CENTRAL STANDARDIZING LABORATORY

The central laboratory responsible for standardizing high-dose measurements not only establishes, verifies and maintains primary standard instruments such as calorimeters or ionization chambers through formal international and national intercomparison exercises but it also makes those standards available through calibration services.

The United States National Bureau of Standards (NBS) offers on a fee-schedule basis dosimetry calibration and test services at high doses (10^1 to 10^6 Gy) using electron sources in the energy range 0.05 to 10 MeV and X-rays and gamma rays in the energy range 0.01 to 5 MeV. These services are as follows:

(1) User irradiates NBS-supplied transfer dosimeter assemblies (generally packaged calibrated radiochromic dosimeters) and returns them to the NBS for readings of absorbed dose in a material of interest
(2) Samples of users' transfer dosimeters are sent to the NBS for irradiation to standard doses
(3) Transfer dosimeters are exchanged for round-robin dosimetry intercomparisons between laboratories
(4) The NBS irradiates or receives irradiated dosimeters and tests them in terms of stability, environmental effects during storage, readout by standardized spectrophotometry, thickness measurement, etc.
(5) The NBS tests environmental effects during and after irradiation on dosimeter response (e.g. temperature and relative humidity)
(6) The NBS makes detailed statistical evaluation of the reproducibility of dosimeter response characteristics
(7) The NBS personnel make custom dose distribution or dose-extreme determinations on location (commissioning dosimetry).

By such periodic checking of the calibration of a proven transfer dosimeter, measurement quality assurance can be maintained in radiation processing and testing to the satisfaction of regulatory bodies, management and the consumer alike.

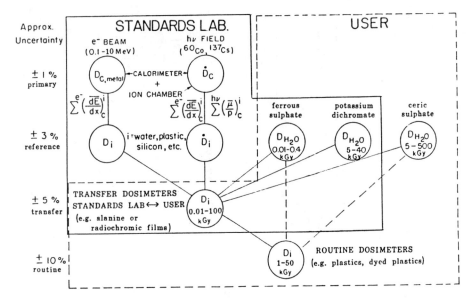

FIG.1. *Chart of traceability in the calibrations and standardization of high-dose dosimeters* [13]. *On the left are estimations of uncertainty for the four levels of traceability from routine (bottom) to primary (top) dosimeters. The standards laboratory deals with systems within the solid-line block and the radiation user within the dashed-line block. Examples of primary, reference, transfer and routine dosimeters are indicated.*

4. TRACEABILITY THROUGH TRANSFER DOSIMETERS

A chart of typical dosimetry traceability to primary standards is illustrated in Fig.1 [13]. Such traceability can be achieved only if reliable high-dose transfer dosimeters are available. They may be specially designed calorimeters for electron beams [23, 24], and can also serve as primary standards. Transfer dosimeters may also consist of certain well-established chemical or physical measurement systems having sufficient stability and ruggedness to withstand the rigours of extreme environments and transportation between laboratories. The latter types of dosimeters are generally used for X-ray and gamma-ray high-dose standardization.

Transfer dosimeter candidates, some of which have been tested in recent IAEA high-dose intercomparison exercises [15–18], are listed in Tables I and II along with their methods of analysis and nominal absorbed dose ranges and sample references [25–41].

The favourable properties of transfer dosimeters are listed in Table III. None of the dosimeters listed in Tables I and II completely satisfies all the ideals listed

TABLE I. LIQUID CHEMICAL DOSIMETERS FOR HIGH DOSES

Solute	Solvent	Methods of analysis	Approximate usable range of absorbed dose (Gy)	Reference
Ferrous sulphate	Aerated aqueous H_2SO_4	UV spectrophotometry	$2 \times 10 - 2 \times 10^2$	[25]
Ceric-cerous sulphate	Aerated aqueous H_2SO_4	UV spectrophotometry; electrochemical potentiometry	$10^3 - 10^6$	[26]
Potassium dichromate	Aerated aqueous perchloric acid + Ag^+ ion	UV, visible spectrophotometry	$5 \times 10^3 - 4 \times 10^4$	[27]
Chlorobenzene	Aqueous ethanol	Titration with indicator; High frequency oscillometry; conductivity	$10 - 10^6$	[28, 29]
Radiochromic leuco dyes	Various	Visible spectrophotometry	$10 - 10^4$	[30]

TABLE II. SOLID CHEMICAL DOSIMETERS FOR HIGH DOSES

Material used	Method of analysis	Approximate usable dose range (Gy)	Reference
Solid state			
Amino acids	ESR	$1 - 10^5$	[31]
Amino acids	Lyoluminescence	$10^3 - 10^6$	[32]
Glucosides	Lyoluminescence	$1 - 10^3$	[33]
Glucosides	Optical rotation	$10^4 - 10^6$	[34]
LiF	TLD or spectrophotometry	$10^{-5} - 10^5$ $10^2 - 10^9$	[35, 36] [35, 36]
Plastics			
Cellulose triacetate	UV spectrophotometry	$10^4 - 10^6$	[37]
Polymethyl methacrylate	UV spectrophotometry	$10^3 - 10^5$	[38]
Dyed plastics			
Polymethyl methacrylate	Visible spectrophotometry	$10^3 - 5 \times 10^4$	[39]
Polyamide (nylon)	Visible spectrophotometry	$10^2 - 10^5$	[40]
Polyvinyl butyral	Visible spectrophotometry	$10^2 - 10^5$	[41]
Cellulose triacetate	Visible spectrophotometry	$10^3 - 10^5$	[40]

in Table III. Nevertheless, the following steps are being taken by the IAEA towards solving this problem [16–18]:

(1) Conduct systematic international dosimetry intercomparisons
(2) Develop calorimeters as high-dose primary standards
(3) Improve knowledge of radiation spectra and energy-dependence data
(4) Establish improved quality control and ease of use of transfer dosimeters
(5) Develop improved routine dosimeters.

TABLE III. FAVOURABLE PROPERTIES
OF TRANSFER DOSIMETERS

Long shelf-life

Easily calibrated

Stability   Mailability
Ruggedness   Intercomparability
Portability

Broad range of doses

Energy independence

Insensitivity to environmental extremes

Correctable systematic errors

Mass produced in reproducible lots

Relatively inexpensive

Simple and reproducible readout

5. DOSIMETRY INTERCOMPARISONS

During the period 1977 to 1983 the IAEA conducted intercomparison studies that tested candidate transfer dosimeter systems over long distance mailing and under severe climate conditions. In the 1983 to 1984 period the Agency started a Pilot Dose Assurance Service, using as its transfer system the alanine dosimeter administered by the Gesellschaft für Strahlen- und Umweltforschung (GSF) in Neuherberg, Federal Republic of Germany [25]. This initial service consisted of mailing a group of calibrated transfer alanine dosimeters to 15 worldwide radiation processing plants for a series of specified irradiations to a given high dose, using the plants' regular ^{60}Co sources. Irradiations were made along with the product in a routine processing situation. The irradiated dosimeters were returned to the GSF by mail for dose assessments. The mean dose readings and error bars in Fig.2 show that, with one exception, the mean dose readings were within 10% of the target administered dose in the plants, and with the exception of four plants the reproducibility of administering the target dose was within ± 10%.

In 1984 an annually scheduled dosimetry intercomparison study was conducted between the NBS and the United Kingdom National Physical Laboratory (NPL) [42]. In this intercomparison ferrous sulphate (Fricke) solution dosimeter ampoules [25] and potassium dichromate solution dosimeter ampoules [27] prepared at the NPL were sent to the NBS for irradiation with ^{60}Co gamma

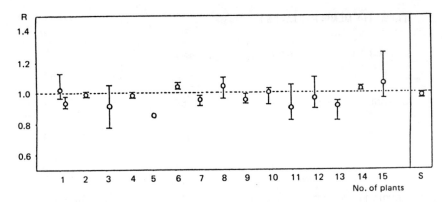

FIG.2. *For the 1983 IAEA Pilot Dose Assurance Service, mean values of the ratio (R) of nominal dose administered by 15 irradiation plants (^{60}Co sources) to the assessed dose by the central administering laboratory (GSF) by means of calibration of the transfer dosimeter (alanine ESR). The laboratory indicated by 'S' was the participating national standards laboratory (NPL). The error bars indicate the extreme limits of ratios of dose values.*

TABLE IV. NPL/NBS DOSE INTERCOMPARISON (1984)

Fricke (ferrous sulphate solution dosimeter)

Dose (kGy)		Mean ratio (NPL/NBS)	Standard deviation 2σ (%)
0.075		1.015	0.9
0.100		1.008	0.7
0.125		1.010	1.0
0.150		0.998	1.0
0.200		0.992	0.9
	Overall	1.005	1.2

Potassium dichromate solution dosimeter

Dose (kGy)		Mean ratio (NPL/NBS)	Standard deviation 2σ (%)
7.5		1.000	0.3
15.0		0.996	1.0
25.0		1.000	0.8
30.0		1.001	0.6
40.0		1.003	0.6
	Overall	1.000	0.7

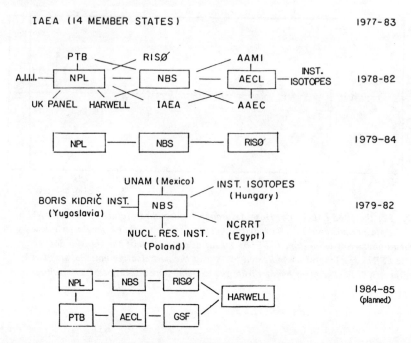

FIG.3. Chart of recent international intercomparison studies for testing the performance of reference and transfer high-dose dosimeters irradiated by gamma radiation from ^{60}Co sources.

rays under standardized conditions at the same time as dosimeter samples were irradiated with ^{60}Co gamma rays at the NPL. The dosimeters irradiated at the NBS were then returned to the NPL for dose assessment. The results are shown in Table IV and reveal reasonable agreement between NBS-administered absorbed doses in water and dose readings and calibration at the NPL, considering the long-term transit times between countries (about 10 days in each direction) and the relatively short irradiation times of Fricke dosimeters (45.5 to 122.5 s).

In addition to these two intercomparisons, other dosimetry studies involving the NBS during the past six years are diagrammed in Fig.3. These generally involve either mailing of calibrated transfer dosimeters between laboratories, or hand-carrying such dosimeters and readout apparatus.

6. SELECTION AND USE OF ROUTINE DOSIMETERS

Routine dosimeters are generally placed at strategic locations in irradiated materials being processed or tested. The main requirements for dosimeters used for routine dose monitoring in radiation processing and testing are availability in

TABLE V. SOURCES OF UNCERTAINTY (ROUTINE DOSIMETERS)

Reproducibility of calibration functions within a given dosimeter batch

Shelf-life stability (repeatability of response function with age)

Stability of response function after irradiation (archival properties)

Dose-rate dependence of response

Spectral energy dependence of response

Response function curve shape, non-linearity, saturation effects

Environmental effects a response function (temperature, humidity, gases, light, etc.)

Uncertainties in radiation-induced signal analysis

---

large reproducible batches and capability of being calibrated, with a radiation response function that is suitably free of random error and correctable for systematic error due to environmental effects. They should also have a relatively long shelf-life and be reasonably compact, rugged and have a radiation signal that covers the dose range and dose-rate range of interest and can be readily analysed in a precise manner. Sources of uncertainty of the dose assessment by calibrated routine dosimeters are listed in Table V.

Most useful routine dosimeters for high-dose applications, including those listed in Tables I and II, have nominal uncertainty limits ranging between about 2 and 10%. Dosimeter selection is then based on this value as well as on convenience for the purpose at hand, such as matching the dosimeter to the radiation absorption properties of the material of interest. The routine dosimeter size and shape are also key factors, especially when detailed mapping of absorbed dose distributions is required. It is a continuous challenge to develop and improve routine dosimeters and dosimetry procedures for the many high-dose applications in industry, agriculture, medicine and materials testing.

Besides the effects of dose-rate dependence and of ambient environmental changes on routine dosimeter response, often neglected sources of error are the effects of self-attenuation by the dosimeter assembly and of radiation scatter and spectral degradation due to surrounding materials [43, 44]. For example, when an assembly of calibrated routine dosimeters is placed in or on an irradiated package corrections are necessary for differences in attenuation by the surrounding materials and by the dosimeter itself, if the quantity of interest is the position-sensitive value of absorbed dose had the dosimeter assembly not been there. If the response of the dosimeter is known to depend on the spectral energy at a given location in the package, some rough estimation of that spectrum may be recommended so that energy-dependence corrections can be accomplished. Solutions to these difficulties lie in matching the geometry of practical irradiation with that of calibration and matching as closely as possible the material of the dosimeter with that of the packaged substance of interest.

7. DISCUSSION

The ability to provide accurate and precise measurement of large absorbed doses in radiation processing and testing of materials is based on establishing and maintaining measurement quality assurance (MQA) of routine dosimeter performance. In this way, the radiation user can comply with regulations that impose certain limits on administered dose to products. Also provided is documentation that products have been given sufficient levels of dose to effect a process or test, and irradiation that does not exceed the prescribed upper limits of dose. Thus, it is through the periodic standardization and traceability procedures that quality control may be achieved in radiation processing and testing.

The critical elements of a reliable high-dose MQA system include: (1) primary radiation measurement standards that serve as a reference base against which routine measurements can be compared by interested radiation industries; (2) a set of reference dosimeters that are capable of achieving appropriate accuracy and whose measurement performances are sufficiently independent of changes in environmental factors such as temperature, humidity, pressure, light or spectral energy of the radiation being used; (3) a set of portable radiation measurement devices (transfer dosimeters) that can be calibrated against primary standards and then be transported or mailed to private-sector radiation facilities to calibrate industrial measurements for processing or testing, large radiation sources, beams and fields, or to check performance of private calibration or test facilities; (4) an 'agreed upon' set of routine measurement devices and MQA procedures and documentation by means of periodic performance checks, calibrations and field-level measurements in radiation processing or testing facilities.[1]

REFERENCES

[1] McLAUGHLIN, W.L., "Radiation measurements and quality control", Radiation Processing (Trans. 1st Int. Meeting, Puerto Rico, 1976) (SILVERMAN, J., VAN DYKEN, A., Eds), Radiat. Phys. Chem. **9** (1977) 147–181.
[2] MILLER, A., CHADWICK, K.H., NAM, J.W., "Dose assurance in radiation processing plants", Radiation Processing (Trans. 4th Int. Meeting Dubrovnik, 1982) (MARKOVIĆ, V., Ed.), Vol. 1, Radiat. Phys. Chem. **22** (1983) 31–40.

---

[1] 'Agreed upon' here means decided by each laboratory or facility to achieve and maintain quality control according to its own particular requirements.

[3] McLAUGHLIN, W.L., MILLER, A., URIBE, R.M., "Radiation dosimetry for quality control of food preservation and disinfestation", Radiation Processing (Trans. 4th Int. Meeting Dubrovnik, 1982) (MARKOVIĆ, V., Ed.), Vol. 1, Radiat. Phys. Chem. **22** (1983) 21–29.

[4] CHADWICK, K.H., EHLERMANN, D.A.E., McLAUGHLIN, W.L., Manual on Food Irradiation Dosimetry, Technical Reports Series No. 178, IAEA, Vienna (1977).

[5] CHADWICK, K.H., "Dosimetry techniques for commissioning a process", Sterilization by Ionizing Radiation (Proc. 1st Symp. Vienna, 1974) (GAUGHRAN, E.R.L., GOUDIE, A.J., Eds), Vol. 1, Multiscience, Montreal (1974) 285–298.

[6] CHADWICK, K.H., "Facility calibration, the commissioning of a process, and routine monitoring practices", Radiosterilization of Medical Products (Proc. Symp. Bombay, 1974), IAEA, Vienna (1975) 69–81.

[7] CHADWICK, K.H., "Radiation measurements and quality control", Advances in Radiation Processing (Trans. 2nd Int. Meeting Miami, 1978) (SILVERMAN, J., Ed.), Radiat. Phys. Chem. **14** (1979) 203–212.

[8] CHADWICK, K.H., "Precision and accuracy in radiation processing and sterilization", Ionizing Radiation Metrology (CASNATI, E., Ed.), Editrice Compositori, Bologna (1977) 475–487.

[9] McLAUGHLIN, W.L., JARRETT, R.H., OLEJNIK, T.A., "Dosimetry", Preservation of Food by Ionizing Radiation (JOSEPHSON, E.S., PETERSON, M.S., Eds), Vol. 1, CRC Press, Boca Raton, Florida (1981) 189–245.

[10] McLAUGHLIN, W.L., "Dosimetry standards for industrial radiation processing", National and International Standardization of Radiation Dosimetry (Proc. Symp. Atlanta, 1977), Vol. 1, IAEA, Vienna (1978) 89–106.

[11] ELLIS, S.C., "The provision of national standards of absorbed dose for radiation processing, the role of NPL in the United Kingdom", High-Dose Measurements in Industrial Radiation Processing, Technical Reports Series No.205, IAEA, Vienna (1981) 7–16.

[12] McLAUGHLIN, W.L., "A national standardization programme for high-dose measurements", ibid., 17–32.

[13] McLAUGHLIN, W.L., HUMPHREYS, J.C., MILLER, A., "Dosimetry for industrial radiation processing", Traceability in Ionizing Radiation Measurements (Proc. Conf. NBS Gaithersburg, 1980), NBS Special Publication 609, National Bureau of Standards, Gaithersburg (1981) 171–178.

[14] ELLIS, S.C., BARRETT, J.H., MORRIS, W.T., "Radiation standards and dosimetry for radiation processing", Radiation Processing for Plastics and Rubber (Proc. 2nd Int. Conf. Canterbury, 1984) (in press).

[15] CHADWICK, K.H., "An international service of dose assurance in radiation technology", IAEA Bulletin **24** (1982) 3.

[16] CHADWICK, K.H., "IAEA Advisory Group Meeting on Dosimetry for High Doses Employed in Industrial Radiation Processing, Nov. 1980, Summary Report", Biomedical Dosimetry: Physical Aspects, Instrumentation, Calibration (Proc. Symp. Paris, 1980), IAEA, Vienna (1981) 573–578.

[17] NAM, J.W., "The high-dose standardization and intercomparison for industrial radiation processing programme of the International Atomic Energy Agency", High Dose Dosimetry in Industrial Radiation Processing (Lecture IAEA Seminar, Risø National Laboratory, Roskilde, 1982), IAEA, Vienna (1982).

[18] NAM, J.W., Assuring good irradiation practice, IAEA Bulletin **25** 1 (1983) 37.

[19] ASSOCIATION INTERNATIONALE D'IRRADIATION INDUSTRIELLE, Code of Practice of AIII, Radiosterilization of Medical Products by Electron Beams and Gamma Rays, AIII, Paris (1975).

[20] ASSOCIATION FOR THE ADVANCEMENT OF MEDICAL INSTRUMENTATION, AAMI Recommended Practice, Process Control Guidelines for Gamma Radiation Sterilization of Medical Devices (proposed), AAMI, Arlington (1984).

[21] INTERNATIONAL ATOMIC ENERGY AGENCY, "Code of practice for radiosterilization of medical products", Radiosterilization of Medical Products (Proc. Symp. Budapest, 1967), IAEA, Vienna (1967) 423–431.

[22] VAS, K., BECK, E.R.A., McLAUGHLIN, W.L., EHLERMANN, D.A.E., and CHADWICK, K.H., "Dose limits versus dose range", Food Irradiat. Newsl. 7 2 (1978) 343–349.

[23] HUMPHREYS, J.C., McLAUGHLIN, W.L., private communication (1984).

[24] MILLER, A., Dosimetry for Electron Beam Applications, Risø National Laboratory, Roskilde, Rep. M-2401 (1983).

[25] ELLIS, S.C., "The dissemination of absorbed dose standards by chemical dosimetry, mechanism, and use of the Fricke dosimetry", Ionizing Radiation Metrology (CASNATI, E., Ed.), Editrice Compositori, Bologna (1977) 163–180.

[26] BJERGBAKKE, E., "The ceric sulfate dosimeter", Manual on Radiation Dosimetry (HOLM, N.W., BERRY, R.J., Eds), Marcel Dekker, New York (1970) 323–326.

[27] SHARPE, P.H.G., BARRETT, J.H., BERKLEY, A.M., Use of dichromate solutions as reference dosimeters in the 10–40 kGy range, Int. J. Appl. Radiat. Isot. **36** (in press).

[28] RAŽEM, D., ANDELIĆ, L., DVORNIK, I., Paper IAEA-SM-272/13, these Proceedings.

[29] KOVÁCS, A., STENGER, V., FÖLDIÁK, G., LEGEZA, L., Paper IAEA-SM-272/33, these Proceedings.

[30] RATIVANICH, N., RADAK, B.B., MILLER, A., URIBE, R.M., McLAUGHLIN, W.L., "Liquid radiochromic dosimetry", Trends in Radiation Processing (Trans. Int. Meeting Tokyo, 1980) (SILVERMAN, J., Ed.), Vol. III, Radiat. Phys. Chem. **18** (1981) 1001–1110.

[31] REGULLA, D., DEFFNER, U., "Dosimetry by ESR spectroscopy of alanine", Trends in Radiation Dosimetry (McLAUGHLIN, W.L., Ed.), Pergamon Press, Oxford; Int. J. Appl. Radiat. Isot. **33** (1982) 1101–1114.

[32] THWAITES, D.E., BUCHAN, G., ETTINGER, K.V., MALLARD, J.R., TAKAVAR, A., A new sensitive technique for study of radiation effects in amino acids, Int. J. Appl. Radiat. Isot. **27** (1976) 663–664.

[33] ETTINGER, K.V., PUITE, K.J., "Lyoluminescence dosimetry", Trends in Radiation Dosimetry (McLAUGHLIN, W.L., Ed.), Pergamon Press, Oxford (1982); Int. J. Appl. Radiat. Isot. **33** (1982) 1115–1159.

[34] SUCHNY, O., FRITTUM, H., EISENLOHR, H., "Radiation dosimetry by measurement of the change in rotation of optically active substances", Solid-State and Chemical Radiation Dosimetry (Proc. Symp. Vienna, 1967), IAEA, Vienna (1967) 317–330.

[35] CAMERON, J., "Lithium fluoride thermoluminescent dosimetry", Manual on Radiation Dosimetry (HOLM, N.W., BERRY, R.J., Eds), Marcel Dekker, New York (1970) 405.

[36] McLAUGHLIN, W.L., LUCAS, A.C., KAPSAR, B.M., MILLER, A., "Electron and gamma-ray dosimetry using radiation-induced color centers in LiF", Advances in Radiation Processing (Trans. Int. Meeting Miami, 1978) (SILVERMAN, J., Ed.), Vol. II, Radiat. Phys. Chem. **14** (1979) 467–480.

[37] TANAKA, R., MITOMO, S., TAMARA, N., Factors influencing the sensitivity of cellulose triacetate dosimeters upon irradiation with electron beams and gamma rays, Radiat. Phys. Chem. **23** (in press).

[38] CHADWICK, K.H., "Solid state dosimetry at high doses", Ionizing Radiation Metrology (CASNATI, E., Ed.), Editrice Compositori, Bologna (1977) 195–211.

[39] BARRETT, J.H., "Dosimetry with dyed and undyed acrylic plastic", Trends in Radiation Dosimetry (McLAUGHLIN, W.L., Ed.), Pergamon Press, Oxford (1982); Int. J. Appl. Radiat. Isot. **33** (1982) 1177–1187.

[40] McLAUGHLIN, W.L., URIBE, R.M., MILLER, A., "Megagray dosimetry, or monitoring of very large radiation doses", Radiation Processing (Trans. 4th Int. Meeting, Dubrovnik, 1981) (MARKOVIĆ, V., Ed.), Vol. II, Radiat. Phys. Chem. 22 (1982) 333–362.

[41] MILLER, A., McLAUGHLIN, W.L., "Evaluation of radiochromic dye films and other plastic dose meters under radiation processing conditions", High-Dose Measurements in Industrial Radiation Processing, Technical Reports Series No. 205, IAEA, Vienna (1981) 119–138.

[42] SHARPE, P.H.G., BARRETT, J.H., private communication (1984).

[43] McLAUGHLIN, W.L., "The measurement of absorbed dose and dose gradients", Radiation Sterilization of Plastic Medical Devices (Proc. Seminar Lowell, Massachusetts, 1979) (MANN, H.K., Ed.), Radiat. Phys. Chem. 15 (1980) 9–30.

[44] MILLER, A., McLAUGHLIN, W.L., "Calculation of the energy-dependence of dosimeter response to ionizing photons", Trends in Radiation Dosimetry (McLAUGHLIN, W.L., Ed.), Pergamon Press, Oxford; Int. J. Appl. Radiat. Isot. 33 (1982) 1299–1310.

IAEA-SM-272/6

CALIBRATION AND INTERCOMPARISON OF RED 4034 PERSPEX DOSIMETERS

K.M. GLOVER, M. KING, M.F. WATTS
UKAEA Atomic Energy Research Establishment,
Harwell, Didcot, Oxfordshire,
United Kingdom

Abstract

CALIBRATION AND INTERCOMPARISON OF RED 4034 PERSPEX DOSIMETERS.

Accurate dosimetry measurements are essential for the successful commercial application of radiation processing where the safety of the product is a function of the absorbed dose, as in sterilization and food processing. To achieve a reproducible response to radiation, red 4034 Perspex dosimeters are conditioned for several months in moist air at an elevated temperature. During processing effective quality control is essential to ensure that dosimeters within a batch have a reproducible response to radiation and meet the specified accuracy of ±2% for any particular batch. The dosimeters are calibrated in a standard ^{60}Co cell at Harwell, the dose rate of which has been intercompared through Fricke dosimetry with the National Physical Laboratory, thus establishing traceability to a National Standard. In addition, each batch of red 4034 dosimeters is subject to a rigorous intercomparison exercise with NPL before being released for sale. Response curves and intercomparison data will be presented to demonstrate that red 4034 dosimeters consistently achieve the accuracy specified.

1. INTRODUCTION

Red 4034 Perspex dosimeters were originally developed in support of the UKAEA investigations into the possible industrial applications of radiation processing, which led to the design and operation of the first ^{60}Co package irradiation plant in 1960. The requirement was for a rugged dosimetry system sensitive in the range 5-50 kGy and capable of accurately measuring an integral dose of 25 kGy which had then been established as that gamma radiation dose which would safely sterilise medical dressings and equipment [1]. A wide range of polymers and polymer dye systems were studied from which red 4034 emerged as an empirical but suitable system with the required working range and a radiation response at 25 kGy which was measurable reproducibly, with minimum post-irradiation fading [2]. On irradiation to doses in excess of 1 kGy red 4034 darkens due to the formation of a new absorption band in the 600-700 nanometer range with a peak at 615 nanometers. The darkening is related to absorbed dose which is measured in terms of optical density at 640 nanometers, where post-irradiation fading is at a minimum. The radiation-induced colour change is measured using a conventional spectrophotometer and a radiation response curve is obtained.

Since 1960 demand for red 4034 Perspex dosimeters has increased steadily in line with the world-wide expansion in radiation sterilisation, particularly since the mid-seventies.

From the establishment of radiation sterilisation in the sixties, national regulatory bodies such as the Department of Health and Social Security (DHSS) in the UK have applied controls to ensure that the correct sterilising dose is given and measured using well established dosimetry systems. In a continuous gamma irradiation plant the dose absorbed at a particular location in a package depends on the source strength, the residence time and the density of the materials passing through the plant. Routine in-plant dosimeters are used to ensure that every unit pack receives at least the required minimum dose and that excessive doses which might damage the product are not given. The rapid expansion in radiation sterilisation has drawn the attention of plant operators to the need for accurate dosimetry systems for routine process verification not only to fulfil the requirements of regulatory bodies but also to optimise plant performance and cost-effectiveness. In this context all Harwell dosimeters used for routine process verification in gamma-irradiation plants, specifically red 4034, have been subjected to stricter quality control procedures. Before release to the market each batch is subject to an intercomparison exercise with the National Physical Laboratory (NPL). Additionally, when opportunities arise, intercomparisons are undertaken with other organisations concerned with radiation processing.

2. RED 4034 PERSPEX DOSIMETERS

Red 4034 Perspex is manufactured in batches in the form of cast sheet 3 ± 0.55 mm thick, from which dosimeters 30mm x 11mm are machined, washed and subjected to a conditioning treatment extending over several months. During the conditioning period the radiation response of each batch of dosimeters is regularly monitored until the required reproducibility is achieved throughout the batch. The batch response is then measured in the Harwell calibration cell.

3. HARWELL CALIBRATION FACILITY

Fig. 1 is a schematic diagram of the Harwell calibration cell. This cell contains four ^{60}Co sources, each of strength

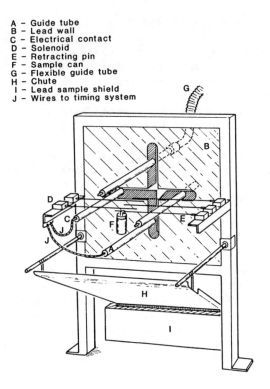

A — Guide tube
B — Lead wall
C — Electrical contact
D — Solenoid
E — Retracting pin
F — Sample can
G — Flexible guide tube
H — Chute
I — Lead sample shield
J — Wires to timing system

FIG.1. Harwell irradiation assembly.

presently around 7.40×10^{13} Bq. In the safe position the sources are housed in dry store in a sand-filled pit enclosed by four foot thick concrete walls. Four electric motors drive the sources by means of teleflex cables from the store through stainless steel guide tubes to the irradiate position. When the sources reach the irradiate position electrical contacts are made which start an electronic timing system. Irradiations are made in aluminium isotope cans 35mm nominal diameter and 70mm long, with screw-on lids. These are hung, prior to exposure of the sources, on steel pins attached to five independently operated solenoids, one controlling each irradiation position. The latter are controlled by the external timing system to cover pre-set times from 1 to 999 999 seconds to within ± 1 second. After the pre-set time for each irradiation position, the pin is released remotely allowing the can to fall into the shielded enclosure shown in Fig. 1. All the cans are retrieved on completion of the irradiation.

4. NATIONAL PHYSICAL LABORATORY FACILITY

The Chemical & Solid State Dosimetry Group of the Radiation Science & Acoustics Division of NPL operates three ^{60}Co irradiators for the calibration of dosimeters at high and medium dose levels [3]. The measurements in these ^{60}Co fields are traceable to National Standards maintained at NPL. A Gammacell 220 irradiator has been used for dosimeter intercomparisons with Harwell. This contains an annular ring of ^{60}Co pencils within a lead shield. The irradiation compartment, a cylindrical aluminium tube of three litres capacity, is driven into the middle of this ring for an accurately timed period.

5. DOSE RATE MEASUREMENTS IN THE HARWELL CALIBRATION CELL

In calibrating a batch of dosimeters which are to be used routinely to measure absorbed dose at different locations in an irradiated product, it is essential to ensure that the calibration is traceable to an accepted reference standard which in turn is traceable to a national standard. Ferrous ammonium sulphate solution (Fricke) is used as the reference dosimeter for measuring the dose rates of the five irradiation positions in the Harwell calibration cell. The Fricke solutions used are prepared according to procedures described elsewhere [4]. Irradiations are carried out in polythene ampoules filled immediately prior to irradiation.

In the range 40 – 400 Gy the ferric ion yield is proportional to the radiation dose and is determined from the height of the ferric ion absorption peak at 304 nanometers, measured spectrophotometrically in a cell thermostatted to 25 ± 0.1°C. The radiation dose is calculated from the formula

$$\text{Dose (rads)} \; \phi = k \frac{D}{\ell}$$

where k is a constant, D is the optical density and ℓ the optical path length.*

The ferric ion extinction coefficient was determined for each spectrophotometer on a range of standard solutions of varying concentrations, prepared by making dilutions from stock solutions.

During July and August 1981 the dose rates of the five positions in the Harwell calibration cell were re-measured

---

* 1 rad = 1.00 x 10^{-2} Gy.

using Fricke. The extinction coefficient of the Pye Unicam SP 1700 spectrophotometer used for optical density measurements was found to be 2137 ± 6 (1σ), the mean of twelve measurements on four dilutions from each of three stock solutions, giving a value for k of 28457.

For the calibration cell dose rate measurements the polytainers are irradiated for twenty five different time intervals to give a range of doses between 0.1 - 0.4 kGy. Eight Fricke solutions were measured; the results are shown in Table I. The errors quoted on the mean dose rate values are one sigma.

6. HARWELL AND NPL INTERCOMPARISONS OF IRRADIATION FACILITIES AND DOSIMETRY SYSTEMS

6.1 Fricke

6.1.1 Fricke 1981

During the latter half of 1981 the irradiation facilities at Harwell and NPL were intercompared using Fricke dosimeters prepared by both establishments. On 27.8.81 Harwell Fricke solutions were irradiated in the NPL Gammacell at a dose rate of approximately 0.8 $Gy.s^{-1}$. A dose rate in good agreement with the value expected from NPL Fricke was obtained from the NPL optical density measurements made on 27.8.81, from which the ratio Harwell/NPL = 1.01 with a standard error of 0.01 was obtained [3]. The Harwell measurements made on 28.8.81 gave a dose rate of 48.82 $Gy.min^{-1}$ compared to the NPL quoted value of 50.32 $Gy.min^{-1}$ a discrepency for which there was no satisfactory explanation, although exposure to sun during transport after irradiation and subsequent temperature fluctuations were thought to have been contributing factors.

The experiment was repeated on 24.9.81. The results are shown in Table II [3]. Good agreement was obtained for the spectrophotometric response and for the Gammacell dose rate. On 12.11.81 Fricke dosimeters supplied by NPL were irradiated in the Harwell cell at a nominal dose rate of 1.6 $kGy.h^{-1}$ for optical density measuremnts at both laboratories to determine the dose rate in all five irradiation positions of the Harwell cell. Three sets of dosimeters were irradiated for each laboratory to doses not exceeding 0.1 kGy. The derived dose rates are shown in Table III [3]. The data collected during the intercomparison showed that Fricke dosimeters prepared by both NPL and Harwell had a similar response and agreement to within 1% for the estimates of dose in both facilities was achieved.

TABLE I. Dose rate measurements using Harwell Fricke in Harwell calibration cell July 1981. Data corrected to 5.8.81

Irradi-ation Position	Dose Rates Gy.s^{-1} corrected to 5.8.81 Fricke Solutions								Mean Values Gy.s^{-1} ± 1σ
	6	7	8	9	10	11	12	13	
1	0.4723	0.4803	0.4721	0.4656	0.4723	0.4743	0.4598	0.4611	0.4697 ± 0.007
2	0.4802	0.4843	0.4902	0.4804	0.4896	0.4806	0.4728	0.4781	0.4820 ± 0.006
3	0.4571	0.4542	0.4607	0.4513	0.4394	0.4579	0.4623	0.4453	0.4510 ± 0.008
4	0.4779	0.4814	0.4778	0.4735	0.4831	0.4738	0.4688	0.4703	0.4758 ± 0.005
5	0.4661	0.4678	0.4778	0.4724	0.4558	0.4619	0.4587	0.4611	0.4652 ± 0.007

TABLE II. Response of NPL and Harwell spectrophotomers and Gammacell dose rates derived from Harwell Fricke [3]

Measuring Laboratory	Spectrophotometer response O.D. (cm^{-1})	Gammacell Dose rate Gy.min^{-1}
Harwell	0.1746 (0.0007)	49.77
NPL	0.1757 (0.0006)	49.82 *
Harwell/NPL	0.994 (0.0005)	0.999

* Dose rate based on measurements with NPL Fricke used as a transfer from the national exposure standards; it is not derived from Harwell Fricke. The NPL and Harwell estimates of dose rate in the Gammacell are in very good agreement.

TABLE III. Comparison of NPL and Harwell measured dose rates on the Harwell facility using NPL Fricke, with Harwell calculated dose rates based on Harwell Fricke [3]

Cell Position	Dose Rate Gy.s^{-1}			Ratio measured/calc.	
	Harwell Calculated	Harwell Measured	NPL Measured	Harwell	NPL
1	0.4534	0.4540	0.4574	1.001	1.009
2	0.4652	0.4719	0.4678	1.014	1.006
3	0.4353	0.4382	0.4436	1.007	1.019
4	0.4592	0.4573	0.4609	0.996	1.004
5	0.4490	0.4524	0.4554	1.009	1.014

6.1.2 Fricke 1983

In July 1983 the Harwell calibration cell was re-loaded with four sources each of approximately 7.40×10^{13} Bq. After installation the dose rates of the five positions were measured using Harwell Fricke solution and then intercompared by NPL and Harwell using Fricke solutions from both laboratories. The Pye Unicam SP 1700 and a new PU 8800 were both used in this

TABLE IVA. Dose rate measurements using Harwell Fricke in Harwell calibration cell corrected to 18.7.83

Cell Position	Mean dose rates $Gy \cdot s^{-1}$		
1	1.650	±	0.011
2	1.679	±	0.007
3	1.562	±	0.012
4	1.679	±	0.012
5	1.625	±	0.011

TABLE IVB. Comparison of NPL and Harwell measured dose rates $Gy \cdot s^{-1}$ in all five positions of the Harwell cell using Harwell and NPL Fricke solutions corrected to 8.8.83

Cell Position	Harwell Fricke		NPL Fricke	
	Measured Harwell	Measured NPL	Measured Harwell	Measured NPL
1	1.6595	1.6603	1.6323	1.6471
2	1.6639	1.6679	1.6823	1.6640
3	1.5538	1.5503	1.5738	–
4	1.6442	1.6629	1.6717	1.6590
5	1.6442	1.6238	1.6532	1.6296

intercomparison. The extinction coefficients were measured and k values of 28167 and 28076, respectively, were found for the two spectrophotometers. The Harwell dose rate values are shown in Table IVA. They are mean values of three solutions measured in both instruments and the errors quoted are 1σ. The dose rates obtained in the intercomparisons are shown in Table IVB. Good agreement was obtained between the two laboratories.

6.2 Red 4034

6.2.1 Red 4034 N

Following the Fricke intercomparison in December 1981 4034 N dosimeters were irradiated at Harwell and NPL to doses of 10,

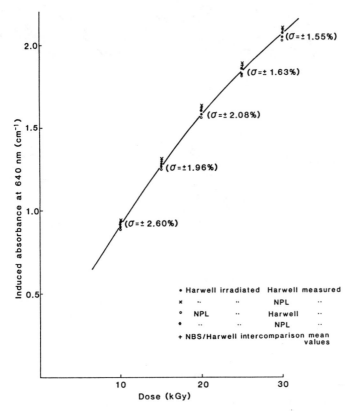

FIG.2. *Intercomparison of red 4034 N dosimeters between the NPL and Harwell (3.12.81).*

15, 20, 25 and 30 kGy. To minimise errors due to fading, optical density measurements were made in both laboratories within twenty four hours from completion of the irradiations. The procedure at Harwell involves irradiating four dosimeters in their sachets at ambient temperature in each of the five irradiation positions.

At NPL dosimeters are placed in the centre of the irradiation volume surrounded by a sufficient thickness of polystyrene to achieve electronic equilibrium. The temperature during irradiation is controlled to ± 1°C.

The results of the red 4034 N intercomparison are shown in Fig. 2. A correction for a mean blank value, obtained from ten dosimeters, was applied to the data, which are plotted as induced absorbance (cm^{-1}) against delivered dose in kGy. These

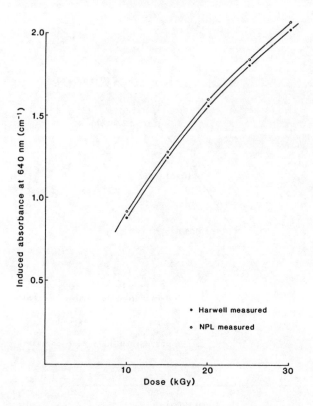

FIG.3. *Red 4034 N dosimeters irradiated at NPL and measured at Harwell and NPL.*

data show agreement between the two laboratories within the quoted accuracy (± 2%) of the red 4034 system.

However, a plot of Harwell and NPL measurements on dosimeters irradiated at NPL (Fig. 3) gives two approximately parallel curves. The Harwell value is 3.2% lower than the NPL value for induced absorbance at 25 kGy, indicating a difference in spectrophotometric read-out of the same samples. A plot of dosimeters irradiated by both laboratories and measured by NPL similarly results in two approximately parallel curves (Fig. 4) from which the Harwell estimate of delivered dose is higher by 3.3% than that estimated by NPL.

The differences identified in Figs. 3 & 4 are approximately equal in magnitude and opposite in sign and are therefore cancelled out in the overall results shown in Fig. 2.

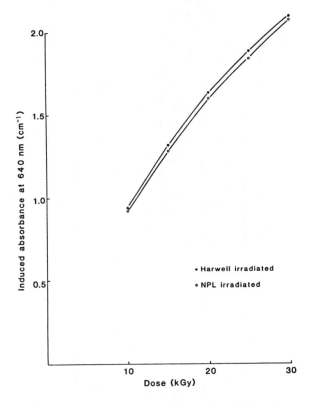

FIG.4. *Red 4034 N dosimeters irradiated at Harwell and NPL and measured by NPL.*

The discrepancy between the NPL Varian 2300 and the Harwell Pye Unicam SP 1700 was investigated using NPL absorbance filters. These again showed higher absorbance values measured at NPL. The measured difference, 1-2%, accounts in part for the 3.2% difference observed during the 4034 N intercomparison. It was concluded that the SP 1700 had a systematic error.

6.2.2 Red 4034 R

As part of an on-going programme batches of red 4034 dosimeters are regularly subjected to intercomparison between NPL and Harwell. Table V shows the data obtained on batch R, which was intercompared on 18.8.82. The Harwell measurements were made using the Pye Unicam SP 1700 corrected against NPL absorbance filters.

TABLE V. Red 4034 R intercomparison results

Dates	Dose kGy	Irradiated Harwell Dose Rate 1.47 kGy.h^{-1}		Irradiated NPL Dose Rate 2.5 kGy.h^{-1}	
		Harwell O.D. (cm^{-1})	NPL O.D. (cm^{-1})	Harwell O.D. (cm^{-1})	NPL O.D. (cm^{-1})
Irradiated 18-19.8.82 Measured 19.8.82	10	1.156 ± 0.007	1.1605 ± 0.0124	1.168 ± 0.004	1.1622 ± 0.0052
	15	1.527 ± 0.009	1.5293 ± 0.0135	1.521 ± 0.003	1.5151 ± 0.0052
	20	1.854 ± 0.022	1.8607 ± 0.0206	1.836 ± 0.015	1.8246 ± 0.0115
	25	2.098 ± 0.017	2.0960 ± 0.0123	2.085 ± 0.006	2.0749 ± 0.0062
	30	2.334 ± 0.019	2.3330 ± 0.0155	2.294 ± 0.012	2.2940 ± 0.0146

Each quoted result is the mean of four dosimeter readings.
Errors quoted 1σ.

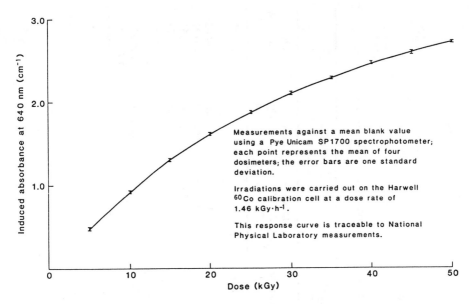

FIG.5. Response curve, induced absorbance versus dose for red 4034 Perspex dosimeters, batch R (Aug. 1982).

A typical response curve (4034 R) as supplied to customers is shown in Fig. 5. These data are plotted as induced absorbance against dose after correcting for a mean blank value derived from ten dosimeters. Each point with the exception of the response at 25 kGy is the mean of eight dosimeters from each of two separate irradiations. The value at 25 kGy is derived from four separate irradiations, the mean of sixteen dosimeters. Error bars are one sigma. Response curves are supplied as either absorbance against air or induced absorbance after correcting for a mean blank.

6.2.3 4034 T and V

In April 1983 red 4034T was intercompared with NPL. Two spectrophotometers were used at Harwell, the Pye Unicam SP 1700 and the new PU 8800. At NPL dosimeters were irradiated in the Gammacell at a dose rate of 2.4 kGy·h^{-1} and in the Hot Spot at a dose rate of 5.5 kGy·h^{-1}, to look for possible dose rate effects. The dose rate in the Harwell cell was 1.3 kGy·h^{-1}. The results are shown in Table VIA. The responses of the spectrophotometers are compared in Table VIB. Good agreement was obtained between the Harwell PU 8800 and the NPL Varian 2300, but the SP 1700 was again discrepant by 1-2%.

TABLE VIA. Intercomparison of spectrophotometers and dose rates Harwell/NPL Batch 4034 T

Dose kGy	Harwell Irradiated Dose rate 1.3 kGy.h⁻¹			NPL Irradiated Gammacell Dose rate 2.4 kGy.h⁻¹			NPL Irradiated Hot Spot Dose rate 5.5 kGy.h⁻¹		
	Harwell SP 1700 O.D. (cm⁻¹)	Harwell PU 8800 O.D. (cm⁻¹)	NPL Varian 2300 O.D. (cm⁻¹)	Harwell SP 1700 O.D. (cm⁻¹)	Harwell PU 8800 O.D. (cm⁻¹)	NPL Varian 2300 O.D. (cm⁻¹)	Harwell SP 1700 O.D. (cm⁻¹)	Harwell PU 8800 O.D. (cm⁻¹)	NPL Varian 2300 O.D. (cm⁻¹)
10	1.086 ±0.017	1.101 ±0.014	1.1092 ±0.0179	1.092 ±0.09	1.104 ±0.008	1.1032 ±0.0127	1.076 ±0.005	1.088 ±0.006	1.0899 ±0.0031
15	1.452 ±0.014	1.469 ±0.014	1.4683 ±0.0131	1.447 ±0.010	1.459 ±0.015	1.4593 ±0.0039	1.423 ±0.005	1.440 ±0.008	1.4443 ±0.0133
20	1.762 ±0.023	1.779 ±0.025	1.7857 ±0.0316	1.744 ±0.011	1.764 ±0.011	1.7639 ±0.0124	1.730 ±0.003	1.746 ±0.008	1.7457 ±0.0017
25	1.996 ±0.013	2.015 ±0.014	2.0280 ±0.0134	1.988 ±0.010	2.007 ±0.009	2.0118 ±0.0126	1.971 ±0.010	1.993 ±0.007	1.9882 ±0.004
30	2.225 ±0.020	2.244 ±0.031	2.2449 ±0.0213	2.209 ±0.003	2.233 ±0.009	2.2330 ±0.0119	2.176 ±0.009	2.202 ±0.014	2.2071 ±0.0192

TABLE VIB. Comparison of spectrophotometer response Harwell/NPL, 4034 T

Irradiator	SP1700/PU 8800	Varian 2300/PU 8800
Harwell 1.3 kGy	0.991	1.006
NPL 2.4 kGy	0.991	1.002
NPL 5.5 kGy	0.989	0.998

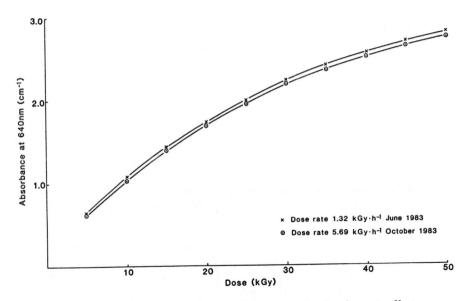

FIG. 6. Response curves for red 4034 V dosimeters showing dose rate effect.

Comparison of the PU 8800 and Varian 2300 estimated dose at the three dose rates, Table VIA, suggests that there may be a dose rate effect. The estimated dose at 5.5 kGy·h^{-1} is 1-2% lower for 25 kGy than that at 1.3 kGy·h^{-1}. A plot of the dosimeter response for 4034 V at two different dose rates, 1.32 kGy·h^{-1} and 5.69 kGy·h^{-1}, in the Harwell cell shows two distinct curves, Fig. 6.

TABLE VIIA. Harwell-NBS intercomparison response of 4034 N irradiated at Harwell, 1982

Dose kGy	Harwell Measurements 4.2.82		NBS Measurements 3.2.82		
	Irradiation Date	Mean O.D. (cm^{-1})	Irradation Date	Mean O.D. (cm^{-1})	Elapsed time (days)
15	25.1.82	1.505	24.1.82	1.52	10
20	"	1.82	"	1.83	"
25	"	2.05	"	2.11	"
30	"	2.30	"	2.35	"
35	"	2.57	"	2.59	"
15	28.1.82	1.50	27.1.82	1.54	7
20	"	1.81	"	1.84	"
25	"	2.07	"	2.12	"
30	"	2.27	"	2.36	"
35	"	2.47	"	2.51	"
15	30.1.82	1.49	29.1.82	1.53	5
20	"	1.82	"	1.86	"
25	"	2.07	"	2.13	"
30	"	2.30	"	2.32	"
35	"	2.49	"	2.55	"

TABLE VIIB. Response of 4034 N irradiated at NBS

Dose kGy	Elapsed time (days)	Irradiated date	Measurement date 9.2.82 O.D. (cm^{-1})	
			Harwell	NBS
20	14	26.1.82	1.82	
30	13	27.1.82	2.27	
25	12	28.1.82	2.06	
15	11	29.1.82	1.48	
35	11	29.1.82	2.46	
20.5	11	29.1.82	1.86	
25.3	8	1.2.82	2.09	2.115
20.1	7	2.2.82	1.81	1.825
25	6	3.2.82	2.07	2.09
15	5	4.2.82	1.49	1.51
35.1	4	5.2.82	2.47	2.495
30	4	5.2.82	2.28	2.31
10	4	5.2.82	1.11	1.14

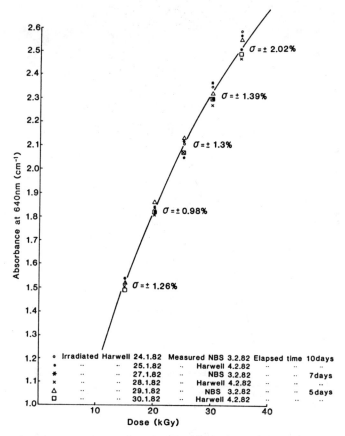

FIG.7. *Intercomparison by Harwell and NBS of red 4034 N dosimeters irradiated at Harwell and read by both laboratories (1982).*

7. HARWELL AND NATIONAL BUREAU OF STANDARDS (NBS) INTERCOMPARISON OF RED 4034 N DOSIMETERS

In February 1982 red 4034 N dosimeters were intercompared by Harwell and NBS.

Six sets of 4034 N dosimeters were irradiated at Harwell in six separate irradiations carried out on January 24, 25, 27, 28, 29 and 30. Three of the sets of dosimeters were retained at Harwell and three sets were taken by hand to NBS. Each set of the dosimeters retained at Harwell, and of those taken to NBS, was read at a specified time from the end of the irradiation; the same time in both laboratories in order to minimise errors due to fading. 4034 N dosimeters were

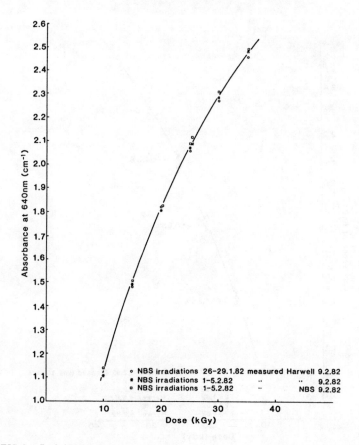

FIG.8. *Red 4034 N irradiated at NBS and measured at Harwell and NBS (1982).*

irradiated in the NBS facility during the period 26.1.82 - 5.2.82. Some were retained at NBS and some were taken by hand to Harwell for measurement. Again dosimeters were read at a specified time from the end of irradiation. The lapsed times from end of irradiation to measurement for the dosimeters irradiated at Harwell were 5, 7 and 10 days. In all cases the dosimeters were irradiated and stored in sealed sachets until read. The results of measurements by Harwell and NBS on dosimeters irradiated at Harwell are shown in Table VIIA. The mean optical densities (cm^{-1}) are obtained from four dosimeters in each case.

The data are plotted in Fig. 7. The graph shows that there is no significant difference between dosimeters measured

5, 7 and 10 days from the end of irradiation. Agreement with the quoted accuracy of the system (± 2%) was achieved. However, there is a difference between the estimates of absorbed dose, the Harwell Pye Unicam SP 1700 giving a lower value by about 1.5%, a systematic error similar to that observed in the intercomparisons with NPL. The data obtained from the NBS irradiations are shown in Table VIIB and plotted in Fig. 8. The NBS values are the mean of two dosimeters, the Harwell results are the mean of four. A comparison of the irradiators did not show a significant trend. When the NBS/Harwell intercomparison data are corrected for a mean blank to induced absorbance, the data are consistent with those from the Harwell-NPL intercomparison and are included in Fig. 8.

Post-irradiation fading was not a significant error source in the Harwell-NBS intercomparison. Post-irradiation fading measurements continued, on those dosimeters measured at Harwell, for a further twenty six days during which time the extent of fading was approximately 3%. The dosimeters were returned to their sachets for storage between readings.

8. INTERCOMPARISONS WITH ATOMIC ENERGY OF CANADA LTD (AECL)

8.1 Red 4034 N

In April 1982 the response of red 4034 N dosimeters was intercompared by Harwell and AECL. Two sets of dosimeters were irradiated to doses of 5, 10, 15, 20, 25, 30, 35, 40 and 45 kGy, with four dosimeters at each dose per set. One set was read at Harwell seven days after completion of the irradiation and the other set was taken by hand to AECL for measurement, also seven days from the end of the irradiation, so that fading corrections would not be necessary. Unirradiated dosimeters were also taken to AECL for irradiation and reading seven days from the end of irradiation at AECL. The results are shown in Table VIIIA and plotted in Fig. 9. These again show agreement within the quoted accuracy of the system.

8.2 Ceric-cerous

The Harwell, NPL and AECL irradiators were also compared in April 1982 through the ceric-cerous dosimetry system operated by AECL. Ceric-cerous dosimeters were supplied by AECL for irradiation at NPL and Harwell. The irradiated dosimeters were again taken by hand to Ottawa for measurements by AECL. The results are shown in Table VIIIB. The delivered and measured doses were in good agreement in all cases (Fig. 10).

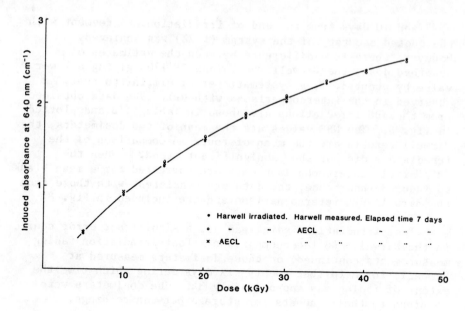

FIG. 9. Harwell-AECL intercomparison response of red 4034 N irradiated and measured by both laboratories (1982).

TABLE VIIIA. 1982 Harwell-AECL intercomparison response of 4034 N dosimeters irradiated and measured by both laboratories

Dose given kGy	Harwell Irradiated		AECL Irradiated and measured O.D. (cm^{-1})
	Harwell measured O.D. (cm^{-1})	AECL measured O.D. (cm^{-1})	
5	0.43	0.45	0.45
10	0.89	0.91	0.92
15	1.24	1.31	1.29
20	1.55	1.61	1.60
25	1.85	1.90	1.86
30	2.05	2.11	2.06
35	2.28	2.30	
40	2.44	2.41	2.43
45	2.58	2.56	
50.7			2.67

TABLE VIIIB. 1982 Intercomparison of NPL, Harwell and AECL
irradiators using ceric-cerous dosimeters

Dose given kGy	Indicated Doses Measured by AECL		
	AECL Irradiated	NPL Irradiated	Harwell Irradiated
5			4.6
5			4.2
10			9.8
10			9.5
15	14.5		15.0
15	14.7		14.7
15	14.8		
15	14.7		
15	14.7		
15	14.8		
15	15.1		
15	14.9		
20		19.8	19.8
20		20	19.8
20		20.1	20.5
25		20	20.1
25		25.2	25.1
25		25.1	24.9
25		25.2	25.1
25		25.2	25.0
25			24.8
25			25.0
30		30.1	30.4
30		30.3	30.3
30		30.3	
30		30.3	
35			35.3
35			35.3
40			40.9
40			40.8

The ceric-cerous intercomparison was valuable in that traceability could be established between red 4034 and a chemical dosimeter sensitive over the same working range, 5-50 kGy.

9. CONCLUSIONS

Well regulated intercomparisons of dosimetry systems can give rise to reliable data which provide background information

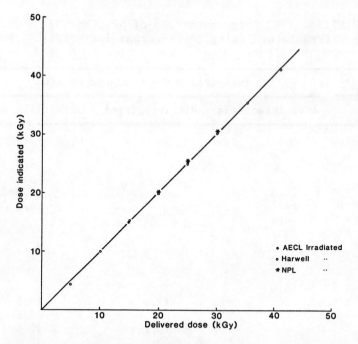

FIG.10. Intercomparison of NPL, Harwell and AECL irradiators using AECL ceric-cerous dosimeters. All dosimeters read by AECL.

on the performance of the systems intercompared. This information provides a base line which can be valuable in helping to reduce problems and systematic errors which may arise when such systems are used in plants for routine process verification.

Under carefully controlled conditions specific problems can be identified and corrected for, even though the response data may be within the quoted accuracy of the system. In the intercomparisons of 4034 N between Harwell and NPL discrepancies (~ 3%) were observed between the spectrophotometers and irradiators, approximately equal and opposite in sign. A similar, although smaller, discrepancy was observed (~ 1.5%) in the Harwell and NBS intercomparison of 4034 N. The Harwell SP 1700 spectrophotometer then used was confirmed as having a systematic error and has since been replaced. There is now good agreement between Harwell and NPL on spectrophotometer response and dose rates in the two facilities.

Consideration of all the data from the Harwell/NPL and Harwell/NBS intercomparisons on 4034 N shows agreement within the quoted accuracy, ± 2%, of the system. The intercomparison, using the AECL ceric-cerous system, of the irradiators at NPL, Harwell and AECL, was valuable because traceability of 4034 N was established with a chemical system sensitive over the same working range.

There are many potential sources of error in the use of dosimetry systems for routine process verification but these can be controlled by care and attention to detail in use.

A potential source of error in red 4034, for example, is that the minimum fading wavelength, 640 nm, does not lie on the peak of the absorption spectrum, but on the slope and is therefore susceptible to wavelength error in reading, unless care is taken. Other factors influencing the precision of dose measurements are dose rate, temperature and humidity, and to achieve the quoted accuracy for 4034 it is important that dosimeters are irradiated and stored in sealed sachets until read in order to minimise fading which is of the order of 2% in three weeks in sachets.

It must be stressed that to achieve the accuracy of which red 4034 is capable, the calibration conditions should duplicate as closely as possible the conditions in use. Plant operators should obtain response curves which relate to their own operating conditions, making use of the services provided by standards laboratories such as NPL and NBS and also Harwell for its own systems. Sets of pre-irradiated dosimeters can be provided for checking spectrophotometer responses and plant irradiated dosimeters can be read on request.

ACKNOWLEDGEMENTS

The authors thank their colleagues at NPL, Dr S.C. Ellis, J.H. Barrett and Dr P.H.G. Sharpe, without whom much of this work would not have been possible, for their continuing support and interest. We also thank Dr W.L. McLaughlin of NBS and Dr R. Chu of AECL for their collaboration and interest.

REFERENCES

[1] WHITTAKER, B. Radiation Dosimetry Technique using Commercial Red Perspex. AERE-R3360 1964.
[2] WHITTAKER, B. Recent Developments in Poly (Methyl Methacrylate)/Dye systems for Megarad Dosimetry. Proceedings of a Symposium on Radiation Dose & Dose Rate Measurements in the Megarad Range, NPL 1970.
[3] BARRETT, J.H., SHARPE, P.H.G. NPL Private Communication.
[4] WHITTAKER, B. The G value and the reproducibility of the Ferrous-Ferric Dosimeter. AERE-R3073, 1963.

IAEA-SM-272/9

ENERGY DEPENDENCE OF RADIOCHROMIC DOSIMETER RESPONSE TO X- AND γ-RAYS*

W.L. McLAUGHLIN
Center for Radiation Research,
National Bureau of Standards,
Gaithersburg, Maryland,
United States of America

A. MILLER
Accelerator Department,
Risø National Laboratory,
Roskilde, Denmark

R.M. URIBE
Instituto de Fisica,
Universidad Nacional Autonoma de Mexico,
Mexico City, Mexico

S. KRONENBERG
United States Army Electronic Technology
 and Devices Laboratory,
Fort Monmouth, New Jersey,
United States of America

C.R. SIEBENTRITT
Federal Emergency Management Agency,
Washington, D.C.,
United States of America

Abstract

ENERGY DEPENDENCE OF RADIOCHROMIC DOSIMETER RESPONSE TO X- AND γ-RAYS.
    Liquid, solid and liquid-core 'fibre-optics' radiochromic dosimeters were studied for their spectral sensitivity to ionizing photons in the energy range 10 keV to 100 MeV. By comparison of ratios of mass energy-absorption coefficients and mass collision stopping powers of water and biological tissues (fat, muscle and bone), approximate errors due to energy dependence for typical ^{60}Co γ-ray scattered spectra and to rough simulation of tissues by means of certain additives to radiochromic dosimeters could be estimated. Design of approximate tissue-simulating dosimeters

---

* Work carried out under NBS-IAEA Research Agreement 3061/CF and with the support of the United States Federal Emergency Management Agency. Commercial products are identified here for technical purposes only, which does not denote endorsement or recommendation.

is accomplished by comparing experimental and computational results for various radiochromic films, liquid solutions and liquid-core waveguides. Several experimental tests of energy dependence were made using X- and γ-rays. For water-, fat-, muscle- or bone-simulation over this photon energy range, chlorides, bromides, triethyl phosphate or dimethyl sulphoxide or a combination of these are added in appropriate concentrations. For simulating other biological tissues, it is possible to test immediately the energy dependence of the response of several material combinations by using suitable approximations, and thereby to design nearly energy-independent, photon-sensitive radiochromic dosimeters for use over wide ranges of absorbed dose (~10^{-2} to 10^6 Gy). Errors due to the presence of dosimeter probe materials placed in biological tissues and irradiated with broad γ-ray spectra are thus diminished.

1. INTRODUCTION

At present calibrated radiochromic systems are used for routine dosimetry of ionizing photons and electrons over a wide range of absorbed doses (10^{-2} to 10^6 Gy) and absorbed dose rates (up to ~10^{13} Gy·s^{-1}) [1–4]. The radiochromic dosimeter forms and their dose ranges are:

Thin films	10^3–10^6 Gy
Thick films and gels	10^1–10^4 Gy
Liquid solutions	10^1–10^3 Gy
Liquid-core waveguides	10^{-2}–10^3 Gy

In the case of X- and γ-ray dosimetry, the dose rates range generally up to ~10^4 Gy·s^{-1}, and if atmospheric conditions and moisture content (for solid systems) and oxygen and acid content (for liquid systems) are properly controlled, there is very little rate dependence of response to photons [3, 5–9]. Temperature dependence of response and stability for different storage conditions are also well established [10–12]. The effects of different atmospheres on response and stability have also been studied [12, 13]. A matter of concern, however, is variation of response with changes in photon spectral energy, namely 'energy dependence' of response, especially at photon energies <100 keV.

In radiation processing dosimetry applications for the sterilization of medical devices, food preservation and disinfestation, waste recycling, etc., and for medical, environmental and radiation protection applications, the main interest is in measuring absorbed dose in biological tissues and plastics. Dose is often expressed in terms of absorbed dose in a polymeric system (e.g. polyethylene or polystyrene) or in water, graphite, muscle, adipose tissue (fat), bone, etc. The spectral sensitivity of conventional radiochromic dosimeters to photons, especially in the energy range 10 to 100 keV and >10 MeV, is appreciably different from that of water and biological tissues [13, 14]. In fact, for a hydrocarbon solvent or a plastic (e.g. nylon) radiochromic dosimeter, the responses to ~30 keV photons are generally lower

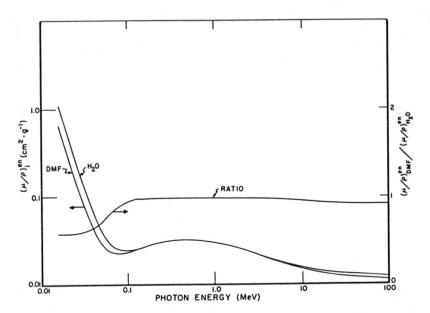

FIG.1. Left ordinate: Calculated mass energy-absorption coefficients of N,N-dimethyl formamide (DMF) and water versus photon energy. Right ordinate: Ratio of the mass energy-absorption coefficient versus photon energy.

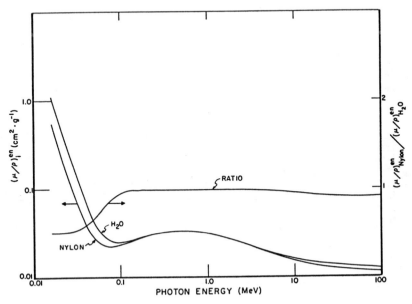

FIG.2. Left ordinate: Calculated mass energy-absorption coefficients of nylon and water versus photon energy. Right ordinate: Ratio of the mass energy-absorption coefficients versus photon energy.

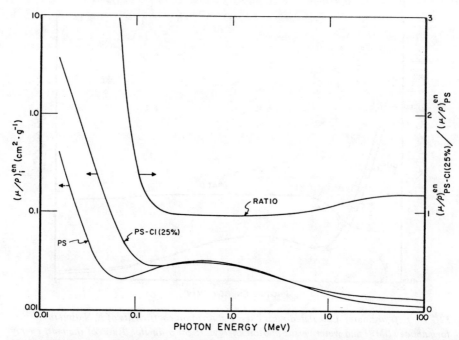

FIG.3. Left ordinate: Calculated mass energy-absorption coefficients of polystyrene (PS) and polystyrene containing 25% chlorine by weight (PS-Cl (25%)) versus photon energy. Right ordinate: Ratio of the mass energy-absorption coefficients versus photon energy.

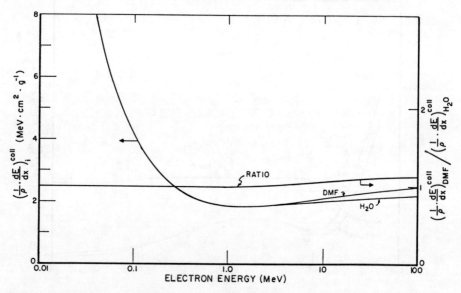

FIG.4. Left ordinate: Calculated collision mass stopping powers of N,N-dimethyl formamide (DMF) and water versus electron energy. Right ordinate: Ratio of the collision mass stopping powers versus electron energy.

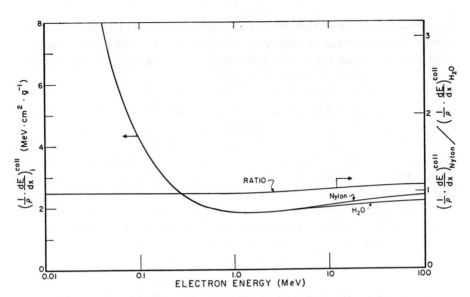

FIG.5. Left ordinate: Calculated collision mass stopping powers of nylon and water versus electron energy. Right ordinate: Ratio of the collision mass stopping powers versus electron energy.

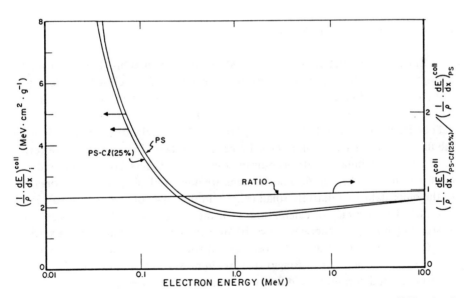

FIG.6. Left ordinate: Calculated collision mass stopping powers of polystyrene (PS) and polystyrene containing 25% chlorine (PS-Cl (25%)). Right ordinate: ratio of the collision mass stopping powers versus electron energy.

by more than a factor of two, because of differences in photoelectron absorption effects. The differences are illustrated in Figs 1 and 2 in terms of mass energy-absorption coefficient ratios of N,N-dimethyl formamide (C_3H_6NOH)-to-water and nylon $[C_6H_{11}NO]_n$-to-water, respectively, over the photon energy range 10 keV to 100 MeV. Figure 3 shows a different trend for a chlorinated plastic, the ratio of mass energy-absorption coefficients, polychlorostyrene (25% Cl)-to-polystyrene becoming very large at photon energies <100 keV and somewhat larger at >10 MeV. Figures 4 to 6 show, on the other hand, that ratios of collision mass stopping powers of these materials for electrons are near unity over the entire electron spectrum of interest. The calculations [14] were made by a simple program for a desk-top calculator [15].

This paper gives comparisons of ratios of both mass energy-absorption coefficients $(\mu/\rho)^{en}$ for photons and collision mass stopping powers $(1/\rho)(dE/dx)^{coll}$ for electrons, in the energy range 10 keV to 100 MeV, and for solid, liquid and liquid-core 'waveguide' radiochromic dosimeters, with and without additives designed to improve their energy dependence of response relative to biological tissues. Special concern is shown for degraded γ-ray spectral effects in the energy range 10 to 1250 keV. Experimental studies of a few sample modified dosimeters over this energy range are given for comparison. Approximate errors due to energy dependence of waveguide radiochromic dosimeters for a typical degraded ^{60}Co γ-ray spectrum are estimated when calibrated as Bragg-Gray 'cavities' under approximate electron equilibrium conditions.

2. CALCULATION PROCEDURES

The methods already described by Miller and McLaughlin [14], involving a program in BASIC applied to a desk-top calculator [15], are used to calculate and plot values of $(\mu/\rho)^{en}$ and $(1/\rho)(dE/dx)^{coll}$ and their ratios, for given elements, and mixtures of elements. Values of $(\mu/\rho)^{en}$ for photons are based on the tabulated data of Hubbell [16], and those of $(1/\rho)(dE/dx)^{coll}$ for electrons are based on the tabulated data of Pages et al. [17] or Berger and Seltzer [18]. Although newer, more accurate tabulated data accounting for density effects, compounds and mixtures and certain empirical corrections are now available [19, 20], errors due to the older data are relatively small (<2%), in the range of energies 15 to 1250 keV [14, 20]. The accuracy of calculations is of the order of ±3%, but for calculations of stopping power for energies above 10 MeV the uncertainty may increase to ±10%. Except for energies very close to photoelectric absorption edges, errors due to the use of the additivity (Bragg) rule are, in most cases, within the +3% uncertainty in determining μ/ρ for compounds and mixtures, i.e.

$$\frac{\mu}{\rho} = \sum_i k_i \left(\frac{\mu}{\rho}\right)^i$$

where k_i is the weight fraction of each element, i [14, 21]. The errors are greatest in the case of nearly equal weight fractions of high- and low-Z elements irradiated at high photon energies [21].

The program uses Burlin's general cavity theory [22] to calculate approximate errors due to the energy dependence of response for certain sensitive probe materials (dosimeter) with thin walls surrounded by another irradiated material of interest, for a given degraded ^{60}Co γ-ray spectrum.

3. EXPERIMENTAL PROCEDURES AND RESULTS

3.1. Irradiations

Dosimeters were irradiated with collimated beams of gamma radiation from 'teletherapy' ^{60}Co and ^{137}Cs sources, with the dosimeters being held between layers of dosimeter material 4.5 and 2.0 mm thick, respectively, in order to maintain approximately electron equilibrium conditions. The calibrations at the positions of irradiation (97.9 cm from the sources) were determined by standard graphite ionization chamber measurements [23]. Collimated X-ray beams were supplied by a tungsten-target industrial X-ray machine equipped with metal filters, and gave fairly narrow photon spectra having effective energies of 39, 72, and 120 keV [24]. In these cases the dosimeters were held in thin polyethylene (~70 μm) envelopes, at 31 cm from the X-ray target, and calibrations of dose rates at this distance were made by calibrated thimble ionization chambers. In addition, irradiations were made at 7.5 keV using K-fluorescence from a nickel target in the Henke X-ray chamber [25]. Absorbed dose rates in this system were measured by a Xe-filled ionization chamber [26]. Table I lists absorbed dose rates in water, which were

TABLE I. EFFECTIVE PHOTON ENERGIES AND ABSORBED DOSE RATES (IN WATER) FOR THE DIFFERENT PHOTON IRRADIATIONS USED FOR ENERGY-DEPENDENCE MEASUREMENTS

Photon source	Effective energy (keV)	Absorbed dose rate at calibration position (Gy/min)
^{60}Co γ-rays	1250	1.45
^{137}Cs γ-rays	662	3.40
150 kVp X-rays	120	1.26
100 kVp X-rays	72	1.06
50 kVp X-rays	39	0.80
Henke X-rays (Ni)	7.5	2.70

TABLE II. RADIOCHROMIC DOSIMETERS

Dosimeter	Ingredients	Approximate chemical formula	Approximate light pathlength (mm)	Approximate dose range (Gy)
		Film systems		
HPR-CN[a]	Nylon	$[C_6H_{13}NO]_n$	0.05	$10^3 - 10^5$
HPR-CN	Polyvinyl butyral	$[C_6H_{13}O_2]_n$	0.05	$5 \times 10^3 - 10^5$
NF-CN[b]	Trogamid[f]	$[C_{22}H_{26}NO_4]_n$	0.06	$10^4 - 10^6$
MG-CN[c]	Polyvinyl chloride	$[C_2H_3Cl]_n$	0.07	$10^4 - 5 \times 10^5$
MG-OHC$_3$[d]	Polychlorostyrene	$[(C_8H_8)(25\% \text{ Cl})]_n$	0.07	$10^3 - 10^5$
		Liquid systems		
PR-CN[e]	2-methoxy ethanol	$C_3H_8O_2$	10.0	$10^1 - 10^3$
HPR-CN	N,N-dimethyl formamide	C_3H_7NO	10.0	$10^1 - 2 \times 10^3$
HPR-CN	Triethyl phosphate	$(C_2H_5)_3PO_4$	10.0	$10^1 - 4 \times 10^3$
HPR-CN	Dimethyl sulphoxide	$(CH_3)_2SO$	10.0	$10^1 - 3 \times 10^3$
		Liquid core waveguide system[g]		
HPR-CN	33% 2 methoxy ethanol 66% dimethyl sulphoxide 1% polyvinyl butyral			

[a] HPR-CN = hexa (hydroxyethyl) pararosaniline cyanide.
[b] NF-CN = new fuchsin cyanide.
[c] MG-CN = malachite green cyanide.
[d] MG-OCH$_3$ = malachite green methoxide.
[e] PR-CN = pararosaniline cyanide.
[f] A polyamide consisting of aromatic diacetamide interspersed with methylated n-hexane (Dynamit Nobel).
[g] Commercial product referred to as 'Opti-chromic' dosimeter (Far West Technology, Inc.).

TABLE III. TISSUE-EQUIVALENT RADIOCHROMIC DOSIMETERS

Dosimeter	Ingredients	Simulated tissue [19, 20]	Approximate light pathlength (mm)	Approximate dose range (Gy)
		Solid systems		
HPR-CN	Nylon	Adipose tissue	0.05	$10^3 - 10^5$
MG-OCH$_3$	Polychlorostyrene (1.5% Cl)	Adipose tissue	0.10	$10^3 - 10^5$
MG-OCH$_3$	Polychlorostyrene (5% Cl)	Striated muscle	0.08	$10^3 - 10^5$
MG-OCH$_3$	Polybromostyrene (5% Br)	Cortical bone	0.07	$10^3 - 10^5$
HPR-CN	Polymer mixture[a] (8.2% PVC)	Striated muscle	1.0 or 6.0	$10^1 - 10^3$
HPR-CN	Polymer mixture[a] (43% PVC$_2$)	Cortical bone	1.0 or 6.0	$10^2 - 10^4$
		Liquid systems		
HPR-CN	N,N-dimethyl formamide	Adipose tissue	10.0	$10^1 - 2 \times 10^3$
HPR-CN	74% N,N-dimethyl formamide 26% triethyl phosphate	Striated muscle	10.0	$10^1 - 5 \times 10^3$
HPR-CN	86% dimethyl sulphoxide 14% triethyl phosphate	Compact bone	10.0	$10^1 - 10^4$
HPR-CN	33% 2-methoxy ethanol 1% polyvinyl butyral 66% dimethyl sulphoxide	Cortical bone	60.0	$10^1 - 10^4$

[a] The polymer mixture consists of 50% polyvinyl pyrrolidone, 25% polymethyl methacrylate, and 25% polyvinyl isobutyl ether, with the indicated per cent by weight polyvinyl chloride (PVC) or polyvinylidene chloride (PVC$_2$) (see Ref. [13]).

obtained from measured values of exposure, by appropriate conversion from exposure (in röntgen) to absorbed dose in water or other compounds (in gray) at these photon energies.

3.2. Radiochromic dosimeters and design of tissue equivalence

Radiochromic dosimeters consist of colourless organic dye precursors dissolved at a concentration of a few per cent by weight in solid systems (plastic films, gels, etc.), liquid systems (polar organic solvents) and liquid-core waveguide plastic tubes with polished glass spheres or rods at each end for transmitting light for analysis. Upon irradiation, the radiochromic medium becomes coloured, and a spectrophotometer or densitometer is used to measure the increase in optical density, which has a functional relationship with absorbed dose as determined by calibration. Some radiochromic dyes (and their approximate wavelengths at the maximum of the radiation-induced absorption band) are the leucocyanides or leucomethoxides of the following triphenyl methane dyes: pararosaniline (550 nm); new fuchsin (560 nm); hexa(hydroxyethyl) pararosaniline (600 nm); and malachite green (630 and 430 nm) [27]. Table II lists some of the films, liquids and waveguides that are commercially available.

Solid and liquid tissue substitutes are described in the literature [28–31], and values of mass energy-absorption coefficients and mass attenuation coefficients (for photons) and collision mass stopping powers (for electrons) of water, biological tissues and other compounds are also available over wide photon and electron energy ranges [19, 20, 32–35]. In the present studies, several simple additions to the conventional radiochromic systems listed in Table II are used to simulate water and tissues. The combinations are listed in Table III. Response curves for ^{60}Co γ-ray irradiations are shown in Figs 7 and 8 for the solid dosimetry systems, and in Fig.9 for the liquid dosimeters.

Plots of mass energy-absorption coefficients for the dosimeters and tissues listed in Table III (and their ratios) are shown as a function of photon energy in Figs 10 to 20. Some dosimeter samples were irradiated with X- and γ-rays, using the sources and absorbed dose rates listed in Table I. The points represent experimental dosimeter sensitivity normalized to the value for ^{60}Co γ-rays (1.25 MeV). By this normalization of the relative responses at the different photon energies to the response at 1.25 MeV energy (note: the absorbed doses are given in terms of energy imparted to the tissues of interest), energy-dependence data are plotted and compared with the curves of the ratios of mass energy-absorption coefficients, dosimeter material-to-tissue (see Figs 10 to 20). In comparing the calculated curves for energy-absorption coefficient ratios with the experimental data, it should also be noted that the ratio is relevant only if the measurements are made under electron equilibrium conditions, that is, if the effective dosimeter size is large relative to the range of secondary electrons.

Text cont. on p. 420.

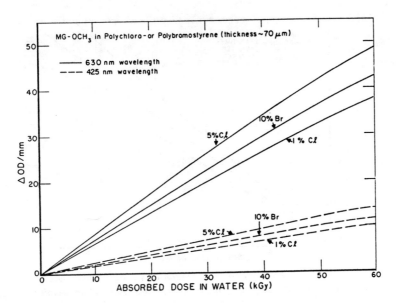

FIG.7. Experimental response curves of thin polyhalostyrene films containing different amounts of halogens and measured at two optical wavelengths at the peaks of the two absorption bands of malachite green (MG). The ordinate is expressed in terms of OD/mm, the increase in optical density per unit thickness of the films.

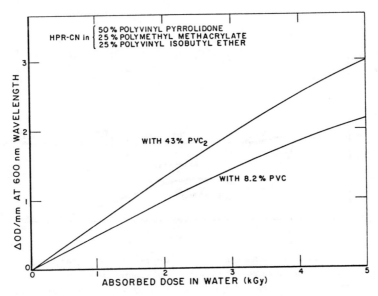

FIG.8. Experimental response curves of thick film (1.0 mm) dosimeters simulating muscle (containing 8.2% PVC) and cortical bone (containing 43% PVC_2), measured at the peak of the absorption band of hexa(hydroxyethyl) pararosaniline (HPR).

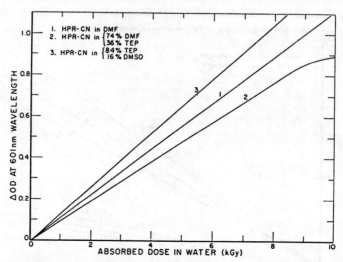

FIG.9. *Experimental response curves of liquid radiochromic dosimeters (pathlength 10 mm), measured at the peak of the absorption band of hexa(hydroxyethyl) pararosaniline (HPR).*

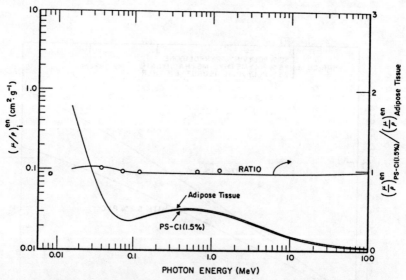

FIG.10. *Left ordinate: Calculated mass energy-absorption coefficients of polystyrene containing 1.5% chlorine by weight (PS-Cl(1.5%)) and adipose tissue versus photon energy. Right ordinate: Ratio of the mass energy-absorption coefficients versus photon energy. The experimental points are plotted in terms of the right ordinate and represent the response of the PS-Cl(1%) film to X- and γ-rays of various photon energies relative to response to 1.25 MeV photons, when the absorbed dose is expressed in units of gray in adipose tissue.*

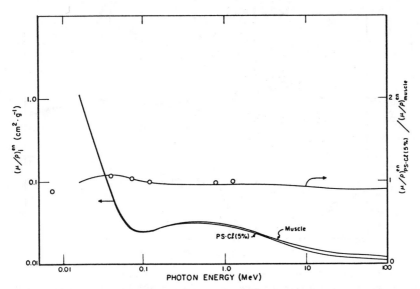

FIG.11. Left ordinate: Calculated mass energy-absorption coefficients of polystyrene containing 5% chlorine by weight (PS-Cl(5%)) and of striated muscle tissue versus photon energy. Right ordinate: Ratio of the mass energy-absorption coefficients versus photon energy. The experimental points are plotted in terms of the right ordinate and represent the response of the PS-Cl(5%) film to X- and γ-rays of various photon energies relative to the response to 1.25 MeV photons, when the absorbed dose is expressed in units of gray in muscle.

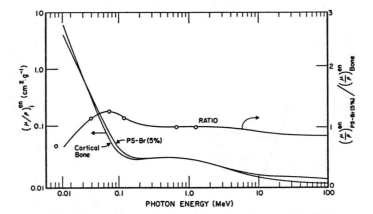

FIG.12. Left ordinate: Calculated mass energy-absorption coefficients of polystyrene containing 5.0% bromine by weight (PS-Br(5.0%)) and cortical bone versus photon energy. Right ordinate: Ratio of the mass energy-absorption coefficients versus photon energy. The experimental points are plotted on the right ordinate and represent the responses of the PS-Br(10%) film to X- and γ-rays of various photon energies relative to the response to 1.25 MeV photons, when the absorbed dose is expressed in units of gray in bone.

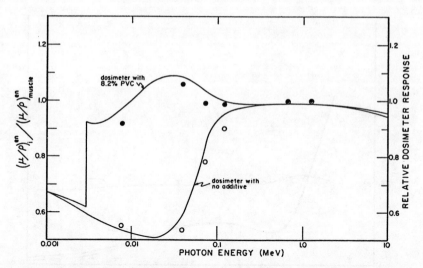

FIG.13. *Curves are plotted on the left ordinate and represent ratios of the calculated mass energy-absorption coefficients of the polymer dosimeter (50% PVP, 25% PMMA, 25% PVI) containing no additive and 8.2% PVC. The experimental points are plotted in terms of the right ordinate and show the responses of the two dosimeter forms to X- and γ-rays of various photon energies relative to the response to 1.25 MeV photons, when the absorbed dose is expressed in units of gray in muscle.*

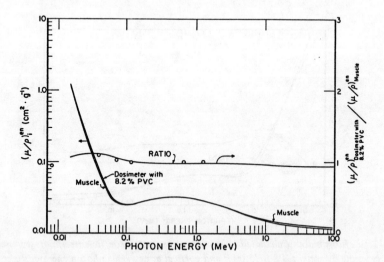

FIG.14. *Left ordinate: Calculated mass energy-absorption coefficients of the polymer dosimeter (50% PVP, 25% PMMA, 25% PVI) containing 8.2% PVC and striated muscle tissue versus photon energy. Right ordinate: Ratio of the mass energy-absorption coefficients versus photon energy.*

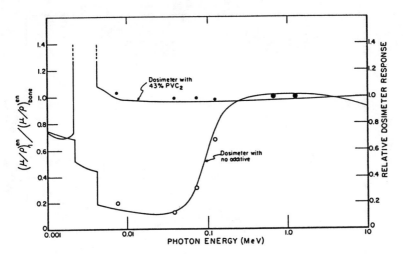

FIG.15. Curves are plotted on the left ordinate and represent ratios of the calculated mass energy-absorption coefficients of the polymer dosimeter (50% PVP, 25% PMMA, 25% PVI) containing no additive and 43% PVC_2. The experimental points are plotted on the right ordinate and show the responses of the two dosimeter forms to X- and γ-rays of various photon energies relative to the response to 1.25 MeV photons, when the absorbed dose is expressed in units of gray in bone.

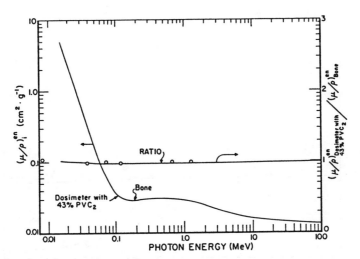

FIG.16. Left ordinate: calculated mass energy-absorption coefficients of the polymer dosimeter (50% PVP, 25% PMMA, 25% PVI) containing 43% PVC_2 and cortical bone versus photon energy. Right ordinate: Ratio of the mass energy-absorption coefficients versus photon energy. The experimental points are plotted in terms of the right ordinate and represent the response of the polymer containing 43% PVC_2 to X- and γ-rays of various photon energies relative to the response to 1.25 MeV photons, when the absorbed dose is expressed in units of gray in bone.

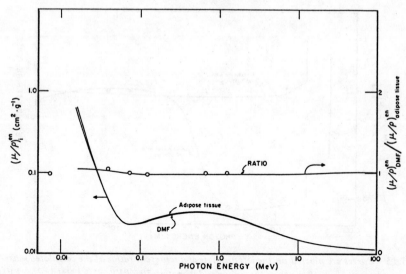

FIG.17. *Left ordinate: Calculated mass energy-absorption coefficients of N,N-dimethyl formamide (DMF) and adipose tissue versus photon energy. Right ordinate: Ratio of the mass energy-absorption coefficients versus photon energy. The experimental points are plotted in terms of the right ordinate and represent the response of the liquid radiochromic dosimeter (HPR-CN in DMF) to X- and γ-rays of various photon energies relative to the response to 1.25 MeV photons, when the absorbed dose is expressed in units of gray in adipose tissue.*

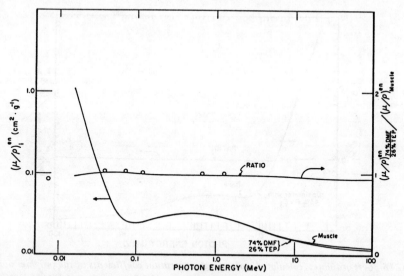

FIG.18. *Left ordinate: Calculated mass energy-absorption coefficients of 74% DMF plus 26% TEP and striated muscle versus photon energy. Right ordinate: Ratio of the mass energy-absorption coefficients versus photon energy. The experimental points are plotted in terms of the right ordinate and represent the response of the liquid radiochromic dosimeter (HPR-CN in 74% DMF plus 26% TEP) to X- and γ-rays of various photon energies relative to the response to 1.25 MeV photons, when the absorbed dose is expressed in units of gray muscle.*

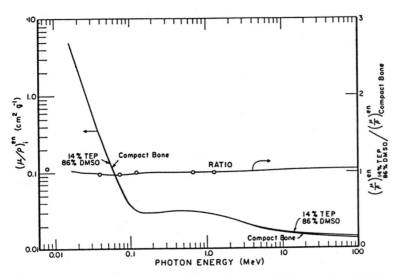

FIG.19. *Left ordinate: Calculated mass energy-absorption coefficients of 86% DMSO plus 14% TEP and compact bone versus photon energy. Right ordinate: Ratio of the mass energy-absorption coefficient versus photon energy. The experimental points are plotted in terms of the right ordinate and represent the response of the liquid radiochromic dosimeter (HPR-CN in 86% DMSO plus 14% TEP) to X- and γ-rays of various photon energies relative to the response to 1.25 MeV photons, when the absorbed dose is expressed in units of gray in bone.*

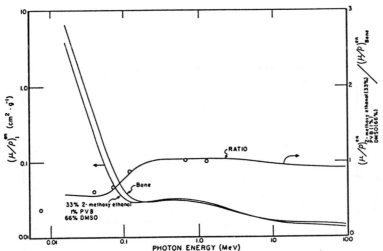

FIG.20. *Left ordinate: Calculated mass energy-absorption coefficients of liquid-core solution used in the waveguide dosimeter (33% 2-methoxy ethanol, 1% polyvinyl butyral, and 66% DMSO) and cortical bone versus photon energy. Right ordinate: Ratio of the mass energy-absorption versus energy. The experimental points are plotted in terms of the right ordinate and represent the response of the liquid dosimeter solutions (33% 2-methoxy ethanol, 1% PVB, 66% DMSO) to X- and γ-rays of various photon energies relative to the response to 1.25 MeV photons, when the absorbed dose is expressed in units of gray in bone.*

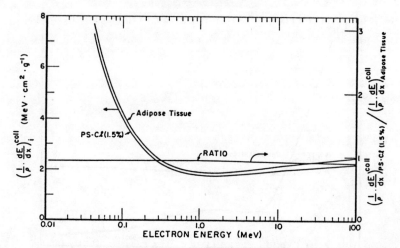

FIG.21. Left ordinate: Calculated collision mass stopping powers of polystyrene containing 1.5% chlorine by weight (PS-Cl(1.5%)) and adipose tissue versus electron energy. Right ordinate: Ratio of the collision mass stopping powers versus electron energy.

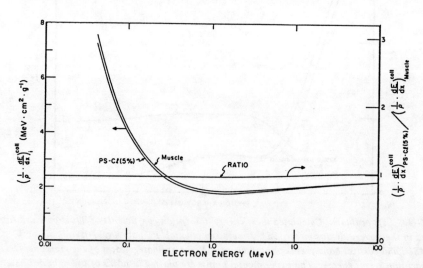

FIG.22. Left ordinate: Calculated collision mass stopping powers of polystyrene containing 5% chlorine by weight (PS-Cl(5%)) and striated muscle versus electron energy. Right ordinate: Ratio of the collision mass stopping powers versus electron energy.

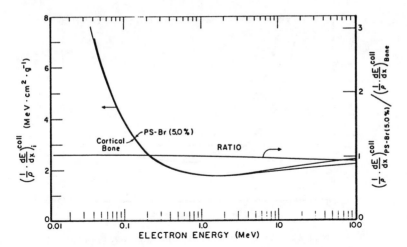

FIG.23. Left ordinate: Calculated collision mass stopping powers of polystyrene containing 5.0% bromine (PS-Br (5%)) and cortical bone versus electron energy. Right ordinate: Ratio of the collision mass stopping powers versus electron energy.

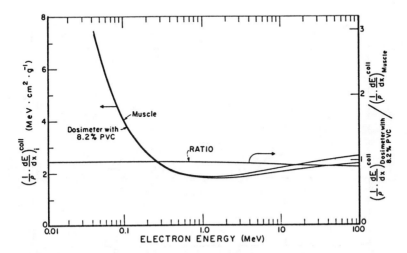

FIG.24. Left ordinate: Calculated collision mass stopping powers of the polymer dosimeter (50% PVP, 25% PMMA, 25% PVI) containing 8.2% PVC and striated muscle tissue versus electron energy. Right ordinate: Ratio of the collision mass stopping powers versus electron energy.

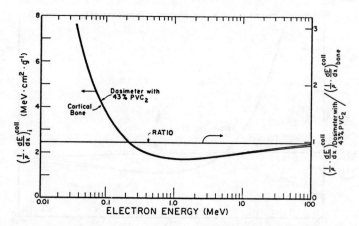

FIG.25. Left ordinate: Calculated collision mass stopping powers of the polymer dosimeter (50% PVP, 25% PMMA, 25% PVI) containing 43% PVC_2 and cortical bone versus electron energy. Right ordinate: Ratio of the collision mass stopping powers versus electron energy.

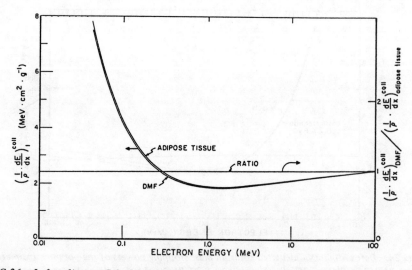

FIG.26. Left ordinate: Calculated collision mass stopping powers of N,N-dimethyl formamide (DMF) and of adipose tissue versus electron energy. Right ordinate: Ratio of the collision mass stopping powers versus electron energy.

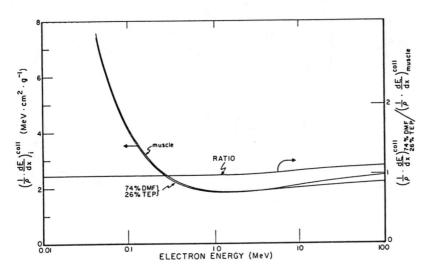

FIG.27. Left ordinate: Calculated collision mass stopping powers of 74% DMF plus 26% TEP and striated muscle versus electron energy. Right ordinate: Ratio of the collision mass stopping powers versus electron energy.

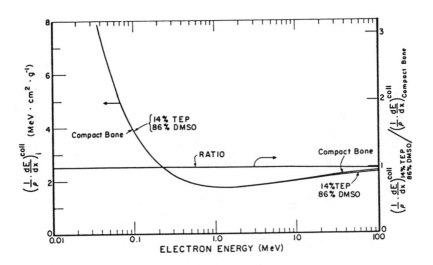

FIG.28. Left ordinate: Calculated collision mass stopping powers of 86% DMSO plus 14% TEP and compact bone versus electron energy. Right ordinate: Ratio of the compact mass stopping powers versus electron energy.

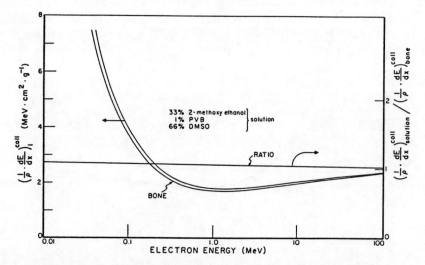

FIG.29. *Left ordinate: Calculated collision mass stopping powers of liquid-core solution used in waveguide dosimeter (33% 2-methoxy ethanol, 1% polyvinyl butyral, 66% DMSO) and cortical bone versus electron energy. Right ordinate: Ratio of the collision mass stopping powers versus electron energy.*

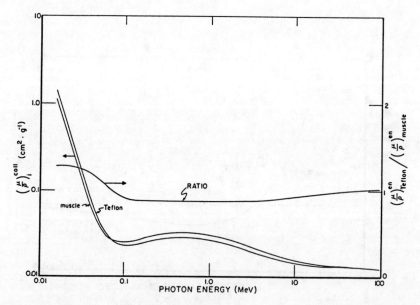

FIG.30. *Left ordinate: Calculated mass energy-absorption coefficients of polytetrafluoroethylene (Teflon) and striated muscle versus photon energy. Right ordinate: Ratio of the mass energy-absorption coefficients versus photon energy.*

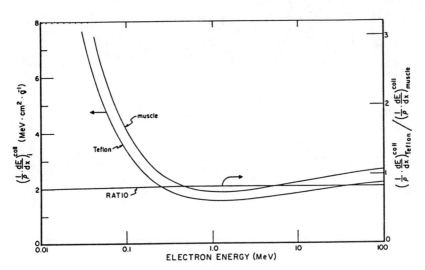

FIG.31. Left ordinate: Calculated collision mass stopping powers of Teflon and striated muscle versus electron energy. Right ordinate: Ratio of the collision mass stopping powers versus electron energy.

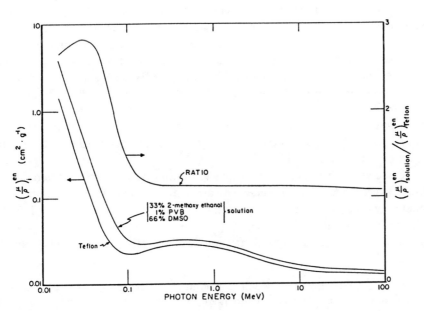

FIG.32. Left ordinate: Calculated mass energy-absorption coefficients of the core solution in the 'Optic-chromic' dosimeter and Teflon versus photon energy. Right ordinate: Ratio of the mass energy-absorption coefficients versus photon energy.

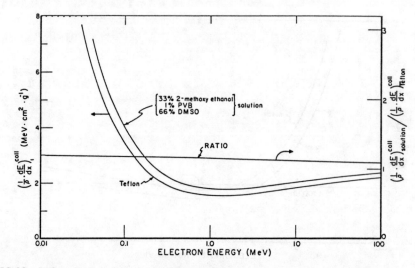

FIG.33. *Left ordinate: Calculated collision mass stopping powers of the core solution in the 'Optic-chromic' dosimeter and Teflon versus electron energy. Right ordinate: Ratio of the collision mass stopping powers versus electron energy.*

The results show that within the uncertainty of the measurements[1], the simulated-tissue dosimeter materials have essentially energy-independent responses to photons over the energy range of interest, as long as the measurement is in terms of absorbed dose in the tissue itself or in the 'tissue-equivalent' dosimeter materials.

Figures 21 to 29 are plots of collision mass stopping powers for the simulated-tissue dosimeter materials and tissues listed in Table III (and their ratios) as a function of electron energy. Here it is seen, as in Figs 4 to 6, that there is very little departure from unity of ratios of electron stopping powers for these materials in this energy range, except at electron energies > 10 MeV. There is some uncertainty in the calculated values at these high energies, since errors due to the 'density effect' are greater in this region [14, 20].

Calculations based on Bragg-Gray and Burlin cavity theory considerations [14, 22] were made for the liquid-core waveguide dosimeter (consisting of 33% 2-methoxy ethanol, 1% polyvinyl butyral, and 66% dimethyl sulphoxide liquid radiochromic solution held in Teflon (polytetrafluoroethylene tube), all surrounded by striated muscle (water-equivalent), and irradiated with degraded γ-ray photons

---

[1] The estimated uncertainties of measurement of optical-density response functions and thus of dose readings by these dosimeters are ±10% and, with the exception of the points at 7.5 and 662 keV, which are essentially monoenergetic, and 1250 keV, which is the mean of 1170 and 1330 keV, the spread of energies of X-ray bands at 39, 72 and 120 keV is about 25% full width at half maximum [13].

from a ^{60}Co plaque source. Different photon spectra were used for primary ^{60}Co γ-rays after scattering due to penetration to depths of 2, 5, 10 and 20 cm in a semi-infinite tissue phantom. The degraded photon spectrum at 10 cm depth in tissue is shown in Fig.1 in Ref. [36].

In these cavity theory calculations, the dosimeter must be considered 'intermediate' in size (0.3 mm inner diameter of the tube containing the liquid dosimeter material, and 1 mm wall thickness of the Teflon tube and 6 cm length). The size of the dosimeter suggests that some secondary electrons arising in the surrounding medium may reach the dosimeter solution in the core of the tube. This fraction may be determined experimentally [37, 38], but for the purposes of these calculations it is assumed that the secondary electrons absorbed in the dosimeter solution in the core of the waveguide tube originate either in the solution or in the wall, and that the electrons interacting in the sensing medium (waveguide core) originate either in the wall or the surrounding medium (muscle tissue), the cavity theory [22] correction factor, f, in the relation

$$D_m = \frac{1}{f} D_d$$

(D_m being the absorbed dose in the surrounding medium and D_d the absorbed dose in the dosimeter) can be expressed as

$$f = [d_2 S_m^w + (1 - d_2) \mu_m^w][d_1 S_w + (1 - d_1) \mu_w^d]$$

where S_m and S_w are respectively the ratio of electron collision mass stopping powers $(1/\rho)(dE/dx)^{coll}$ wall-to-medium and dosimeter solution-to-wall averaged over the secondary electron spectrum shown by Fig.8(b) of Ref. [14]; μ_m^w and μ_w^d are respectively the ratios of photon mass energy-absorption coefficients $(\mu/\rho)^{en}$ to medium and dosimeter solution-to-wall, averaged over the degraded photon spectrum shown in Fig.1 of Ref. [36] for 10 cm depth in water (similar to the muscle); d_1 and d_2 are the weighting factors

$$d_{1,2} = \frac{1 - e^{-\beta g}}{\beta g}$$

β being the effective mass attenuation coefficient for electrons and g the average pathlength (in g/cm²) through the dosimeter [22], so that when calculated approximately for a typical liquid-core waveguide dosimeter, $d_1 = 0.84$ for a 0.3 mm inner diameter of the Teflon tube holding the sensitive dosimeter solution, and $d_2 = 0.19$ for a 1 mm wall thickness, having a density of $\rho = 2.2$ g·cm^{-3}. For the typical scattered electron spectrum for ^{60}Co γ-ray interaction, shown in Fig.8(b) in Ref. [14], $S_m = 0.823$, $S_m = 1.165$, and for the degraded ^{60}Co γ-ray spectrum shown in

Fig.1 of Ref. [36], $\mu_m^w = 0.872$ and $\mu_w^d = 1.187$. Substituting these values in the relation for f above

$$f = [(0.19 \times 0.823) + (1 - 0.19) \times 0.872] [(0.84 \times 1.165) + (1 - 0.84) \times 1.187] = 1.01$$

Thus, the estimated error in determining absorbed dose to water (or muscle) from the liquid-core dosimeter irradiated with a typical degraded ^{60}Co gamma-ray spectrum at 10 cm depth in water (or muscle), is predicted to be about 1%. Even at 20 cm depth, the error increases to only about 1.4%.

The plots of mass energy-absorption coefficients and collision mass stopping powers and their ratios are shown in Figs 30 to 33 for Teflon-to-muscle and for dosimeter solution mixture-to-Teflon. Therefore, despite the rather large energy dependence in photon interaction with dosimeter solution and Teflon wall material relative to water or muscle, the combinations of effects compensate for each other, so that the overall effect is a relatively small energy dependence, at least for the stated waveguide dosimeter dimensions and the typical degraded γ-ray spectra used in radiation processing and medical applications.

4. SUMMARY

Certain chemical dosimeters (e.g. ceric sulphate solutions or halogenated hydrocarbons) [14, 39] and physical dosimeters (e.g. thermoluminescence dosimetry) [14, 40] show energy dependence of response when absorbed doses due to irradiation with degraded photon spectra from ^{60}Co and ^{137}Cs sources are expressed in terms of absorbed dose in water or biological soft tissues (e.g. muscle or fat). It has been shown here that, by combining certain ingredients with several forms of radiochromic dosimeters, a relatively energy-independent response relative to several different biological systems can be achieved. In addition, specially designed radiochromic dosimetry systems have energy absorption properties over wide photon and electron spectral energies similar to those of many absorbing media and covering wide ranges of absorbed dose and dose rate (~10^{-2} to 10^6 Gy).

ACKNOWLEDGEMENT

The authors are grateful to A. Hadjinia-Lailibadi, of the Atomic Energy Organization of Iran who, during his IAEA Fellowship at the Risø National Laboratory, Denmark, calculated the photon energy-absorption coefficients and electron stopping powers and their energy-absorption coefficients and ratios for different materials.

REFERENCES

[1] MILLER, A., McLAUGHLIN, W.L., "Evaluation of radiochromic dye films and other plastic dose meters under radiation processing conditions", High-Dose Measurements in Industrial Radiation Processing, IAEA Technical Reports Series No. 205, IAEA, Vienna (1981) 119.

[2] RADAK, B.B., "Radiation process control and acceptance of dosimetric methods", ibid, 101.

[3] RADAK, B.B., McLAUGHLIN, W.L., The gamma-ray response of Opti-chromic dosimeters, Radiat. Phys. Chem. 23 (1984).

[4] HUMPHREYS, K.C., WILDE, W.O., KANTZ, A.D., "An opti-chromic dosimetry system for radiation processing of food", Radiation Processing (Trans. 4th Int. Meeting Dubrovnik, 1982) (MARKOVIĆ, V., Ed.), Vol. 2, Radiat. Phys. Chem. 22 (1983) 291.

[5] McLAUGHLIN, W.L., HUMPHREYS, J.C., RADAK, B.B., MILLER, A., OLEJNIK, T.A., "The response of plastic dosimeters to gamma rays and electrons at high absorbed dose rates", Advances in Radiation Processing (Trans. 2nd Int. Meeting Miami, 1978) (SILVERMAN, J., Ed.), Vol. 2, Radiat. Phys. Chem. 14 (1979) 535.

[6] GEHRINGER, P., ESCHWEILER, H., PROKSCH, E., Dose rate and humidity effects on the gamma radiation response of nylon-base radiochromic film dosimeters, Int. J. Appl. Radiat. Isot. 31 (1980) 595.

[7] McLAUGHLIN, W.L., HUMPHREYS, J.C., LEVINE, H., MILLER, A., RADAK, B.B., RATIVANICH, N., "The gamma-ray response of radiochromic dye films at different absorbed dose rates", Trends in Radiation Processing (Trans. 3rd Int. Meeting Tokyo, 1980) (SILVERMAN, J., Ed.),Vol. 3, Radiat. Phys. Chem. 18 (1981) 987.

[8] KOSANIĆ, M.M., NENADOVIĆ, M.T., RADAK, B.B., MARKOVIĆ, V.M., McLAUGHLIN, W.L., Liquid radiochromic dye dosimetry for continuous and pulsed radiation field over a wide range of energy flux densities, Int. J. Appl. Radiat. Isot. 28 (1977) 313.

[9] MILLER, A., Dosimetry for Electron Beam Applications, Risø National Laboratory, Rep. M-2401 (1984).

[10] MILLER, A., BJERGBAKKE, E., McLAUGHLIN, W.L., Some limitations to the use of plastic and dyed plastic dosimeters, Int. J. Appl. Radiat. Isot. 28 (1977) 313.

[11] LEVINE, H., McLAUGHLIN, W.L., MILLER, A., "Temperature and humidity effects on the gamma-ray response and stability of plastic and dyed plastic dosimeters", Radiation Processing (Trans. 2nd Int. Meeting Miami, 1978) (SILVERMAN, J., Ed.), Radiat. Phys. Chem. 14 (1979) 571.

[12] CHEN Wenxiu, HUMPHREYS, J.C., McLAUGHLIN, W.L., The response of radiochromic film dosimeters to gamma rays in different atmospheres, (in preparation).

[13] McLAUGHLIN, W.L., ROSENSTEIN, M., LEVINE, H., "Bone- and muscle-equivalent solid chemical dosimeters for photon and electron doses above 1 kilorad", Biomedical Dosimetry (Proc. Symp. Vienna, 1975), IAEA, Vienna (1975) 267.

[14] MILLER, A., McLAUGHLIN, W.L., "Calculation of the energy dependence of dosimeter response to ionizing photons", Trends in Radiation Dosimetry (McLAUGHLIN, W.L., Ed.), Pergamon Press, Oxford (1982); Int. J. Appl. Radiat. Isot. 33 (1982) 1299.

[15] CHRISTENSEN, E.B., MILLER, A., A Program in BASIC for Calculation of Cavity Theory Corrections, Risø National Laboratory, Rep. M-2345 (1982).

[16] HUBBELL, J.H., Photon Cross-Sections, Attenuation Coefficients, and Energy Absorption Coefficients from 10 keV to 100 GeV, Superintendent of Documents, U.S. Government Printing Office, Washington, DC, Rep. NSRD NBS 29 (1969).

[17] PAGES, L., BERTEL, E., JOFFRE, H., SKLAVENITIS, L., Energy loss, range, and bremsstrahlung yields for 10 keV to 100 MeV electrons in various elements and chemical compounds, At. Data **4** (1972) 1.

[18] BERGER, M.J., SELTZER, S.M., Tables of Energy Losses and Ranges of Electrons and Positrons, National Technical Information Service, Springfield, Virginia, Reps NASA SP 3012 and NASA SP 3036 (1964, 1966).

[19] HUBBELL, J.H., "Photon mass attenuation and energy-absorption coefficients from 1 keV to 20 MeV", Trends in Radiation Dosimetry (McLAUGHLIN, W.L., Ed.), Pergamon Press, Oxford (1982); Int. J. Appl. Radiat. Isot. **33** (1982) 1269.

[20] SELTZER, S.M., BERGER, M.J., Improved procedure for calculating the collision stopping power of elements and compounds for electrons and positrons, Int. J. Appl. Radiat. Isot. **34** (1984) (in press).

[21] ATTIX, F.H., Energy-absorption coefficients for γ-rays in compounds and mixtures, Phys. Med. Biol. **29** (1984) 869.

[22] BURLIN, T.E., A general theory of cavity ionization, Br. J. Radiat. **39** (1966) 727.

[23] LOFTUS, T.P., WEAVER, J.T., Standardization of ^{60}Co and ^{137}Cs gamma-ray beams in terms of exposure, J. Res. NBS **78A** (1974) 465.

[24] LAMPERTI, P.T., WYCKOFF, H.O., NBS free-air chamber for measurement of 10-60 kV X-rays, J. Res. NBS **69C** (1965) 39.

[25] HENKE, B.L., "Microanalysis with ultrasoft X-radiation", Advances in X-ray Analysis, Vol. 5, Plenum Press, New York (1961) 285.

[26] LYONS, P.B., BARAN, J.A., McGARRY, J.H., A Total Absorption Ionization Chamber for 1.5 to 10 keV X-rays, Los Alamos Scientific Laboratory of the University of California, Rep. LA 4568 (1974).

[27] McLAUGHLIN, W.L., "Radiochromic dye-cyanide dosimeters", Manual on Radiation Dosimetry (HOLM, H.W., BERRY, R.J., Eds), Marcel Dekker, New York (1970) 377.

[28] GOODMAN, L.J., A modified tissue equivalent liquid, Health Phys. **16** (1969) 763.

[29] FRIGERIO, N.A., COLEY, R.F., SAMSON, M.J., Depth dose determinations. I. Tissue equivalent liquids for standard man and muscle, Phys. Med. Biol. **17** (1972) 792.

[30] WHITE, D.R., A survey of tissue substitutes and phantoms in 90 European and non-European centres, Br. J. Radiol. **50** (1970) 603.

[31] WHITE, D.R., Tissue substitutes in experimental radiation physics, Med. Phys. **5** (1978) 467.

[32] ATWATER, H.F., Evaluation of mass energy-absorption coefficients of bone and muscle, Health Phys. **23** (1972) 739.

[33] WHITE, D.R., FITZGERALD, M., "Calculated attenuation and energy absorption coefficients for ICRP reference man (1975) organs and tissues, Health Phys. **33** (1977) 73.

[34] WHITE, D.R., The photon attenuation and absorption properties of clear and white polystyrene, Br. J. Radiol. **51** (1978) 397.

[35] WHITE, D.R., PEAPLE, L.H.T., CROSBY, T.J., Measured attenuation coefficients at low photon energies (9.88–57.32 keV) for 44 materials and tissues, Radiat. Res. **84** (1980) 239.

[36] MILLER, A., Paper IAEA-SM-272/27. these Proceedings.

[37] LEMPERT, G.D., NASH, R., SCHULTZ, R.J., Fraction of ionization from electrons arising in the wall of an ionization chamber, Med. Phys. **10** (1983) 1.

[38] ALMOND, D.P., SWENSSON, H., Acta Radiol., Ther., Phys., Biol. **16** (1977) 177.

[39] McLAUGHLIN, W.L., The measurement of absorbed dose and dose gradients, Radiat. Phys. Chem. **15** (1980) 9.

[40] ATTIX, F.H., Basic γ-ray dosimetry, Health Physics **15** (1968) 49.

IAEA-SM-272/27

CALCULATION OF ENERGY DEPENDENCE OF SOME COMMONLY USED DOSIMETERS*

A. MILLER,
Accelerator Department,
Risø National Laboratory,
Roskilde, Denmark

Abstract

CALCULATION OF ENERGY DEPENDENCE OF SOME COMMONLY USED DOSIMETERS.
 The cavity theory correction factors according to Burlin's general cavity theory are calculated for several dosimeters used in routine and reference dosimetry for gamma-radiation processing. A simplified computer code suitable for a small desk-top computer was applied to calculate different degraded energy spectra. The results show that although there is a marked material dependence of the correction factor, only the energy dependence is limited.

1. INTRODUCTION

 The radiation quantity of interest in radiation processing is the dose absorbed by the product. This quantity is usually measured by dosimeters placed in or around the product. Only in very few cases is the dose measured by observing physical or chemical changes in the product itself. The dosimeter and the product are thus often different materials with different radiation absorbing properties. The dose absorbed by the dosimeter may therefore not equal the dose absorbed by the product, even if the dosimeter and the product are irradiated under equal conditions. This problem is treated by the cavity theories, by means of which the dose in the product can be calculated if the dose in the dosimeter is known. To perform the calculation, knowledge of certain parameters must be available, including the energy distribution of the photons and electrons, radiation absorption properties such as the mass energy absorption coefficient and the mass collision stopping power, and their variation with energy. Which radiation absorbing property is the most important for photon irradiations is influenced by the size and to some extent the shape of the dosimeter.
 Calculation of mass collision stopping powers and mass energy absorption coefficients is rather time consuming and tabulations of these parameters are often used when cavity theory calculations are carried out. The most recent tables are by Hubbell [1] and Berger and Seltzer [2]. Recently Seltzer and Berger provided a simplified calculation of mass collision stopping powers [3] which uses one

---

* Research carried out with the support of the IAEA under Research Contract No. 2883/RB.

set of energy-dependent parameters and one set of material-dependent parameters. Using this method, mass collision stopping powers for all elements and 180 compounds and mixtures can be calculated rather easily at specific energies. Miller and McLaughlin [4] in their approach use the sum of weighted fractions to calculate the stopping powers of compounds. This method will give rise to errors, particularly at higher energies, but it may still be the only easy way to estimate the stopping powers for compounds not listed in Ref. [3]. Miller and McLaughlin further calculate the cavity theory correction factors according to Burlin's general cavity theory [5] for specified dosimeters surrounded by various materials. We have now included Seltzer and Berger's simplified calculation in our calculation of cavity theory correction factors, and making use of this procedure have carried out calculation of correction factors and their energy dependence for six dosimeters (Perspex, radiochromic dye film, ceric-cerous, ethanol monochlorobenzene, potassium dichromate, alanine) irradiated with ^{60}Co photons.

2. PHOTON ENERGY SPECTRUM

Photons are emitted from a radionuclide source as a line spectrum, e.g. for ^{60}Co at 1.17 and 1.33 MeV. These photons are attenuated exponentially through the passage of matter and simultaneously photons of lower energy are created primarily by the Compton scatter process, thereby giving rise to a continuous spectrum from the maximum energy down towards lower energies. Since the photoelectric effect is dominant at the lowest energies, fewer scattered photons are found here.

Degraded photon energy spectra have been calculated by Bruce and Johns [6] and Goldstein and Wilkins [7] using different source arrangements and several absorbing materials. More recently, calculations on specific geometries such as test gamma cells have been carried out [8]. In the calculations of cavity theory correction factors we have used photon spectra in 2, 5, 10 and 20 cm depth in water. We assume that spectra in other low atomic number absorbers are closely similar. The fractions of energy absorbed from the primary photon beam and the degraded spectrum are taken from Tables 1 to 4 in Ref. [9]. In our calculations an approximation of the spectrum is used in the form of a histogram, as shown in Fig. 1 [6].

3. SECONDARY ELECTRON ENERGY SPECTRUM

During photon irradiation, secondary electrons are created by the photoelectric and Compton processes. The maximum energy of the electrons is almost equal to the maximum energy of the photons and the minimum energy is close to zero. Information on the shape of secondary electron energy spectra is a little

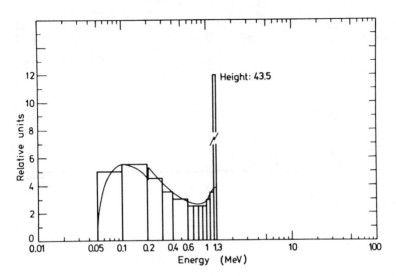

FIG.1. *Relative fluence spectrum at $10 \, g \cdot cm^{-2}$ depth in the product irradiated by an infinite area plane parallel ^{60}Co source. The smooth curve is from Ref. [6], while the histogram is the approximation used in these calculations.*

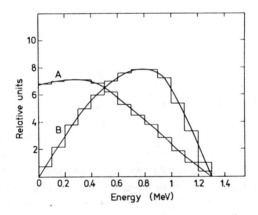

FIG.2. *Relative fluence spectra for secondary electrons, created by ^{60}Co photon irradiation. Spectra A and B are used in these calculations to demonstrate the influence of different electron spectra on the cavity theory correction factor.*

sparse, but some examples of spectra in different materials are given in Ref. [10]. Irradiation geometry can give rise to changes in the spectrum; the low energy part will be increased if the number of backscattered electrons is increased. We have chosen to make calculations for two spectra with average energies at high and low energies, respectively (see Fig. 2).

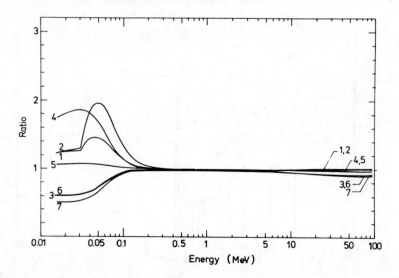

FIG.3. Variation with energy of mass energy absorption coefficient ratios for several dosimeter materials versus water. Elemental composition ((*) Ref. [3]; (°) see Table III): 1. ceric sulphate*; 2. ceric-cerous°; 3. Perspex*; 4. ethanol monochlorobenzene°; 5. potassium dichromate°; 6. alanine/wax°; 7. radiochromic dye film°.

4. EVALUATION OF RATIOS

For evaluation of the response of a dosimeter irradiated with photons, it may be an advantage to visualize the variation with energy of the ratios of the mass collision stopping powers, S_M^C, and mass energy absorption coefficients, μ_M^C.

The ratios should be taken for the dosimeter material and the material in which it is irradiated. In Figs.3 and 4 the ratio μ_M^C is shown for several dosimeter materials versus water and for several product materials also versus water. Table I supplements Fig.4 where values of μ_M^C exceed the boundaries of the computer plot. Although several materials are almost equivalent to water at medium energies, considerable differences occur at low energies.

Figure 5 shows the variations of S_M^C with energy for the same product materials as in Fig.4. A figure for the dosimeter materials is not included, since the stopping power ratio is uniform and close to one over the energy range of 0.01 to 100 MeV, as seen in Table II. The spectral distribution of the secondary electrons may therefore be expected to have only a minor influence on the cavity theory correction factor.

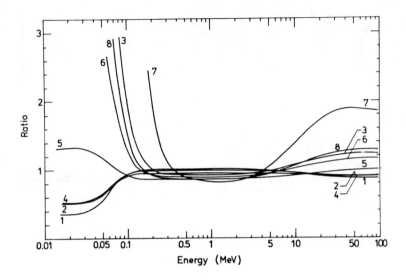

FIG.4. *Variation with energy of mass energy absorption coefficient ratios for several typical product materials versus water. Elemental composition is taken from Ref. [3]: 1. polypropylene; 2. nylon 6; 3. polyvinylchloride; 4. polycarbonate; 5. Teflon (PTFE); 6. glass; 7. copper; 8. aluminium.*

TABLE I. MASS ENERGY ABSORPTION COEFFICIENT RATIOS FOR SOME MATERIALS VERSUS WATER. THE TABLE INCLUDES THE MATERIALS MENTIONED IN FIG. 4 FOR WHICH THE RATIOS EXCEED THE BOUNDARIES OF THE FIGURE

Energy (MeV)	μ^M_{water}			
	PVC	Aluminium	Copper	Glass
0.015	7.333	5.535	44.568	3.566
0.02	7.774	5.769	53.282	3.673
0.03	7.958	5.697	61.918	3.685
0.04	7.444	5.228	61.822	3.427
0.06	4.900	3.467	41.284	2.412
0.08	2.892	2.128	21.827	1.635
0.1	1.909	1.486	11.675	1.254
0.15			3.733	
0.2			1.957	

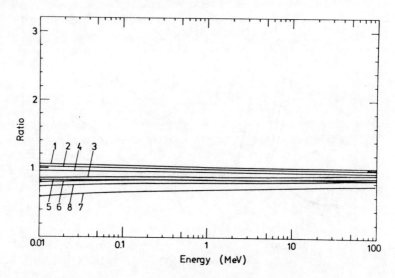

FIG.5. Variation with energy of mass collision stopping power ratios for the same materials as in Fig. 4 versus water.

5. DOSIMETER WALL

The dosimeter is often surrounded by a wall that may influence the response of the dosimeter, because it may change the fluence of secondary electrons in the dosimeter. The wall may be considered so thin that it has a vanishing influence, or it may be so thick that it 'isolates' the dosimeter from the surrounding medium, i.e. no secondary electrons generated in the medium will reach the dosimeter. In either case the cavity theory treatment is straightforward [4]. If the wall has an intermediate thickness, an experimentally determined correction factor may have to be applied [11]. For monodirectional and monoenergetic photon beams the effect of the wall may be considerable, but for the highly scattered irradiation field encountered in radiation processing the effect of the wall on the absorbed dose in the dosimeter may only reach a few tens of microns into the dosimeter [12], and may therefore often be ignored. Under certain irradiation conditions, e.g. a glass wall surrounding a small liquid dosimeter, the wall influence cannot be ignored.

It may be difficult to predict the actual influence from the wall, and we have therefore chosen to calculate cavity theory correction factors for individual dosimeters surrounded by water and glass, these two materials being the likely extremes for materials surrounding the dosimeter, both as a wall and as products, e.g. to be radiation sterilized.

TABLE II. STOPPING POWER RATIOS FOR SEVERAL DOSIMETER MATERIALS VERSUS WATER

Energy (MeV)	S^C_{water}					
	1	2	3	4	5	6
0.01	0.993	0.991	0.966	0.974	1.018	0.968
0.05	0.994	0.992	0.965	0.974	1.011	0.967
0.1	0.994	0.993	0.964	0.973	1.009	0.967
0.5	0.995	0.993	0.963	0.973	1.004	0.961
1	0.995	0.994	0.957	0.970	0.995	0.957
5	0.996	0.995	0.949	0.967	0.985	0.952
10	0.996	0.995	0.948	0.966	0.983	0.955
50	0.996	0.995	0.949	0.966	0.982	0.962
100	0.996	0.995	0.949	0.966	0.983	0.964

1. Ferrous sulphate.
2. Ceric sulphate.
3. Glutamine.
4. Perspex.
5. Nylon.
6. Ethanol monochlorobenzene.

6. CORRECTION FACTORS FOR DOSIMETERS

For calculation of cavity theory correction factors we use Burlin's general cavity theory [5]. We realize that the validity of this theory has been disputed by several authors (see, for example, Ref. [13]), but there is still no agreement on which modification of the theory should be used.

In Burlin's general cavity theory the correction factor, f, may be calculated as

$$D_M = \frac{1}{f} D_C \tag{1}$$

$$f = d\, S^C_M + (1-d)\, \mu^C_M \tag{2}$$

$$d = \frac{1 - e^{-\beta g}}{\beta g} \tag{3}$$

where D_M and D_C are the doses in the product, M, and the dosimeter, C, S^C_M is the ratio of mass collision stopping powers and μ^C_M is the ratio of mass energy absorption coefficients for these two materials. Both ratios must be evaluated over the energy

TABLE III. VALUES FOR THE CORRECTION FACTOR, f, CALCULATED FOR SIX DOSIMETERS IRRADIATED WITH ^{60}Co PHOTONS

TABLE III(a) PERSPEX DOSIMETER

Density: 1.19 g·cm^{-3}
Red Perspex: 10 × 30 × 3 mm^3, d = 0.201
Clear Perspex: 10 × 30 × 1 mm^3, d = 0.427

Medium: water	S_M^C, spectrum A	0.973
	S_M^C, spectrum B	0.972
	Used in calculation of f:	0.973
Medium: glass	S_M^C, spectrum A	1.165
	S_M^C, spectrum B	1.154
	Used in calculation of f:	1.160

	f			
	Red Perspex		Clear Perspex	
Depth (g·cm^{-2})	Water	Glass	Water	Glass
2	0.972	1.054	0.972	1.030
5	0.971	1.047	0.971	1.025
10	0.970	1.033	0.971	1.015
20	0.969	1.022	0.970	1.007

TABLE III(b) RADIOCHROMIC DYE FILM, BASE MATERIAL NYLON

Density: 1.14 g·cm^{-3}, area 10 × 10 mm^2
Thin film, 0.05 mm, d = 0.968
Thick film, 1 mm, d = 0.444

Medium: water	S_M^C, spectrum A	1.005
	S_M^C, spectrum B	1.001
	Used in calculation of f:	1.003
Medium: glass	S_M^C, spectrum A	1.212
	S_M^C, spectrum B	1.202
	Used in calculation of f:	1.207

	f			
	Thin film		Thick film	
Depth (g·cm^{-2})	Water	Glass	Water	Glass
2	1.002	1.203	0.994	1.143
5	1.002	1.203	0.994	1.138
10	1.002	1.202	0.993	1.128
20	1.002	1.202	0.992	1.120

TABLE III(c) CERIC-CEROUS DOSIMETER

Density: 1.03 g·cm^{-3}
Composition: 15mM Ce(SO$_4$)$_2$, 15mM Ce$_2$(SO$_4$)$_3$, 0.8 N H$_2$SO$_4$

1 mL ampoule, d = 0.132
2 mL ampoule, d = 0.102
5 mL ampoule, d = 0.075

Medium: water S_M^C, spectrum A	0.991	
S_M^C, spectrum B	0.992	
Used in calculation of f:	0.992	
Medium: glass S_M^C, spectrum A	1.182	
S_M^C, spectrum B	1.192	
Used in calculation of f:	1.187	

	f			
	Water	Water	Water	Glass
Depth (g·cm^{-2})	2 mL	1 mL	5 mL	2 mL
2	1.004			1.119
5	1.009			1.117
10	1.019			1.114
20	1.028	1.028	1.030	1.112

spectrum of photons and electrons at the position of the dosimeter. The average pathlength, g, of the secondary electrons penetrating the dosimeter is determined as four times the dosimeter volume divided by its surface area. The effective mass attenuation coefficient, β, is determined as

$$\beta = \frac{16}{(E_{max} - 0.036)^{1.4}} \tag{4}$$

where E_{max} is the maximum electron energy (equal to the maximum photon energy). Calculations in Table III (a–f) [14, 15] have been carried out for ^{60}Co photon irradiations, and the E_{max} was chosen to be 1.3 MeV.

The values of f in Table III show that the energy variations, which might be expected during industrial ^{60}Co irradiation, will only give rise to minor changes of the dose response. Irradiation with photons of lower energy, e.g. from ^{137}Cs, may give larger variations of response. Changing the material directly surrounding the dosimeter may produce significant correction factors. In general, a small dosimeter will be less energy dependent because the response depends on stopping power, while a large dosimeter, depending more on energy absorption coefficients, is more likely to show a change of response as the energy is changed.

TABLE III(d) ETHANOL MONOCHLORO-BENZENE DOSIMETER

Density: 0.88 g·cm^{-3}
Composition: volume %, 24% monochlorobenzene, 4% water, 0.04% benzene, 0.04% acetone, 71.92% ethanol

2 mL ampoule, d = 0.102

Medium: water	S_M^C, spectrum A	0.984
	S_M^C, spectrum B	0.984
	Used in calculation of f:	<u>0.984</u>
Medium: glass	S_M^C, spectrum A	1.183
	S_M^C, spectrum B	1.173
	Used in calculation of f:	<u>1.178</u>

	f	
Depth (g·cm^{-2})	Water	Glass
2	0.992	1.096
5	0.995	1.091
10	0.999	1.082
20	1.002	1.074

TABLE III(e) DICHROMATE DOSIMETER, DENSITY 1.00 g·cm^{-3} [15]

Composition: 2mM $K_2Cr_2O_7$, 0.5mM $Ag_2Cr_2O_7$, 0.1M $HClO_4$ in aqueous solution

2 mL ampoule, d = 0.102

Medium: water	S_M^C, spectrum A	0.998
	S_M^C, spectrum B	0.998
	Used in calculation of f:	<u>0.998</u>
Medium: glass	S_M^C, spectrum A	1.201
	S_M^C, spectrum B	1.191
	Used in calculation of f:	<u>1.196</u>

	f	
Depth (g·cm^{-2})	Water	Glass
2	0.999	1.115
5	1.000	1.107
10	1.000	1.093
20	1.000	1.082

TABLE III(f) ALANINE DOSIMETER [15]

80% by weight alanine, 20% paraffin
Density: 1.3 g·cm^{-3}
Pellets: size, 4 mm diameter, 7.5 mm length
d = 0.122

Medium: water	S_M^C, spectrum A	0.984
	S_M^C, spectrum B	0.984
	Used in calculation of f:	0.984
Medium: glass	S_M^C, spectrum A	1.183
	S_M^C, spectrum B	1.173
	Used in calculation of f:	1.178

	f	
Depth (g·cm^{-2})	Water	Glass
2	0.983	1.099
5	0.983	1.091
10	0.981	1.076
20	0.980	1.062

Estimation of the correction factor may be made, based on data such as in Figs 3 to 5, but by having a small computer and this program available the hard work and/or guesswork are removed from estimation of the cavity theory correction factor.

REFERENCES

[1] HUBBELL, J.H., Int. J. Appl. Radiat. Isot. **33** 11 (1982) 1269.
[2] BERGER, M.J., SELTZER, S.M., Stopping Powers and Ranges of Positrons and Electrons, National Bureau of Standards, Rep. NBSIR 82-2550-A (1982).
[3] SELTZER, S.M., BERGER, M.J., Int. J. Appl. Radiat. Isot. **33** 11 (1982) 1189.
[4] MILLER, A., McLAUGHLIN, W.L., ibid., 1299.
[5] BURLIN, T.E., Br. J. Radiol. **39** (1966) 727.
[6] BRUCE, W.R., JOHNS, H.E., The spectra of X-rays scattered in low atomic number materials, Br. J. Radiol., Suppl. No. 9 (1960) 37.
[7] GOLDSTEIN, H., WILKINS, J.W., Calculations of the Penetration of Gamma Rays, USAEC New York Operations Office, Rep. NYO-3075 (1954).
[8] WOOLF, S., BURKE, E.A., (accepted for publication in IEEE Trans. Nucl. Sci. **NS-31** 6 1984).
[9] INTERNATIONAL ATOMIC ENERGY AGENCY, Manual of Food Irradiation Dosimetry, Technical Reports Series No. 178, IAEA, Vienna (1977).

[10] ST. GEORGE, F., ANDERSON, D.W., McDONALD, L., Radiat. Res. **75** (1978) 453.
[11] ALMOND, P.R., SVENSSON, H., Acta Radiol., Ther., Phys., Biol. **16** (1977) 177.
[12] McLAUGHLIN, W.L., "A national standardization programme for high-dose measurements", High-Dose Measurements in Industrial Radiation Processing, Technical Reports Series No. 205, IAEA, Vienna (1981).
[13] HOROWITZ, Y.S., MOSCOVITCH, M., Radiat. Prot. Dosim. **6** 1–4 (1984) 37.
[14] SHARPE, P.H.G., BARRETT, J.H., BERKLEY, A.M., (accepted for publication in Int. J. Radiat. Isot.).
[15] REGULLA, D.F., DEFFNER, U., TUSCHY, H., "A practical high-level dose meter based on tissue-equivalent alanine", High-Dose Measurements in Industrial Radiation Processing, Technical Report Series No. 205, IAEA, Vienna (1981).

CHAIRMEN OF SESSIONS

Session 1	D.F. REGULLA	Federal Republic of Germany
	S. ONORI	Italy
Session 2	B. WHITTAKER	United Kingdom
	H.M. BÄR	German Democratic Republic
	W.L. McLAUGHLIN	United States of America
	R. TANAKA	Japan
Session 3	A. MILLER	Denmark
	A. KOVÁCS	Hungary

SECRETARIAT OF THE SYMPOSIUM

Scientific Secretary:	J.W. NAM	Division of Life Sciences, IAEA
Administrative Secretary:	E. PILLER	Division of External Relations, IAEA
Editor:	P. HOWARD KITTO	Division of Publications, IAEA

LIST OF PARTICIPANTS

Bär, H.M.	Central Institute for Isotope and Radiation Research, Academy of Sciences of the German Democratic Republic, Permoserstrasse 15, DDR-7050 Leipzig, German Democratic Republic
Barthe, J.R.	CEA, Centre d'études nucléaires (IPSN/DPI/SIDR), BP 6, F-92260 Fontenay-aux-Roses, France
Bartolotta, A.	Laboratorio di Fisica, Istituto Superiore de Sanità, Viale Regina Elena 299, I-00161 Rome, Italy
Beck, J.A.	Johnson & Johnson, Ethicon, Inc., Somerville, NJ 08876-0151, United States of America
Brinkman, G.A.	National Institute of Nuclear and High Energy Physics (NIKHEF-K), P.O. Box 4395, NL-1009 AJ Amsterdam, Netherlands
Cance, M.	CEA, Centre d'études nucléaires (LMRI), BP 21, F-91191 Gif-sur-Yvette, France
Chen, L.	Radiation Protection Laboratory, Institute of Atomic Energy, Ministry of Nuclear Industry, P.O. Box 275, Beijing, China
Chen, W.	Beijing Teachers University, Xiaoxitian, Huoko, West District, Beijing, China
Chu, R.	Atomic Energy of Canada Ltd, Radiochemical Company, 413, March Road, P.O. Box 13500, Kanata, Ontario, K2K 1X8, Canada
Debel, C.	Risø National Laboratory, P.O. Box 49, DK-4000 Roskilde, Denmark
de Graaff-Kant, C.	Ministry of Housing, Physical Planning and the Environment, P.O. Box 5811, NL-2280 HV Rijswijk, Netherlands

LIST OF PARTICIPANTS

Dorda de Cancio, E.M.
Comisión Nacional de Energía Atómica,
Avenida del Libertador 8250,
1429 Buenos Aires, Argentina

Drissler, F.E.
Willy Rüsch AG,
Radiation Sterilization Service,
Waiblinger Strasse 4-10,
D-7053 Kernen, Federal Republic of Germany

Du Plessis, T.A.
Iso-Ster (Pty) Ltd,
P.O. Box 3219,
1620 Kempton Park, South Africa

Du Preez, M.L.
National Physical Research Laboratory,
South African Council for Scientific and
  Industrial Research,
P.O. Box 395,
0001 Pretoria, South Africa

Eschweiler, H.
Institut für Chemie,
Österreichisches Forschungszentrum Seibersdorf,
A-2444 Seibersdorf, Austria

Etienne, J.C.
Institut national des radioéléments,
B-6220 Fleurus, Belgium

Fremerye, M.
CEA, Centre d'études nucléaires (DS),
BP 27,
F-94190 Villeneuve-St-Georges, France

Gavahi, M.
Standard Calibration Division,
Radiation Protection Department,
Atomic Energy Organization of Iran,
P.O. Box 41/2663,
Tehran, Islamic Republic of Iran

Gehringer, P.
Institut für Chemie,
Österreichisches Forschungszentrum Seibersdorf,
A-2444 Seibersdorf, Austria

Ghoos, L.
Centre d'étude de l'énergie nucléaire (CEN/SCK),
Département de métrologie nucléaire - dosimétrie,
Boeretang 200,
B-2400 Mol, Belgium

Glover, K.M.
Chemistry Division, B.10.30,
UKAEA Atomic Energy Research Establishment,
Harwell, Didcot,
Oxfordshire OX11 0RA, United Kingdom

Golnik, N.
Institute of Atomic Energy,
Świerk, Poland

LIST OF PARTICIPANTS

Gosh, A.	Department of Bio-Medical Physics and Bio-Engineering, University of Aberdeen, Foresterhill, Aberdeen AB9 2ZD, United Kingdom
Hadjinia-Lielabadi, A.	Gamma Irradiation Centre, Atomic Energy Organization of Iran, P.O. Box 3327, Tehran, Islamic Republic of Iran
Hernalsteen, P.	Tractionel, Rue de la Science 31, B-1040 Brussels, Belgium
Heylen, C.A.M.	Ministère de l'Emploi et du Travail, Rue Belliard 53, B-1040 Brussels, Belgium
Hickman, C.	Technicatome, BP 18, F-9190 Gif-sur-Yvette, France
Hristova, M.G.	Institute of Nuclear Research and Nuclear Energy, Bulgarian Academy of Sciences, Bd Lenin 72, BG-1184 Sofia, Bulgaria
Huber, E.L.	Labour Inspection, Fichtegasse 11, A-1010 Vienna, Austria
Icre, P.	SODETEG/CARIC, Domaine de Corbeville, BP 35, F-91402 Orsay, France
Janovský, I.	Nuclear Research Institute, CS-250 68 Řež, Czechoslovakia
Kálmán, I.	Research Institute for the Plastics Industry, XIV Hungária krt 114, H-1950 Budapest, Hungary
Khan, H.A.	Pakistan Atomic Energy Commission, Islamabad, Pakistan
Klimov, S.V.	All-Union Scientific Research Institute of Radiation Technology, Varshavskoe shosse 46, Moscow, Union of Soviet Socialist Republics
Kovács, A.	Institute of Isotopes, Hungarian Academy of Sciences, P.O. Box 77, H-1525 Budapest 114, Hungary

LIST OF PARTICIPANTS

Kratschmann, H.   Permanent Mission of the Holy See to the IAEA,
Theresianumgasse 31,
A-1040 Vienna, Austria

Kronenberg, S.   CS/TA Laboratory,
Radiac Division DEL-CS-K,
United States Department of the Army,
Fort Monmouth, NJ 07703 5304,
United States of America

Lécuyer, P.   LCC/Thomson-CSF,
BP 71,
F-84500 Bollène, France

Lecat, X.V.A.   Commissariat à l'énergie atomique,
29-31, rue de la Fédération,
F-75015 Paris, France

Link, R.   Isotopenforschung Dr. Sauerwein GmbH,
Bergische Strasse 16,
D-5657 Haan, Federal Republic of Germany

McLaughlin, W.L.   Center for Radiation Research,
National Bureau of Standards,
Gaithersburg, MD 20899,
United States of America

Menessier, P.   Centre d'études nucléaires de Cadarache (DRE/SEN),
BP 1,
F-13115 Saint-Paul-lez-Durance, France

Miller, A.   Accelerator Department,
Risø National Laboratory,
P.O. Box 49,
DK-4000 Roskilde, Denmark

Möhlmann, J.H.F.   Gammaster BV,
P.O. Box 4250,
NL-6716 AH Ede, Netherlands

Mosse, D.   CEA, Centre d'études nucléaires (LMRI/ORIS),
BP 21,
F-91191 Gif-sur-Yvette, France

Musílek, L.   Faculty of Nuclear Science and Physical Engineering,
Technical University of Prague,
Brehová 7,
CS-115 19 Prague 1, Czechoslovakia

Onori, S.   Laboratorio di Fisica,
Istituto Superiore di Sanità,
Viale Regina Elena 299,
I-00161 Rome, Italy

Osvay, M.   Institute of Isotopes, Hungarian Academy of Sciences
P.O. Box 77,
H-1525 Budapest 114, Hungary

LIST OF PARTICIPANTS 443

Ott, F.	Gammaster München, Produktionsveredelung GmbH, Kesselbodenstrasse 7, D-8051 Allershausen, Federal Republic of Germany
Panta, P.P.	Institute of Nuclear Chemistry and Technology, Dorodna 16, PL-03 195 Warsaw, Poland
Pešek, M.	Institute for Research, Production and Application of Radioisotopes, Radiová 1, CS-102 27 Prague 10 - Hostivar, Czechoslovakia
Proksch, E.	Institut für Chemie, Österreichisches Forschungszentrum Seibersdorf, A-2444 Seibersdorf, Austria
Rantanen, E.J.	Finnish Centre for Radiation and Nuclear Safety, P.O. Box 268, SF-00101 Helsinki 10, Finland
Rativanich, N.	Office of Atomic Energy for Peace, Vibhavadee Rungsit Road, Bangkok 10900, Thailand
Ražem, D.	'Ruđer Bošković' Institute, Bijenička česta 54, YU-41000 Zagreb, Yugoslavia
Regulla, D.F.	Gesellschaft für Strahlen- und Umweltforschung mbH München, Ingolstädter Landstrasse 1, D-8042 Neuherberg, Federal Republic of Germany

Romanin, R.	Comitato Nazionale per la Ricerca e per lo Sviluppo
	   dell'Energia Nucleare e delle Energie Alternative
	   (ENEA),
	SAG-ICAS,
	Sicurezza SP Anguillarese,
	I-30100 Rome, Italy

Sauerwein, K.J.	Isotopenforschung Dr. Sauerwein GmbH, Bergische Strasse 16, D-5657 Haan, Federal Republic of Germany

Scarpa, G.	Comitato Nazionale per la Ricerca e per lo Sviluppo
	   dell'Energia Nucleare e delle Energie Alternative
	   (ENEA),
	Località Poggio dei Pini SNC,
	I-00061 Anguillara, Rome, Italy

Schneider, M.K.	Physikalisch-Technische Bundesanstalt, Bundesallee 100, D-3300 Braunschweig, Federal Republic of Germany

LIST OF PARTICIPANTS

Soliman, F.A.S.	Nuclear Materials Corporation, Inchas, Cairo, Egypt
Stenger, V.	Institute of Isotopes, Hungarian Academy of Sciences, P.O. Box 77, H-1525 Budapest 114, Hungary
Szinovatz, W.	Institut für Chemie, Österreichisches Forschungszentrum Seibersdorf, A-2444 Seibersdorf, Austria
Tanaka, R.	Takasaki Radiation Chemistry Research Establishment, Japan Atomic Energy Research Institute, 1233, Kamishinden-machi, Takasaki, Gunma-ken, Japan
Thomassen, J.A.	Institute for Energy Technology, P.O. Box 40, N-2007 Kjeller, Norway
Tranteeva, M.G.H.	Institute of Nuclear Research and Nuclear Energy, Bd Lenin 72, BG-1184 Sofia, Bulgaria
Valdezco, E.M.	Philippine Atomic Energy Commission, P.O. Box 932, Diliman, Quezon City, Philippines
Whittaker, B.	Chemistry Division, B.220, UKAEA Atomic Energy Research Establishment, Harwell, Didcot, Oxfordshire OX11 0RA, United Kingdom
Wiesner, L.	BGS Beta-Gamma-Service, P.O. Box 1128, Fritz-Kotz Strasse, D-5276 Wiehl, Federal Republic of Germany
Zavala, J.L.	Mediterranean Fruit Fly Laboratory, Programa Moscamed, Dirección General de Sanidad Vegetal, Secretaría de Agricultura y Recursos Hidráulicos, Apdo. Postal 496, Tapachula, Chiapas 30700, Mexico
Zuppiroli, L.	CEA, Centre d'études nucléaires (FAR/SESI), BP 6, F-92260 Fontenay-aux-Roses, France

LIST OF PARTICIPANTS

INTERNATIONAL ORGANIZATIONS

ASSOCIATION INTERNATIONALE D'IRRADIATION INDUSTRIELLE (AIII)

Vidal, P.E(). 59, route de Paris,
F-69260 Charbonnières-les-Bains, France

EUROPEAN ORGANIZATION FOR NUCLEAR RESEARCH (CERN)

Baeyens, B. CH-1211 Geneva 23, Switzerland
Schönbacher, H.

AUTHOR INDEX

Numerals refer to the first page of paper(s) by the author concerned.

Agematsu, T.: 317
Ambrosimov, V.K.: 183
Anđelić, L.: 143
Ascaño, L.M.: 31
Baeyens, B.: 275
Bär, H.M.: 79
Bartolotta, A.: 245
Boeykens, W.: 61
Bułhak, Z.: 47
Cabalfin, E.G.: 31
Caccia, B.: 245
Chaise, P.M.: 165
Chen, W.: 345
Coninckx, F.: 275
Deffner, U.: 221
Dorda, E.M.: 193
Du Plessis, T.A.: 13
Dvornik, I.: 143
El-Behay, A.Z.: 255
Eschweiler, H.: 333
Farahani, M.: 109
Fierro, M.M.: 23
Földiák, G.: 135
Gehringer, P.: 333
Generalova, V.V.: 183
Ghoos, L.: 61
Glover, K.M.: 373
Golder, F.: 285
Grinov, M.P.: 183
Guerra, M.: 23
Gurskij, M.N.: 183
Heneghan, M.: 293
Hristova, M.G.: 69
Humphreys, J.C.: 109
Indovina, P.L.: 245
Janovský, I.: 307
Jędrzejewski, K.: 157
Jia, H.: 345
Jiang, D.: 105
Kálmán, I.: 85
Kaneko, H.: 203
Katoh, A.: 203
King, M.: 373
Klimov, S.V.: 183
Kon'kov, N.G.: 183
Kovács, A.: 69, 135

Kronenberg, S.: 397
Krystek, M.: 237
Lecuyer, P.L.: 165
Legeza, L.: 135
Maier, P.: 275
McLaughlin, W.L.: 109, 345, 357, 397
Mellor, S.: 293
Miller, A.: 109, 397, 425
Moriuchi, Y.: 203
Muñoz, S.S.: 193
Navarro, Q.O.: 31
Nikolaev, S.M.: 183
Onori, S.: 245
Orozco, D.H.: 23
Osvay, M.: 285
Panta, P.P.: 47, 157
Pešek, M.: 263
Proksch, E.: 333
Rageh, M.S.I.: 255
Ražem, D.: 143
Regulla, D.F.: 221
Reinhardt, J.: 79
Remer, M.: 79
Roediger, A.H.A.: 13
Romaniuk, R.: 157
Rosati, A.: 245
Schneider, C.C.J.: 237
Schneider, M.K.H.: 237
Schönbacher, H.: 275
Schwarz, A.J.: 23
Siebentritt, C.R.: 397
Singson, C.C.: 31
Soliman, F.A.S.: 255
Stenger, V.: 69, 135
Sunaga, H.: 317
Tamura, N.: 203
Tanaka, R.: 203, 317
Thomassen, J.: 171
Tolkachev, B.V.: 183
Uribe, R.M.: 397
Valdezco, E.M.: 31
Vanyushkin, B.M.: 183
Vidal, P.E.: 3
Watts, M.F.: 293, 373
Whittaker, B.: 293
Wiesner, L.: 95
Zavala, J.L.: 23

TRANSLITERATION INDEX

С.В. Климов	S.V. Klimov
Б.М. Ванюшкин	B.M. Vanyushkin
Н.Г. Коньков	N.G. Kon'kov
С.М. Николаев	S.M. Nikolaev
В.В. Генералова	V.V. Generalova
М.Н. Гурский	M.N. Gurskij
В.К. Амбросимов	V.K. Ambrosimov
Б.В. Толкачев	B.V. Tolkachev
М.П. Гринев	M.P. Grinov

INDEX OF PAPERS BY NUMBER

IAEA-SM-272/	Page	IAEA-SM-272/	Page
1	193	26	47
2	263	27	425
3	13	28	61
5	293	29	69
6	373	30	255
7	31	31	85
9	397	32	285
10	109	33	135
11	275	34	165
12	237	35	171
13	143	36	23
16	95	37	307
17	203	38	333
18	317	39	221
22	345	41	245
23	105	42	3
24	79	44	357
25	157	46	183

HOW TO ORDER IAEA PUBLICATIONS

An exclusive sales agent for IAEA publications, to whom all orders and inquiries should be addressed, has been appointed in the following country:

UNITED STATES OF AMERICA   UNIPUB, P.O. Box 433, Murray Hill Station, New York, NY 10157

In the following countries IAEA publications may be purchased from the sales agents or booksellers listed or through your major local booksellers. Payment can be made in local currency or with UNESCO coupons.

ARGENTINA	Comisión Nacional de Energía Atomica, Avenida del Libertador 8250, RA-1429 Buenos Aires
AUSTRALIA	Hunter Publications, 58 A Gipps Street, Collingwood, Victoria 3066
BELGIUM	Service Courrier UNESCO, 202, Avenue du Roi, B-1060 Brussels
CZECHOSLOVAKIA	S.N.T.L., Spálená 51, CS-113 02 Prague 1 Alfa, Publishers, Hurbanovo námestie 6, CS-893 31 Bratislava
FRANCE	Office International de Documentation et Librairie, 48, rue Gay-Lussac, F-75240 Paris Cedex 05
HUNGARY	Kultura, Hungarian Foreign Trading Company P.O. Box 149, H-1389 Budapest 62
INDIA	Oxford Book and Stationery Co., 17, Park Street, Calcutta-700 016 Oxford Book and Stationery Co., Scindia House, New Delhi-110 001
ISRAEL	Heiliger and Co., Ltd., Scientific and Medical Books, 3, Nathan Strauss Street, Jerusalem 94227
ITALY	Libreria Scientifica, Dott. Lucio de Biasio "aeiou", Via Meravigli 16, I-20123 Milan
JAPAN	Maruzen Company, Ltd., P.O. Box 5050, 100-31 Tokyo International
NETHERLANDS	Martinus Nijhoff B.V., Booksellers, Lange Voorhout 9-11, P.O. Box 269, NL-2501 The Hague
PAKISTAN	Mirza Book Agency, 65, Shahrah Quaid-e-Azam, P.O. Box 729, Lahore 3
POLAND	Ars Polona-Ruch, Centrala Handlu Zagranicznego, Krakowskie Przedmiescie 7, PL-00-068 Warsaw
ROMANIA	Ilexim, P.O. Box 136-137, Bucarest
SOUTH AFRICA	Van Schaik's Bookstore (Pty) Ltd., Libri Building, Church Street, P.O. Box 724, Pretoria 0001
SPAIN	Diaz de Santos, Lagasca 95, Madrid-6 Diaz de Santos, Balmes 417, Barcelona-6
SWEDEN	AB C.E. Fritzes Kungl. Hovbokhandel, Fredsgatan 2, P.O. Box 16356, S-103 27 Stockholm
UNITED KINGDOM	Her Majesty's Stationery Office, Publications Centre P.O. Box 276, London SW8 5DR
U.S.S.R.	Mezhdunarodnaya Kniga, Smolenskaya-Sennaya 32-34, Moscow G-200
YUGOSLAVIA	Jugoslovenska Knjiga, Terazije 27, P.O. Box 36, YU-11001 Belgrade

Orders from countries where sales agents have not yet been appointed and requests for information should be addressed directly to:

Division of Publications
International Atomic Energy Agency
Wagramerstrasse 5, P.O. Box 100, A-1400 Vienna, Austria

85-00699